Foundations of Atmospheric Remote Sensing

Dmitry Efremenko · Alexander Kokhanovsky

Foundations of Atmospheric Remote Sensing

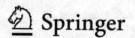 Springer

Dmitry Efremenko
German Aerospace Center (DLR)
Remote Sensing Technology Institute (IMF)
Oberpfaffenhofen, Germany

Alexander Kokhanovsky
Vitrociset Belgium
A Leonardo Company
Darmstadt, Germany

ISBN 978-3-030-66747-4   ISBN 978-3-030-66745-0   (eBook)
https://doi.org/10.1007/978-3-030-66745-0

This Springer imprint is published by the registered company Springer Nature Switzerland AG
The registered company address is: Gewerbestrasse 11, 6330 Cham, Switzerland

Preface

As this book is being written and compiled, there are thousands of satellites in orbit around the Earth, and over 800 satellites are used for Earth observation. It is expected that this number will be increased in the next decade. In recent years, there has been a tremendous increase in the use of data obtained from satellites for solving practical problems such as weather prediction, hazard prevention and development of climatological models. Spaceborne atmospheric composition sensors enabled global and continuous monitoring of air quality. Atmospheric remote sensing has evolved into an interdisciplinary technology and became an integral part of Earth observation science. A big amount of data acquired from satellites has become available to everyone on an open and free-of-charge basis, thus allowing many new services in the public sector to be developed. Just as our daily life is inconceivable without the Internet, modern atmospheric science requires satellite observations.

Atmospheric remote sensing lies on the intersection of three theoretical disciplines, namely electromagnetic scattering, radiative transfer and inversion theory. There are plenty of textbooks which describe each of these topics as independent disciplines. Our goal was to provide the overview of these topics in a common context and compile the data available in the literature and results of our research, as well as fill the gaps, where needed. In this regard, we hope that our book will help students who choose to work in the field of remote sensing to get a broad view of its theoretical background.

Chapters 1 and 2 of this book are devoted to the introduction to remote sensing technology and atmospheric physics, respectively. Essentially, they are designed to convince readers that remote sensing is the best available solution to many of the important problems faced by humanity. Chapter 3 contains basic definitions of quantities used in radiative transfer. The foundation of the radiative transfer theory is discussed. We also briefly outline the electromagnetic scattering theory. Chapter 4 is a review of the radiative transfer theory and computational techniques to solve the radiative transfer equation. Chapter 5 outlines retrieval techniques and regularization principles which are used to extract information about atmospheric constituents from spaceborne measurements. Although the last two chapters have mathematical characters, we often appeal to physical intuition providing various schemes and illustrations. Moreover, Chap. 5 might sound like a collection of general recipes, which the reader can use for solving practical problems he may face. All chapters have extensive lists of references, and we will be happy if the readers will be motivated to refer to those papers after reading our book.

Dmitry Efremenko would like to thank V. P. Afanas'ev, V. P. Budak and A. Doicu for revealing to him the wonderful world of science, as well as for the every-day scientific support. The authors are grateful to J. P. Burrows, S. V. Korkin, D. Loyola, R. Munro, T. Nakajima, V. V. Rozanov, T. Trautmann, W. von Hoyningen-Huene, R. Weichert and E. P. Zege for many stimulating discussions and cooperation. The important insights on various aspects of radiative transfer theory and its theoretical foundations have been provided by Viktor V. Danilov (1951–2020) and Michael I. Mishchenko (1959–2020) who passed away at the zenith of their research activities. The authors would like to dedicate this book to the memory of these two outstanding scientists.

Munich, Germany   Dmitry Efremenko
October 2020   Alexander Kokhanovsky

Contents

Chapter 1
Introduction to Remote Sensing

1.1 Remote Sensing: From the Past to the Present

The history of remote sensing as an independent scientific discipline can be traced back to the sixteenth century. Galileo Galilei was one of the first observational astronomers who employed a telescope to observe objects such as Jupiter's moons, Venus, Saturn, and Neptune, at long distances. Probably, he could not even imagine that the viewing direction would be reversed in the future.

The idea of taking pictures of the Earth by the instrument elevated above the surface was used by balloonists in the nineteenth century. Photography from balloons was used mostly for topographic purposes. At the beginning of the twentieth century, Konstantin Tsiolkovsky in his treatise, *Exploring Space Using Jet Propulsion Devices* [1], described a mental experiment about the Earth observation from space. He wrote: *From the rocket, a huge planet globe is visible in one or another phase, like the moon. It can be seen how the ball turns, as in several hours it shows all its sides in sequence. The closer it is to the rocket, the more enormous it seems, the more concave, spreading across the sky, its fanciful, the more brilliance it gives to its rocket, the latter revolves more likely around its mother - the Earth.... This picture is so majestic, attractive, infinitely varied that I sincerely wish to myself and to you to see it.* These lines gave a prefiguration of what we call now as Earth remote sensing. Later, Herman Potoénik in his book, *The problem of Space Travel—The Rocket Motor* [2], considered how a spacecraft could be used for the detailed observation of the underlying surface.

The precursor of the satellite remote sensing was the first image of Earth taken in 1946 over the New Mexico desert from a camera attached to a V-2 sounding rocket. On October 1957, USSR launched Sputnik I, the first artificial Earth satellite [3]. It has been demonstrated that satellites could be launched from the Earth and operated on a continuous basis. This success was followed by the first U.S. satellites, Explorer 1 in January 1958 and Vanguard 1 in March 1958 [4]. On the 12 April 1961, the first cosmonaut Yuri Gagarin traveled to space and did first visual observations of the Earth's surface. Less than in four months after, Gherman Titov filmed the Earth from space. These breakthroughs heralded the era of instrumental studies of

© Springer Nature Switzerland AG 2021
D. Efremenko and A. Kokhanovsky, *Foundations of Atmospheric Remote Sensing*,
https://doi.org/10.1007/978-3-030-66745-0_1

optically active components of the atmosphere from space and revolutionized the way the mankind thought about what modern technology could do.

At first, the development of remote sensing technology was motivated primarily by military requirements. However, in the next few decades, it evolved into an interdisciplinary technology and became an integral part of Earth observation science. Nowadays, remote sensing technology is actively developed through many international projects. The Earth observation programs provide high-resolution data to everyone on an open and free-of-charge basis, thus allowing many new services in a public sector to be developed. The National Aeronautics and Space Administration (NASA) Space Science Data Coordinated Archive [5] lists more than 1300 spacecrafts launched in human history, which contributed to Earth science. Starting with Sputnik's journey over 60 years ago, the amount of data available for the environmental science has tremendously grown up in just a few decades due to the use of satellites. Clearly, even more advances are expected in the nearest future.

1.2   Satellite Remote Sensing

1.2.1   General Methodology

Formally, *remote sensing* is defined as acquisition of information about the state of a target by a sensor that is not in direct physical contact with it [6]. From the scientific point of view, satellite remote sensing is an interdisciplinary technology based on several fields, including atmospheric physics, optics, celestial mechanics, radiometry, electromagnetism, signal processing, environmental science, high-performance computing, optimization theory and data science. Currently, remote sensing has a lot of applications in atmospheric science, ecology, urban studies, geology, farming, forest cover monitoring, natural hazard prevention and economy.

Despite the various methods developed in remote sensing, it is possible to formulate a single basic scheme of measurement, shown in Fig. 1.1. A signal coming from the source of electromagnetic radiation interacts with the object of interest (atmosphere in our case), and after that it is measured by the instrument. Here we distinguish between *passive* and *active* remote sensing techniques. A passive sensor simply responds to the radiation coming from the natural source (such as the Sun, or thermal emission of the Earth's surface and atmosphere). To a certain degree, we are confronted with passive remote sensing in our daily life when visually observing an object in the daylight. In the active remote sensing technique, the radiation is generated and received by the instrument. The majority of systems operating in the visible and infrared parts of the electromagnetic spectrum are passive. Figure 1.1 shows passive and active remote sensing approaches. In the first scheme, a sensor measures the reflected sunlight. The solar radiation traveling through the atmosphere interacts with it and with the surface. Consequently, the radiation is perturbed, which is then captured by the sensor. In the second scheme, the terrestrial thermal emission

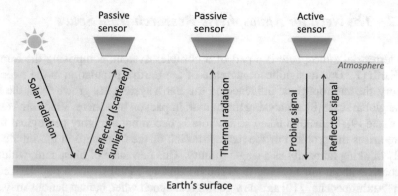

Fig. 1.1 Three schemes of remote sensing approach. Left and middle pictures refer to the passive remote sensing, while the right scheme shows the active remote sensing setup

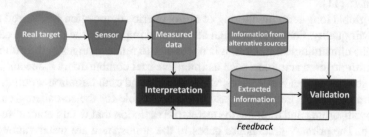

Fig. 1.2 A general scheme of the remote sensing-based approach

acts as a source of radiation, while in the third scheme, an active sensor has its own source of energy. Note that not only the spectrum itself but also the angular distributions of the radiation at a given location contain information about atmospheric constituencies.

Usually, the parameters of interest are not measured directly. The measurements have to be processed and interpreted in order to obtain the information of interest. In general, it is customary to distinguish between data and information. In remote sensing of the atmosphere from space, information about the atmosphere is extracted from satellite data. It is schematically illustrated in Fig. 1.2. Obviously, we need a sort of interpretation block which converts the data into information. In fact, this book is focused on how to perform such a transformation. The extracted information has to be validated against other sources of information, e.g., by comparing with ground-based measurements. The validation process allows to estimate the uncertainty of the obtained results and then to suggest improvements for the interpretation methods and instrumentation.

1.2.2   The Need for Atmospheric Research from Space

The terrestrial atmosphere is a part of a complex system, comprising the Sun and the Earth [7]. Due to significant increase of the Earth's population in the twentieth century, the anthropogenic influence on the Earth system has grown from the local to the global scale [8]. Analyzing the human impact on biosphere, Vladimir Vernadsky wrote [9]: "Mankind taken as a whole is becoming a mighty geological force. There arises the problem of the reconstruction of the biosphere in the interests of freely thinking humanity as a single totality. This new state of biosphere, which we approach without our noticing it, is the noosphere". Paul Crutzen popularized the term "Anthropocene" [10] as a new geological epoch when human actions may have a drastic impact on the Earth. In this regard, an international environmental policy is required to assess and minimize the anthropogenic impact on the environment and the climate [11].

The global long-term monitoring of atmospheric composition is required to control the air quality. For environmental studies and improving the predictive capabilities of the climatological models, it is necessary to relate changes in the atmosphere with anthropogenic activities, such as intensive fuel combustion. Of special interest are the so-called greenhouse gases, which absorb and emit radiation within the thermal infrared range. Greenhouse gases are responsible for the so-called greenhouse effect by absorbing and trapping the terrestrial radiation that would otherwise escape to space. The primary greenhouse gases in the atmosphere are water vapor (H_2O), carbon dioxide (CO_2), methane (CH_4), nitrous oxide (N_2O) and ozone (O_3). The change in their concentrations can cause an imbalance between the energy received and emitted by the Earth. Consequently, the radiative equilibrium between the Sun and the Earth is shifted, resolving in new values of the mean temperature. There is an estimation that without greenhouse gases, the average temperature of Earth's surface would be about $-18\,°C$. One consequence of the average temperature increase is the melting of glaciers. In its turn, it may lead to the global sea level rising and natural disasters.

Remote sensing instruments on-board space-based platforms are extremely useful for measuring regional variations of the atmospheric composition on a global scale. As opposite to airborne and ground-based sounding with limited geographical and temporal coverage, the region of interest can be observed from satellites frequently and repetitively with low costs per unit surface area coverage. The data can be acquired even from the locations that are difficult to access and where it is problematic due to various reasons to maintain observation facilities. Having long-term observation archives, it is possible to estimate the efficiency of climate protection agreements (e.g., the Montreal protocol [12] and the Kyoto protocol [13]). In fact, the spaceborne remote sensing is a key instrument to analyze our planet as a global interdependent system of the Earth "spheres" interacting with each other—atmosphere, hydrosphere, cryosphere, biosphere, magnetosphere and lithosphere. The ability to make observations of the Earth globally facilitated the development of Earth system science [14], which treats the whole planet as an integrated system consisting of physical, chemical, biological and anthropogenic (noosphere) interactions.

Remote sensing provides valuable information for weather forecasting [15], including cloud identification, monitoring of cyclones, air quality, observations of sandstorms and volcanic eruptions. Spaceborne observations are also used to study the melting of glaciers, ice and snow extent and albedo [16]. They play a vital role in systems of real-time monitoring and managing natural disasters [17], thereby contributing to the sustainable development of society and protecting the Earth's environment.

1.2.3 Satellite Orbits

In this section we distinguish between polar/near-polar orbits and geostationary orbits [18]. Suppose that a satellite of mass m travels around the Earth in a circular orbit of radius r. The mass of the Earth is M. Then the satellite experiences a gravitational force of GMm/r^2 with G being the gravitational constant. The gravitational force is responsible for causing the normal acceleration of the satellite, given by $r\omega^2$, where ω is the angular velocity. Thus, Newton's second law of motion is written as

$$G\frac{mM}{r^2} = mr\omega^2. \tag{1.1}$$

Expressing ω from Eq. (1.1), the period of revolution $T = 2\pi/\omega$ can be found:

$$T = 2\pi\sqrt{\frac{r^3}{GM}} \tag{1.2}$$

or, since $r = R_E + h$, where R_E is the Earth radius and h is the height above the surface of the Earth,

$$T = 2\pi\sqrt{\frac{(R_E + h)^3}{GM}}. \tag{1.3}$$

Then, the orbital speed v is computed as

$$v = \frac{2\pi(R_E + h)}{T} = \sqrt{\frac{GM}{R_E + h}}. \tag{1.4}$$

To simplify Eq. (1.4) we note that

$$G\frac{mM}{R_E^2} = mg \tag{1.5}$$

with $g \approx 9.8 \, \text{m/s}^2$ being the free fall acceleration. Hence

$$GM = g R_E^2. \tag{1.6}$$

Substituting Eq. (1.6) into Eq. (1.4) we obtain

$$v = \sqrt{\frac{g R_E^2}{R_E + h}}. \tag{1.7}$$

If $h \ll R_E$, then $v = \sqrt{g R_E} \approx 7.9 \, \text{km/s}$—that is the so-called first orbital speed, i.e., the speed at which the satellite orbits the Earth.

In *a polar orbit*, a satellite passes above the Earth poles with an inclination very close to 90° at the equator. Usually, polar-orbiting satellites are placed at the height h from 800 to 900 km above the surface of the Earth, where the air resistance is negligible. Using Eq. (1.3) it can be estimated that a period T is of about 90–100 min. A satellite in a polar orbit will pass over the equator at a different longitude on each of its orbits due to the Earth's rotation. Polar orbits are often used for Earth observations, during which the Earth is monitored at different locations as time passes. Polar orbits are usually preferred because of global coverage requirements. In near-polar orbits, the satellites travel northwards on one side of the Earth—that is the ascending pass. Then they travel southwards on the second half of the orbit—that is the descending pass. Among near-polar orbits one can distinguish Sun-synchronous orbits, meaning that each successive orbital pass occurs at the same local time of day. In this case, the ascending pass is most likely performed on the shadowed side of the Earth.

The revolution period increases with the altitude. If $h = 35786 \, \text{km}$, then T is exactly one sidereal day (23 h, 56 min, 4 s). Assume we place a satellite at that height. If the satellite moves along the orbit in the equatorial plane and follows the direction of Earth's rotation, then it remains over the same point on the Earth surface and has a stationary footprint on the ground. That is a *geostationary orbit*. Geostationary orbits have an inclination of 0° with respect to the Earth's equator.

A geostationary orbit is a particular type of *geosynchronous orbit*. An object in geosynchronous orbit returns to the same point in the sky at the same time each day. Geosynchronous orbits have varying inclinations.

Other regimes are also used for Earth remote sensing. For instance, the Deep Space Climate Observatory (DSCOVR) with the Earth Polychromatic Imaging Camera (EPIC) on board is at the Sun-Earth L_1 Lagrangian point (that is about 1.5 million km from the Earth) [19]. This is the so-called Lissajous orbit. In Lagrangian points, the satellite maintains its position relative to the Earth and the Sun. Figure 1.3 shows the example of the images captured by the EPIC instrument.

Fig. 1.3 Example of the EPIC/DSCOVR image, lunar transit (*credits* https://epic.gsfc.nasa.gov/galleries/)

1.2.4 Scan Modes, Geometry of Observation and Swaths

The geometry of observations is defined by three angles, as shown in Fig. 1.4. The sunlight strikes Earth's atmosphere and surface at the solar zenith angle (SZA). The satellite receives the signal at the viewing zenith angle (VZA). The relative azimuthal angle (RAA) describes the angle between the plane comprising the incident direction and the normal vector to the surface, and the plane incorporating the viewing direction and the normal vector to the surface. Note that the convention for RAA can differ in different applications (for instance, by π). Moreover, more accurate definition may involve both satellite and sun azimuth angles, and the measurement can be dependent on both angles (and not just on their difference) due to the scene heterogeneity.

There are several types of observational geometries, as shown in Fig. 1.5:

1. In *nadir* geometry, the atmosphere directly under the sensor is probed. This observational mode is used by many spaceborne remote sensing instruments.
2. In *limb* measurement mode, the sensor looks tangentially to the Earth's surface toward the edge of the atmosphere.
3. In solar/lunar occultation mode, the sun/the moon is observed directly through the atmosphere in the limb geometry.

The geometry of observation strongly influences the spatial vertical and horizontal resolution of the measurement. For nadir sounding, the collected signal is influenced by the parameters of the atmosphere along the viewing direction. However, the sensitivity of the measurement is low in the vertical direction, while the instantaneous field of view (IFOV), i.e., the angle subtended by a single detector element on the axis of the optical system, can be small. Typical spatial resolution of nadir measurements can be ∼10 km vertical and ∼10 km horizontal for remote sensing of gases

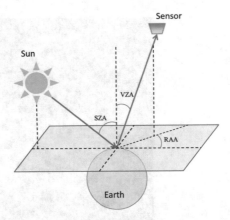

Fig. 1.4 Geometry of the measurement: solar zenith angle (SZA), viewing zenith angle (VZA) and relative azimuth angle (RAA)

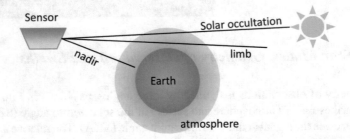

Fig. 1.5 Nadir, limb and solar measurement modes

and ~1 km for aerosol and ~0.1 km for land remote sensing. As opposite to nadir, limb observations are highly sensible to the vertical distribution of the parameters. The spatial resolution of limb measurements can reach 1 km in vertical. However, the signal is collected along the tangent path, which is quite large (~600 km), providing poor spatial resolution in horizontal direction.

The swath is the area "seen" by the sensor. While the satellite travels, the observed area forms the "strips", as shown in Fig. 1.6. Typically for spaceborne instruments, the swath widths vary between tens and hundreds of kilometers. The swath widths for some spaceborne instruments used for remote sensing of the atmosphere are shown in Table 1.1. In the case of polar orbits, the satellite swath covers a new area with each consecutive pass due to Earth's rotation, thereby providing global coverage of the Earth's surface. The regions close to poles can be imaged more frequently than those near to the equator due to the overlap in adjacent swaths of the orbits. The viewing angle of the sensor can be adjusted allowing off-nadir measurements. That option increases the field of view and the revisit capacity, meaning that the time between two consecutive measurements over a certain location is decreased.

Table 1.1 The swath width for several instruments

Instrument	Swath width (km)
SCIAMACHY	960
GOME	960
POLDER	2400
GOME-2	1920
MODIS	2330
OLCI	1270
SLSTR	1400
OMI	2600
TROPOMI	2600

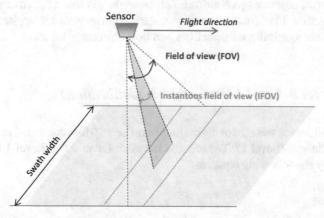

Fig. 1.6 The field of view and the instantaneous field of view. The viewing angle of the sensor is changing, thereby allowing off-nadir viewing

1.3 Data Processing in Remote Sensing

1.3.1 General Scheme

In remote sensing, quantity of interest is not measured directly. The signal received by the instrument goes through several processing stages before the information about a certain parameter is retrieved and can be used in applications. A common set of the so-called "levels" is adopted to describe the stages of data in the processing chain. The latter is broken down into sub-processes with corresponding inputs and outputs. Thus, each process can be considered independently. These levels are the following:

1. Level 0: raw unprocessed instrumental data, just as it is collected at the sensor; typically counts (which are proportional to the number of registered photons by the detector) or volts;

2. Level 1: converted Level 0 data into energy units (typically radiance);
3. Level 2: retrieved geophysical parameters (e.g., trace gas concentrations) at the same resolution and location as Level 1 data;
4. Level 3: Level 2 data mapped onto uniform space/time grid scales, making possible to combine data from different sources;
5. Level 4: data, conclusions, results, derived by using Level 3 data.

The conversion of Level 0 data into Level 1 data is called absolute radiometric calibration of the sensor. Sometimes, several sublevels are distinguished (e.g., Level 1A and Level 1B) describing intermediate corrections applied to the data. For instance, a typical remote sensor has several light detecting devices, which might look at different locations and operate in different wavelength regions. To obtain consistent measurements, these detectors have to be calibrated with their own calibration coefficients (the detector equalization). This procedure is also referred to as the radiometric correction. In addition, several geometrical corrections are applied to remove distortions due to misaligned scan lines and non-uniform pixel sizes.

1.3.2   Level 0–Level 1 (Radiometric Calibration)

The recorded sensor voltages or digital numbers (Level 0) are converted to an absolute scale of radiances (Level 1). The relation between Level 0 and Level 1 data can be described by the following equation:

$$S = k \int_{\lambda_1}^{\lambda_2} L_\lambda r_\lambda d\lambda + O, \tag{1.8}$$

where S is the output of the sensor (counts or volts), k is the instrument peak intensity, r_λ is the relative spectral responsivity of the instrument, L_λ is the spectral radiance, O is the sensor offset, and λ_1 and λ_2 are the lower and upper bandpass limits, respectively, determining the spectral resolution of the instrument. Here it is assumed that the radiance is spatially uniform at the instrument entrance aperture. Effects related to polarization and finite integration time are neglected as well. It follows from Eq. (1.8) that the obtained quantity from the sensor is an integral over a given bandpass, which means fast variations of L_λ along λ will be smoothed and might not appear in the measurements. The average spectral radiance $\bar{L}_\lambda$ over the spectral wavelength bandpass is

$$\bar{L}_\lambda = \frac{\int_{\lambda_1}^{\lambda_2} L_\lambda r_\lambda d\lambda}{\int_{\lambda_1}^{\lambda_2} r_\lambda d\lambda}. \tag{1.9}$$

Introducing the gain as

$$G = k \int_{\lambda_1}^{\lambda_2} r_\lambda d\lambda, \tag{1.10}$$

Equation (1.8) is rewritten as follows:

$$S = G\bar{L}_\lambda + O. \tag{1.11}$$

Thus, to compute $\bar{L}_\lambda$, it is sufficient to know the gain, the offset and the wavelength spectral bandpass. By exposing the sensor to two levels of known spectral radiance, the gain and the offset can be determined by using Eq. (1.11). This procedure is called sensor calibration. Note that this description is simplified. In practice, calibration involves several steps such as prelaunch calibration and on-board measurements. In addition, the measurements over known targets at the Earth's surface can be involved. But the idea behind remains the same, i.e., to use known values of radiances in order to establish calibration coefficients (in practice, there can be more than two coefficients). Note that the conversion from Level 0 to Level 1 data eliminates a hardware-related details and provides fundamental physical quantities (e.g., the spectral radiances or the Stokes vector at the top of the atmosphere). Thus, the hardware-specific features and physical modeling are decoupled in the processing chain.

1.3.3 Level 1–Level 2 (Retrieval)

Level 1 to Level 2 conversion is referred to as retrieval. An atmospheric processor is an algorithm which relates physical parameters of the atmosphere to radiances. In general, a forward model is defined as a quantitative tool for simulation of observables for a given set of model parameters [20]. Here observables refer to the quantities which can be measured. For example, observables could be the measured spectra. Therefore, forward modeling in remote sensing is computing the radiances for a given atmospheric state. Inversion means retrieval of the atmospheric parameters from the spectral radiances.

Forward modeling starts with the conversion of the physical parameters of the atmosphere to the optical parameters through a physical adapter. Usually for this, the atmosphere is discretized into several atmospheric layers. The obtained atmospheric model can be then treated as an optically active system/medium. The boundary conditions should be defined. In most cases, the incident solar irradiance at the top of the atmosphere and the properties of the underlying surface are assumed to be known. Then the radiative transfer model (RTM) is used to compute the scattered radiance field. Mathematically the RTMs are based on the radiative transfer equation, which is considered in detail in Chap. 4. After the radiance field is computed, the post-processing step may be needed to convert the obtained result into quantities which are required in a certain application (e.g., the spectral radiance convolved with the slit function of the instrument). These steps are shown in Fig. 1.7. The clockwise direction corresponds to the forward modeling.

Opposite to forward modeling is inversion, which corresponds to the anticlockwise direction. Here, the spectral radiances are given and used as inputs. By inversing the radiative transfer model, we come from the radiances to the optical parameters of

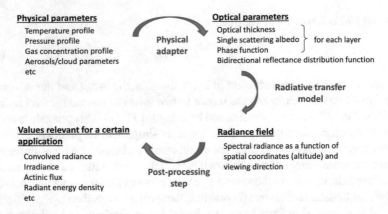

Fig. 1.7 Processing chain in the forward modeling

the atmosphere, and then to its physical parameters. Note that the inversion of the radiative transfer model is not a trivial step, which requires special approaches. They are considered in Chap. 5 of this book. So far, we just note that the inversion procedure is application-specific. The extraction of information about atmospheric constituents and surface properties is called *retrieval*. For remotely sensed imagery it is important to remove the effects of the atmosphere on the signal coming from the surface. Retrieval of surface image from the signal with the atmospheric effects removed is called *atmospheric correction* [21].

1.3.4 Big Data in Remote Sensing

The recent developments in optics, sensor design and measurement techniques significantly improved the characteristics of hyperspectral atmospheric composition sensors, such as the spatial resolution and the signal-to-noise ratio [22]. Table 1.2 shows a comparison between previous generation instruments (Global Ozone Monitoring Experiment (GOME) and GOME-2) and the newest The TROPOspheric Monitoring Instrument (TROPOMI) [23] on-board of the Copernicus Sentinel 5 Precursor (S5P) satellite. The spatial resolution of TROPOMI is two orders of magnitude higher, providing 21 million level-1B spectra per day, i.e., almost 8 billion spectral points, while the signal-to-noise ratio in the UV/Vis channels reaches the values of about 1500. Figure 1.8 shows the example of a retrieved map of tropospheric nitrogen dioxide from S5P measurements (the example of the script for reading S5P data can be found in Appendix 1). Air pollution emitted by big cities and shipping lines is clearly visible. It is possible to detect air pollution over individual cities using high spatial resolution data as well as to locate where pollutants are being emitted, and so, identifying pollution hotspots. Such high spatial resolution satellite remote sensing observations are extremely useful for diagnosing the impact of atmospheric con-

Table 1.2 Characteristics of atmospheric composition sensors

Instrument	GOME	GOME-2	TROPOMI
Platform	ERS-2	MetOp (A, B, C)	Sentinel-5 Precursor
Spatial resolution (km^2)	320×40	80×40	7×3.5
Amount of Level-1 data (TBytes per year)	0.8	4.2	240
Operational	1995–2011	2006-present	2017-present

stituents on a global scale, in particular, allowing detection of small-scale sources, and increasing the fraction of cloud-free observations. However, the high spatial resolution of the state-of-the-art atmospheric composition sensors results in very challenging data volumes to be processed—hundreds of TBytes per year of level-1 data.

The traditional approaches implemented in the current atmospheric processors are inadequate to deal with a large amount of data. As a matter of fact, the amount of satellite data increases faster than the computational power [24]. The remote sensing data is recognized as Big Data [25] since it satisfies Doug Laney's 3V criterion [26]: significant growth in the volume, velocity and variety. Volume is the most natural characteristic of Big Data. It refers to the data scale, i.e., physical amount of data. Velocity means the speed at which the data is created. For instance, near-real-time Level-2 products should be available already two hours after sensing. Variety means different forms of data. Broadly speaking, data about atmospheric composition which is used in applications is not provided by a certain instrument, but rather it is a compilation of data coming from different sources (several satellites and ground stations) equipped with model predictions. A complete analysis of all available data is a challenging task and is at the border of state-of-the-art computational capabilities. Sometimes, 3V criterion goes together with the 4th V, namely veracity, which means uncertainty in the data. Indeed, data coming from different sources is often misaligned in time and space. Some pieces of data might be damaged. The reliability of the data strongly depends on many factors, such as weather conditions, performance and state of the instrument. On top of these Vs is another V, which refers to "value": quantity is transformed into quality. Substantial value can be discovered in remote sensing big data, including understanding the global changes of climate, optimizing processes of agriculture, improving quality of life, etc.

In this regard, new efficient techniques are to be developed for the next-generation atmospheric processors to cope with these high-efficiency requirements. The design of new generation atmospheric processors can be regarded as a disruptive development. In fact, new computational paradigms are required to be able to understand a massive amount of remote sensed data [27] including integration of data from the sources in different locations, distribution of data across massively parallel platforms for processing and providing fast access to remote sensing data to users worldwide.

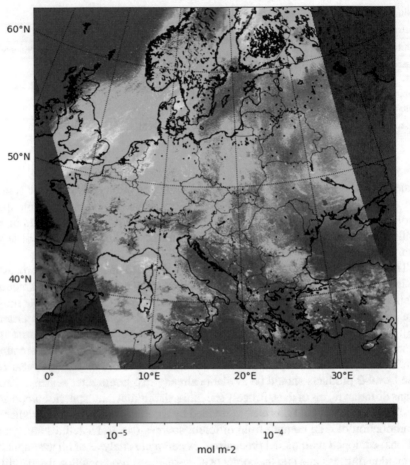

Fig. 1.8 Example of Sentinel-5P tropospheric nitrogen dioxide (NO_2) measurements over Europe (data is freely available at [28])

1.4 Electromagnetic Spectrum

In most remote sensing techniques, the electromagnetic radiation is used as the main information carrier. According to the concept of wave-particle duality, the electromagnetic radiation phenomenon can be partly described by using wave theory, and partly, by particle theory. In the first case, the electromagnetic energy propagates through space in the form of waves, while in the second case the propagation is modeled as the movement of discrete corpuscles called photons. The wave is characterized by its wavelength λ, while the photon possesses the momentum p. The de Broglie hypothesis establishes a link between these two concepts and relate the wave properties, i.e., λ, with corpuscle properties, i.e., p, through the following relation:

Table 1.3 Electromagnetic spectrum regions

Name	Wavelength	Frequency	Photon energy
Gamma-ray	<0.01 nm	>30 EHz	>124 keV
X-ray	0.01–10 nm	30 PHz–30 EHz	1.24 eV–124 keV
Extreme ultraviolet	10–120 nm	2.5 PHz–30 PHz	10.35 eV–124 eV
Far ultraviolet	120–200 nm	1.5 PHz–2.5 PHz	6.21 eV–10.35 eV
Middle ultraviolet	200–300 nm	1.0 PHz–1.5 PHz	4.14 eV–6.21 eV
Near ultraviolet	300–400 nm	750 THz–1.0 PHz	3.1 eV–4.14 eV
Visible	400–700 nm	420 THz–750 THz	1.77 eV–3.1 eV
Near infrared	700–3500 nm	86 THz–420 THz	0.35 eV–1.77 eV
Thermal infrared	3.5–100 μm	3 THz–86 THz	0.012–0.35 eV
Microwaves	1 mm–1 m	300 MHz–300 GHz	1.24 μeV–1.24 meV
Radio waves	1 mm–10000 km	30 Hz–300 GHz	1.24 meV–124 feV
Extremely low frequency	>10000 km	<30 Hz	<12.4 feV

$$\lambda = \frac{h}{p}, \tag{1.12}$$

where h is Planck's constant ($6.626 \cdot 10^{-34}$ J·s). According to the Planck–Einstein relation, the energy E_p carried by the photon can be expressed through its wavelength λ or frequency v as

$$E_p = \frac{hc}{\lambda} = hv, \tag{1.13}$$

where c is the speed of light ($\approx 3 \cdot 10^8$ m/s in vacuum). The standard unit of energy in particle physics is electron volt, 1 eV is approximately equivalent to 1.602 10^{-19} Joule.

The electromagnetic spectrum can be divided into spectral regions listed in Table 1.3. Typically, X-rays and gamma rays are not used in terrestrial remote sensing because of atmospheric opacity. Due to small wavelengths, this part of the electromagnetic spectrum can be used for studying crystal structures and surface state analysis in laboratory measurements or during studies of planetary surfaces with no atmosphere (e.g., the Moon). Gamma radiation with photon energy of 10^{11}–10^{14} eV is called very-high-energy gamma rays [29]. They are used for studies of white dwarfs and neutron stars. X-rays that are present in the solar spectrum have energy higher than the binding energy of any element (including He with the highest ionization energy of 24.59 eV), and so they can ionize all atmospheric species. The extreme ultraviolet is generated by the solar corona (extreme ultraviolet), highly absorbed in air ionizing O_2 and N_2. X-rays and extreme ultraviolet comprises the ionized part of the upper atmosphere, namely the ionosphere. The extreme ultraviolet region ends with the hydrogen Lyman-alpha spectral line at 121.6 nm. The far UV region is absorbed by O_2, while the middle ultraviolet region is mostly absorbed by the ozone layer located around 25 km above the sea level. The near-ultraviolet can reach the Earth's surface and cause burn of the skin. It can also cause cataracts, immune system suppression and genetic damage, resulting in skin cancer. However, moderate sun exposure is important for human health as it stimulates the synthesis of vitamin D in the skin [30]. Far and part of middle ultraviolet regions are considered to be lethal to biosphere.

The part of radiation which is visible to the human eyes belongs to the 400–700 nm spectral range, which is roughly divided into the following wavelength sub-regions:

- violet 400–430 nm;
- indigo 430–450 nm;
- blue 450–500 nm;
- green 500–570 nm;
- yellow 570–590 nm;
- orange 590–610 nm;
- red 610–700 nm.

The visible part is used in imaging techniques, mapping and multispectral photography. The clouds and aerosols have a very strong impact on the spectrum. This part of the electromagnetic spectrum can be used for both passive and active remote sensing techniques.

The main gaseous absorbers of near-infrared radiation are oxygen, water vapor and carbon dioxide. The thermal infrared part can be used for thermal imaging and retrieval of land and surface water temperatures. The emission of thermal radiation is an important mechanism of radiation cooling, and so this part of the spectrum has a strong impact on the global climate.

The microwaves are used by passive and active sensors. One of the advantages of this spectral range is that microwave radiation can penetrate through clouds. It is reflected by the ionosphere. In passive microwave remote sensing, a sensor detects the naturally emitted microwave energy. Note that the amount of energy in this region is small as compared to the optical region. Consequently, the area from which the signal is collected should be large enough. For this reason, the spatial resolution of the microwave sensors is usually lower than those operated in the visible region. Active microwave sensors can be imaging and non-imaging. The imaging sensors are called RADARs (radio detection and ranging). They can be used for imaging at any time, day or night, and in most weather conditions. The time between the transmitted and reflected signals (and the corresponding phase difference) determines the distance to the target. Non-imaging sensors (e.g., Meteorological Temperature Profiler MTP-5) serve the same purpose.

In this book we are focused on the passive remote sensing techniques in the ultraviolet, visible and near-infrared spectral regions.

1.5 Instruments in Remote Sensing

1.5.1 Measurement Principles

We consider two techniques for decomposing the radiation into a spectrum. The first technique relies on the spatial separation of the spectrum, i.e., *the selective filtering*. The second method is *selective modulation*.

It is known that the simplest way to perform the spatial separation is to employ a prism. The refractive index varies with the wavelength (a phenomenon known as dispersion). Since the light of different wavelengths is refracted differently, it escapes from the prism at different angles. More advanced technology is the diffraction grating, in which a plate with a periodic surface modulation is employed. It creates multiple slit diffraction patterns with diffraction maxima associated with a given wavelength. That is the basis for the so-called whisk-broom (i.e., scanning) and push broom (i.e., staring) instrument imaging spectrometers, as shown in Fig. 1.9. The whisk-broom technology uses the scan mirror, so the pixels along the scan direction are measured one by one. Such scanning concept with 1D detector arrays is used in GOME, GOME-2 and SCIAMACHY.

Unlike the whisk-broom, the push-broom technology allows measuring of the spectra from several pixels along the 1D array simultaneously. The staring push-broom concept measures all ground pixels in the swath simultaneously and allows improved spatial resolution. TROPOMI uses an improved push-broom concept based on a 2D charge-coupled device (CCD) sensor technology [31]. That means, the spatial information is resolved over one side of the CCD sensor, while the spectral information is resolved over another side, as shown in Fig. 1.10.

Another principle used in remote sensing measurements is based on the selective modulation of the incoming signal. The most common scheme is based on the Michelson interferometer (see Fig. 1.11), which consists of a light source, a fixed mirror, a translating mirror, a beam splitter and a detector. A beam splitter divides the light into two beams. After reflection by the mirrors, they are merged by the same beam-splitter and sent to the detector. By moving one of the mirrors, one of the reflected beams travels an additional distance corresponding to the so-called optical path difference Δz.

Let us consider the case of a monochromatic source with wavelength λ_0. The intensity L at the detector is given by

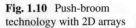

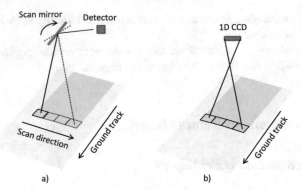

Fig. 1.9 Illustration of the whisk-broom (**a**) and push-broom (**b**) technologies

Fig. 1.10 Push-broom
technology with 2D arrays

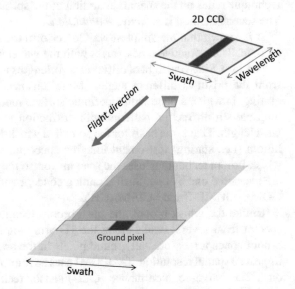

$$L\left(\Delta z\right) = \frac{L_0}{2}\left(1 + \cos\left(\omega_0 x\right)\right),\tag{1.14}$$

where L_0 is the intensity of the source, and $\omega_0 = 2\pi/\lambda_0$. From Eq. (1.14) it is obvious, that if $x = n\lambda_0$ with n being integer number, then the recombined beams will produce constructive interference, while if $x = (n + 0.5)\lambda_0$ the destructive interference is produced and the intensity at the detector is zero. The intensity $L\left(\Delta z\right)$ measured as a function of the optical path difference is called the interferogram. Let the mirror moves from location $-a$ to a. Applying the Fourier transform to Eq. (1.14), one derives:

$$L\left(\omega, L\right) = \frac{L_0}{2\pi}\int_{-a}^{a}\cos\omega_0 x \cdot \cos\omega x dx\tag{1.15}$$

Fig. 1.11 The Michelson
interferometer

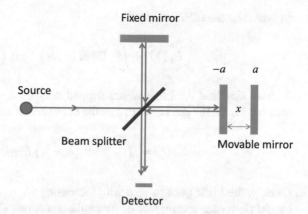

and after rearranging,

$$L\left(\omega, L\right) = \frac{L_0}{2\pi}\left(\frac{\sin\left(\omega - \omega_0\right)a}{\omega - \omega_0} + \frac{\sin\left(\omega + \omega_0\right)a}{\omega + \omega_0}\right). \tag{1.16}$$

The second term in Eq. (1.16) rapidly decays and can be neglected, while the first term produces a sharp peak at $\omega = \omega_0$. Therefore, in the case of monochromatic light, we will see a narrow line at $\omega = \omega_0$. The width of the line decreases with increasing a. Thus, a determines the spectral resolution of the instrument. In the case of multispectral source, the spectrum can be retrieved from the interferogram by applying the inverse cosine Fourier transform expressed by

$$L\left(\omega\right) = \int_{-\infty}^{\infty}\left(L\left(x\right) - \frac{1}{2}L\left(0\right)\right)\cos\omega x dx. \tag{1.17}$$

An important advantage of Fourier spectrometers is due to the fact that they perform simultaneous measurements of all wavelengths (although at the cost of complex math required for analysis)—this is the so-called Fellgett's advantage [32]. In addition, they have fine spectral resolution (of about $0.002\,\text{cm}^{-1}$), making them useful for measurements in the infrared spectral range.

1.5.2 The Instrumental Spectral Response Function

The instrumental spectral response function (ISRF), also referred to as the slit function, describes the spectral response of the detector to a monochromatic signal. Typically, the instrument "sees" the monochromatic signal at the wavelength λ_0 as a certain distribution ISFR (λ, λ_0). The observed spectrum $L_o\left(\lambda\right)$ is equal to the spectrum at the sensor aperture (e.g., collected at the top of the atmosphere (TOA) $-L_{\text{TOA}}$)

weighted by the ISFR:

$$L_o(\lambda) = \int_0^\infty \text{ISFR}(\lambda, \lambda') L_{\text{TOA}}(\lambda') d\lambda'. \tag{1.18}$$

Assuming that the ISFR values depend only on $\lambda - \lambda'$ over the whole spectral interval, Eq. (1.18) can be written as the convolution explicitly

$$L_o(\lambda) = \int_0^\infty \text{ISFR}(\lambda - \lambda') L_{\text{TOA}}(\lambda') d\lambda'. \tag{1.19}$$

Usually, the ISFR goes to zero with increasing $(\lambda - \lambda')$ and can be approximated by the Gaussian, rectangular or triangular functions [33]. Thus, the integration in Eq. (1.19) is performed over the finite wavelength domain. The ISFR is determined by the entry and exit slits, gratings and detector properties of the instrument. The slit function is usually available in technical reports on the instrument.

Although Eq. (1.19) looks similar to Eq. (1.8), they describe different phenomena. Equation (1.19) describes the smoothing of the measured signal due to imperfection of the instrument in the continuous wavelength space, while Eq. (1.8) represents the translation from the continuous wavelength space to the discrete wavelength space through averaging over the bandpass and subsequent conversion to the output signal of the sensor.

1.5.3 Spatial, Temporal and Spectral Resolution

The important characteristics of remote sensing data are spatial, temporal and spectral resolution [34].

Spatial resolution defines the smallest object that can be resolved by the sensor and refers to the number of pixels of the digital image. An image having higher spatial resolution contains a greater number of pixels per unit area than those of lower spatial resolution, thereby providing more information. However, as the number of sensor pixels increase, each pixel receives less light and so the data becomes noisier.

Temporal resolution corresponds to the frequency with which a measurement is performed over the same area. It is determined by the satellite orbit and the sensor design. Modern Sun-synchronous orbit satellites can revisit the same part of the Earth's surface each day.

Spectral resolution is defined by the number and width of spectral bands of the instrument. Hyperspectral sensors take the measurements in narrow spectral bands over a continuous spectral range. These data can be utilized to extract information about atmospheric constituencies with characteristic spectral traces. Multispectral instruments have tens of wider spectral bands, which often correspond to different colors (e.g., blue, green and red). A so-called panchromatic (black and white) band is one wide band. It cannot "distinguish" the colors, so the data can be visualized as the gray-scale image. As the signal is summed up across the wide band instead of being partitioned into spectral bins, the resulting luminosity and a signal-to-noise ratio are much higher compared to multispectral bands enabling much sharper images at higher spatial resolution.

There is also radiometric resolution which refers to the number of different output numbers in the data. It is determined by the number of bits the recorded radiation is represented. For instance, the radiometric resolution of the sensor MODIS is 12 bits, meaning that the measurement value can range from 0 to $2^{12} = 4096$ for each pixel.

1.5.4 Signal-to-Noise Ratio

Noise is a critical factor in remote sensing data processing, as it can "hide" the valuable information of the measurements and even lead to misinterpreted results. Noise can be understood as random fluctuations, which are added to the signal received by a detector. Typically, the so-called additive noise model is adopted:

$$S = S_0 + N, \tag{1.20}$$

where S is the measured signal, S_0 is the ideal, i.e., noise free, signal, while N is the noise term. The noise level is expressed by using the signal-to-noise ratio (SNR) [35]. For simplicity it can be assumed that the noise has zero-mean and is not correlated with the signal. The SNR is an important parameter of the sensor, which characterizes the impact of the noise on the measured signal. Essentially, it tells about the stability of the instrument and the quality of the data. Higher SNR values lead to better results of retrieval algorithms. The SNR parameter can be increased by improving the design of the sensors, e.g., cooling of the detector and increasing the instrument aperture. The design of an instrument with high SNR is not an easy task, as the SNR parameter behaves in opposition to other sensor parameters (e.g., spectral resolution, since less signal is received by the latter's increase). In principle, SNR, IFOV, sampling time, spectral resolution and special resolution have to be balanced. Certainly, such optimization is always application-specific.

1.5.5 Spectrometers and Imagers

The instruments used in remote sensing applications fall into one of two categories, namely *spectrometers (hyperspectral instruments)* and *imagers (multispectral instruments)* [36]. Actually, since every image can be considered as a sort of a spectral image, the distinction between spectrometers and imagers is not fundamental. They both employ the same measurement principles outlined in the previous sections.

Table 1.4 Atmospheric trace gases and corresponding spectral ranges used for retrieval from hyperspectral measurements (data according to [38] and references therein)

Target parameter	Spectral range (nm)
SO_2	315–327
O_3	325–335
BrO	335–350
NO_2	425–450
H_2O	688–700, 2353–2368
CH_4	1627–1671
CO_2	1558–1594
CO	2321–2335

However, the ideas put behind the design of the instruments have some peculiarities, which deserve some words of explanation.

In simple words, a spectrometer acquires the data at multiple narrow and contiguous spectral bands, while an imager captures the image data in a few wide spectral bands within certain spectral intervals. Thus, the data provided by spectrometers can be regarded as a spectrum, while it is more natural to think of data of imagers as an image. The spectrometers are designed to measure certain absorption bands. Here the spectral resolution is a critical parameter. The spectral ranges in which the measurements are performed can be quite narrow. On the contrary, the imagers are designed to capture the signal within broad spectral ranges, at the cost of poorer spectral resolution than that of spectrometers. Several spectral channels can be identified in the data provided by imagers; each channel is characterized by the mean wavelength, the spectral response function, and the width of the channel. Here, the similarity with TV monitors working with blue, green and red channels can be seen.

The fine resolution of the data provided by spectrometers make the measurements very sensible to the gaseous composition of the atmosphere. The optical characteristics of the atmosphere within a given absorption band can vary by several orders of magnitude, thereby making the measurements extremely sensible to the gaseous compounds and at the same time giving the possibility to filter out the influence of other factors. It is also possible to retrieve information about the mineral composition of the surface [37]. The signal-to-noise ratio is extremely important for hyperspectral measurements. In particular, the SNR is expected to be of magnitude of about 1000 for the state-of-the-art instruments. Table 1.4 lists some atmospheric trace gases and corresponding informative spectral ranges.

The multispectral scanners measure the reflected sunlight for several wavelength bands. Each band is related to a specific property of the Earth's surface (the land cover, the snow cover, the vegetation state, the soil type etc.). For example, the infrared channels provide information on vegetation, its type and the fraction of vegetated land. The multispectral data acquired at different channels provide a contrast

Table 1.5 Spectral ranges and parameters, which can be retrieved from multispectral measurements (data is taken from [39])

Spectral range (nm)	Usage
430–450	Coastal and aerosol studies
450–500 (blue)	Distinguish soil from vegetation
530–590 (green)	Emphasizes peak vegetation—useful for assessing plant vigor
640–670 (red)	Vegetation slopes, the detection of vegetated areas
770–900	Biomass content and shorelines
1550–1750	Moisture content of soil and vegetation

image used for a spatial distinction and classification of areas. Some bands of the LANDSAT-8 instrument and their applications are shown in Table 1.5.

In the following sections, we briefly outline some spectrometers and imagers, which played or play a vital role in getting the data about atmospheric composition and surface parameters that became an integral part of the Earth's remote sensing history.

1.5.6 Spectrometers

Spectrometers in atmospheric remote sensing were primarily designed for retrieving the following trace gases: BrO, CH_4, ClO, H_2O, HCHO, N_2O, NO, NO_2, O_3 and SO_2. In this regard, the important characteristics of the instrument are spectral ranges it covers, number of channels and spectral resolution.

The Global Ozone Monitoring Experiment (GOME) instrument is a UV/VIS grating spectrometer. It was launched in 1995 on board the European Space Agency's (ESA) second Earth Remote Sensing Satellite (ERS-2). The instrument GOME is a four-channel spectrometer measuring the scattered solar radiation in the wavelength range of 240–790 nm with a spectral resolution of 0.2–0.4 nm. Also, the instrument has three polarization channels (see Table 1.6). GOME was specifically designed to monitor the total ozone column. In addition, GOME had the capability to study other atmospheric constituents, such as NO_2, OClO, BrO, clouds and aerosols. Since 2011, the instrument is not operational anymore.

The instrument *GOME-2* is a successor of GOME. It is a UV/VIS/NIR grating spectrometer [41]. GOME-2 performs measurements in the ultraviolet and visible part of the spectrum (240–790 nm), at a spectral resolution between 0.2 and 0.4 nm, as shown in Table 1.7. It measures 4096 spectral points in total. GOME-2 is on board the satellites Metop-A (launched on 19 October 2006), Metop-B (launched on 17 September 2012) and Metop-C (launched 7 November 2018). As compared

Table 1.6 Channels and spectral resolution of the GOME spectrometer [40]

Spectral range (nm)	Number of channels	Spectral resolution (nm)
240–295	1024	0.22
290–405	1024	0.24
400–605	1024	0.40
590–790	1024	0.40
292–402	1	110
402–597	1	195
597–790	1	193

Table 1.7 Channels and spectral resolution of the GOME-2 spectrometer

Spectral range (nm)	Number of channels	Spectral resolution (nm)	SNR
240–315	1024	0.24–0.29	7–177
311–403	1024	0.26–0.28	372–3000
401–600	1024	0.44–0.53	4000
590–790	1024	0.44–0.53	2000–4000
312–790	200	2.8–40	100–1000

to GOME, GOME-2 has improved capabilities in registering the polarization of backscattered solar light. It has 200 polarization channels.

SCIAMACHY is a UV/VIS/NIR/SWIR grating spectrometer [42]. It was on board ENVISAT. SCIAMACHY had eight bands with 8192 channels in total. SCIA-MACHY also had seven polarization channels, as shown in Table 1.8. The spectrometer was operated in nadir, limb and Solar occultation modes. SCIAMACHY

Table 1.8 Channels and spectral resolution of the SCIAMACHY spectrometer

Spectral range (nm)	Number of channels	Spectral resolution (nm)	SNR
214–334	1024	0.24	200
300–412	1024	0.26	2300
383–628	1024	0.44	2600
595–812	1024	0.48	2800
773–1063	1024	0.54	1900
971–1773	1024	1.48	1500
1934–2044	1024	0.22	100
2259–2386	1024	0.26	320
310–2380	7	67–137	–

Table 1.9 Channels and spectral resolution of Sentinel-4

Spectral ranges (nm)	Number of channels	Spectral resolution (nm)	SNR
305–400	600	0.5	200–1000
400–500	600	0.5	1400
750–775	600	0.25	600

Table 1.10 Channels and spectral resolution of the TROPOMI instrument

Spectral ranges (nm)	Number of channels	Spectral resolution (nm)	SNR
270–495	1200	0.55	1000
710–775	600	0.55	500
2305–2385	800	0.25	100

had the ability to observe the same location first in limb and thereafter in nadir viewing geometries within 7 min—that was one of the most important and novel features. Since 2012 ENVISAT is not operational.

The Sentinel-4 mission involves an ultraviolet visible near-infrared imaging spectrometer embarked on a geostationary satellite. The mission's objective is to monitor key trace gases and aerosols over Europe at high spatial resolution with almost hourly revisit time. The instrument is a wide-field push-broom hyperspectral imaging spectrometer providing data with a partial resolution of $8 \times 8\,km^2$. Its characteristics are given in Table 1.9.

The newest Tropospheric Monitoring Instrument (*TROPOMI*) is on board the Sentinel-5 Precursor [23, 31]. It was launched in October 2017. TROPOMI uses three grating spectrometers with characteristics shown in Table 1.10. It extends the capabilities of instrumentation mentioned above, in particular, in terms of higher spatial resolution of the measurements.

1.5.7 Imagers

Below we briefly recall some imagers. In terrestrial remote sensing, the data coming from imagers are mostly used for the study of aerosols, biomass, clouds, oceans (in particular, extracting the information about colored dissolved organic matter), surface properties, vegetation, oil spills etc.

MERIS (medium resolution imaging spectrometer) [43] was on board ENVISAT. The measurements have been based on the push-broom scanning technique with a total swath of 1150 km. Its IFOV is about 300 m. As shown in Table 1.11, MERIS has many narrow-bandwidth channels in the visible spectral range, which allow for the detailed study of open ocean as well as coastal zone waters. The channel around 760 nm is used for aerosol and cloud detection and vertical profiling.

Table 1.11 Channels and spectral resolution of MERIS

Central wavelength (nm)	Bandwidth (nm)	SNR
412.5	10	1871
442.5	10	1650
490	10	1418
510	10	1222
560	10	1156
620	10	863
665	10	708
681.25	7.5	589
708.75	10	631
753.75	7.5	486
760.625	3.75	205
778.75	15	628
865	20	457
885	10	271
900	10	211

MODIS (NASA) is a moderate-resolution imaging spectro-radiometer on board the satellites Terra and Aqua (see [44] and references therein). Together they image the entire Earth every 1 to 2 days. It has 36 channels and uses the whisk-broom scanning technology (the characteristics of VIS/NIR channels are shown in Table 1.12). The IFOV is in the range between 250 m and 1000 m for different channels.

Another example of imagers is *VIIRS* [45]—Visible/Infrared Imager Radiometer Suite. The characteristics are shown in Table 1.13. It has 22 channels from 412 nm to 12.01 μm and provides spatial resolution of about 750 m in the visible spectral range. The VIIRS sensor was designed to extend and improve the measurements performed by its predecessors (e.g., MODIS). The VIIRS instrument employs the whisk-broom scanning technique. One of the channels with the central wavelength of 0.7 μm that covers 0.5–0.9 μm spectral range is the so-called Day/Night band. It is a panchromatic (i.e., black and white) band sensitive to visible and near-infrared wavelengths. Using this band VIIRS observes nighttime lights on the Earth's surface with better spatial and temporal resolution compared to its predecessors. Currently, VIIRS instruments are operated on board SNPP and NOAA-20.

Besides single-view imagers, there are multiple view imagers [46]. Multi angle imaging spectro-radiometer (MISR) is one of the first multi angle multispectral space-based imagers. It is on board satellite Terra. MISR has nine cameras at different viewing angles. Each camera has four spectral channels in VIS and NIR regions. Nine cameras look along track at the following viewing angles: nadir, ±26.1°, ±45.6°, ±60.0° and ±70.5°. Thus, a given target is observed under different geometries, thus enabling the studies of reflection properties of the surface (BRDF, see Chap. 3). The push-broom technique is used for cross-track scanning. MISR provides the data

Table 1.12 Channels and spectral resolution of MODIS (only for VIS/NIR channels)

Central wavelength (nm)	Bandwidth (nm)	SNR
645	50	128
858	35	201
469	20	243
555	20	228
1240	20	74
1640	24	275
2130	50	110
412	15	880
443	10	838
488	10	802
531	10	754
551	10	750
667	10	910
678	10	1087
748	10	586
870	15	516
905	30	167
936	10	57
940	50	250
1375	30	150

Table 1.13 Channels and spectral resolution of the VIIRS (for VIS/NIR channels)

Central wavelength (nm)	Bandwidth (nm)	SNR
412	20	352
445	18	380
488	20	416
555	20	362
672	20	242
746	15	199
865	39	215
1240	20	101
1378	15	83
1610	60	342
2250	50	10

Table 1.14 Channels and spectral resolution of the MISR instrument

Central wavelength (nm)	Bandwidth (nm)	SNR
446.4	41.9	300
557.5	28.6	620
671.7	21.9	650
866.4	39.7	740

Table 1.15 Channels and spectral resolution of the POLDER instrument

Central wavelength (nm)	Bandwidth (nm)	SNR
443.5	13.4	200
490.8	16.3	200
563.8	15.4	200
669.9	15.1	200
762.9	10.9	200
762.7	38.1	200
863.7	33.7	200
907.1	21.1	200
1019.6	17.1	200

of spatial resolution of 250 m at nadir. The characteristics of MISR are shown in Table 1.14.

Imagers with a capability to measure the polarization are called *polarimeters*. They are used primarily for aerosol observation and retrieval of microphysical properties of the scene. Polarimetric measurements are also useful for studies of the Earth radiation budget, atmopsheric aerosol, clouds, soil condition and ocean color.

POLDER (POLarization and Directionality of the Earth's Reflectances) is a multispectral imaging radiometer providing measurements of polarization of the solar radiation reflected by the Earth-atmosphere system [47]. It was developed by CNES, the French space agency. POLDER has a camera with a two-dimensional CCD detector array. The instrument (with some modifications) was on board ADEOS, ADEOS 2 and PARASOL satellites. As the satellite moves, POLDER can take 14 successive images of a given target during the same orbit path, which is important for determining the surface properties. In addition, from POLDER measurements it is possible to retrieve the properties of aerosols and ocean color. The characteristics of the POLDER are shown in Table 1.15. The POLDER instrument has pioneered space-based polarimetric remote sensing. It proved the concept of 3MI, which stands for multi angle, multi wavelength and multi polarization imagers.

Another important example of polarimeters is multi viewing multi channel multi polarization imager—*3MI* [48]. Essentially, it is an evolution of POLDER/PARASOL mission with enhanced spatial resolution. It will be on board satellites Metop-SG-A1 (2023–2028), Metop-SG-A2 (2028–2035) and Metop-SG-A3 (2035–2042).

Table 1.16 Channels and spectral resolution of the 3MI instrument

Central wavelength (nm)	Bandwidth (nm)	SNR
410	20	200
443	20	200
490	20	200
555	20	200
670	20	200
763	10	200
765	40	200
865	40	200
910	20	200
1370	40	200
1650	40	200
2130	40	200

Table 1.17 Channels and spectral resolution of the MSI (Sentinel-2A)

Central wavelength (nm)	Bandwidth (nm)	SNR
443	20	129
490	65	154
560	35	168
665	30	142
705	15	117
740	15	89
783	20	105
842	115	174
865	20	72
945	20	114
1375	20	50
1610	90	100
2190	180	100

The characteristics of 3MI are shown in Table 1.16. The information in VIS/NIR channels is captured by 512×512 CCD array. Just like POLDER, each ground scene can be viewed from 14 directions as satellite moves. For SWIR channels, 512×256 CCD array is used providing 12 viewing directions. The swath of 3MI is 2200 km from both CCD arrays.

The multi spectral imager (MSI) for Sentinel-2A is a push-broom instrument designed to perform high-resolution land observation, vegetation, territory management and hazard mitigation. The MSI has 13 channels in VIS/NIR/SWIR spectral

Table 1.18 Channels and spectral resolution of the OLCI/Sentinel-3

Central wavelength (nm)	Bandwidth (nm)	SNR
400	10	2188
412.5	10	2061
442.5	10	1811
490	10	1541
510	10	1488
560	10	1280
620	10	997
665	10	883
673.75	7.5	707
681.25	7.5	745
708.25	10	785
753.75	7.5	605
761.25	2.5	232
764.375	3.75	305
767.5	2.5	330
778.75	15	812
865	20	666
885	10	395
900	10	308
940	20	203
1020	40	151

ranges (see Table 1.17). The IFOV is 10–60 m depending on the channel. The global coverage is done in 10 days.

Sentinel-3 Ocean and Land Colour Instrument (OLCI) is a medium resolution push-broom imaging spectrometer of MERIS heritage. OLCI is a visible imaging push-broom radiometer with more spectral bands, improved signal-to-noise ratio, among other improvements, compared to MERIS. The sampling distance is 1.2 km over the open ocean and 0.3 km for coastal zone and land observations. OLCI has 21 channels with characteristics shown in Table 1.18. The main focuses of the instrument are color dissolved organic matters, ocean chlorophyll as well as ocean suspended sediments concentrations, aerosol properties and vegetation index. The linear size of a ground pixel is about 300 m. The global coverage is achieved in 2 days.

1.6 Concluding Remarks

So, *what is remote sensing*? We started with a simple definition in Sect. 1.2.1. But now we see that it is far more than that. Spaceborne remote sensing is a multidisciplinary field of research and development dealing with data collected by sounding instruments on board the satellites. That includes data acquisition, data processing, analysis of the data and decision making using the data. Remote sensing as a field of industry provides to the community with a lot of innovations and research assets, increasing the quality of life.

Remote sensing is a vital instrument for policymakers to make decisions regarding sustainable development of the society [49, 50]. In [51], nine societal benefit areas were identified in which a global observation system can provide social benefits. These areas are:

1. reducing loss of life and property from natural and human-induced disasters;
2. understanding environmental factors affecting human health and well-being;
3. improving management of energy resources;
4. understanding, assessing, predicting, mitigating and adapting to climate variability and change;
5. improving water resource management through better understanding of the water cycle;
6. improving weather information, forecasting and warning;
7. proving the management and protection of terrestrial, coastal and marine ecosystems;
8. supporting sustainable agriculture and combating desertification;
9. understanding, monitoring, and conserving biodiversity.

And the role of Earth observations, in general, and atmospheric remote sensing, in particular, is expected to be increasing in the next decade [52].

Remote sensing technology is actively developed by means of international collaboration [53]: according to www.scimagojr.com, the number of papers in the high-rank peer-review journals such as Remote Sensing (MDPI), International Journal of Remote Sensing (Taylor & Francis), Remote Sensing of Environment (Elsevier) written by researchers from more than one country, has increased from 10–20% in 1999 to 40% in 2018 from the total number of papers (that is of course a part of a global trend). There is also a tendency to make the remote sensing data available to the community (e.g., TROPOMI data [28]), as well as tools for data reading and processing. All these factors transform remote sensing from being just a technology into a paramount scientific tool needed to understand the state of our planet and its multidimensional global change [54, 55].

Appendix 1: Example of the Plotting Program for the Level-2 Data of S5P

Below there is an example of a python script for plotting the nitrogen dioxide tropospheric column data provided by Sentinel 5 Precursor (the script was used to make

Fig. 1.8). For plotting, the Matplotlib Basemap toolkit is used. Some references data archives and tools for reading the data can be found in section "Useful links" at the end of the book.

```
 1  import numpy as np
 2  from mpl_toolkits.basemap import Basemap
 3  from netCDF4 import Dataset
 4  import matplotlib.pyplot as plt
 5  from matplotlib.colors import LogNorm
 6
 7  # load data from https://s5phub.copernicus.eu/dhus. The data
        file has an extension 'nc'
 8  # Assume that the name of the file is 'file.nc'
 9  nc1 = 'file.nc'
10  # read data
11  fh = Dataset(nc1, mode='r')
12  lons = fh.groups['PRODUCT'].variables['longitude'][:][0,:,:]
13  lats = fh.groups['PRODUCT'].variables['latitude'][:][0,:,:]
14  no2 = fh.groups['PRODUCT'].variables[
        nitrogendioxide_tropospheric_column_precision'][0,:,:]
15  no2_units = fh.groups['PRODUCT'].variables['
        nitrogendioxide_tropospheric_column_precision'].units
16  lon_0 = lons.mean()
17  lat_0 = lats.mean()
18  # use Basemap class
19  m = Basemap(width=3500000,height=3500000, resolution='h',
        projection='stere', lat_ts=40,lat_0=50,lon_0=17)
20  m.bluemarble()
21  xi, yi = m(lons, lats)
22  # Plot Data
23  cs = m.pcolor(xi,yi,np.squeeze(no2),norm=LogNorm(), cmap='jet
        ')
24  # Add Grid Lines
25  m.drawparallels(np.arange(-80., 81., 10.), labels=[1,0,0,0],
        fontsize=10)
26  m.drawmeridians(np.arange(-180., 181., 10.), labels
        =[0,0,0,1], fontsize=10)
27  # Add Colorbar
28  cbar = m.colorbar(cs, location='bottom', pad="10%")
29  cbar.set_label(no2_units)
30  # Add Title
31  m.drawcoastlines()
32  m.drawstates()
33  m.drawcountries()
34  plt.show()
```

References

1. K. Tsiolkovsky, *The Exploration of Cosmic Space by Means of Reaction Devices* (In Russian: Issledovanie mirovih prostranstv reaktivnimi priborami): reissue of works in 1903 and 1911 with some changes and additions (Kaluga, 1926)
2. H. Potocnik, *The Problem of Space Travel - The Rocket Motor* (In German, Das Problem der Befahrung des Weltraums - der Raketen-Motor) (Berlin, 1929)
3. V. Kuznetsov, V. Sinelnikov, S. Alpert, Yakov Alpert: Sputnik-1 and the first satellite iono-spheric experiment. Adv. Space Res. **55**(12), 2833–2839 (2015). https://doi.org/10.1016/j.asr.2015.02.033

4. A. Tatem, S. Goetz, S. Hay, Fifty years of Earth-observation satellites. Am. Sci. **96**(5), 390 (2008). https://doi.org/10.1511/2008.74.390
5. National Aeronautics and Space Administration (NASA) Space Science Data Coordinated Archive, https://nssdc.gsfc.nasa.gov/nmc/SpacecraftQuery.jsp. Accessed 1 Jun 2020
6. A. Cracknell, L. Hayes, *Introduction to Remote Sensing*, 2nd edn. (CRC Press, New York, 2007)
7. J.P. Burrows, A.P.H. Goede, C. Muller, H. Bovensmann, SCIAMACHY – the need for atmospheric research from space, in *SCIAMACHY - Exploring the Changing Earth's Atmosphere* (Springer Netherlands, 2010), pp. 1–17. https://doi.org/10.1007/978-90-481-9896-2_1
8. M. Subramanian, Anthropocene now: influential panel votes to recognize Earth's new epoch. Nature (2019). https://doi.org/10.1038/d41586-019-01641-5
9. V. Vernadsky, The biosphere and the noosphere. Am. Sci. **33**(1), 1–12 (1945)
10. W. Steffen, J. Grinevald, P. Crutzen, J. McNeill, The anthropocene: conceptual and historical perspectives. Philos. Trans. R. Soc. A: Math. Phys. Eng. Sci. **369**(1938), 842–867 (2011). https://doi.org/10.1098/rsta.2010.0327
11. A. Al Gore, *Earth in the Balance: Ecology and the Human Spirit* (Houghton Mifflin, 1992)
12. *Handbook for the Montreal Protocol on Substances that Deplete the Ozone Layer* (2016), www.efcc.eu/media/1079/2016-ods-montreal_protocol-handbook.pdf. Accessed 20 Apr 2019
13. Kyoto protocol to the united nations framework convention on climate change (1998), https://unfccc.int/resource/docs/convkp/kpeng.pdf. Accessed 20 Apr 2019
14. B. Smirnov, Ekologicheskie problemy atmosfery zemli. Uspekhi Fizicheskikh Nauk [in Russian] **117**(10), 313–332 (1975). https://doi.org/10.3367/UFNr.0117.197510d.0313
15. G. Ohring, S. Lord, J. Derber, K. Mitchell, M. Ji, Applications of satellite remote sensing in numerical weather and climate prediction. Adv. Space Res. **30**(11), 2433–2439 (2002). https://doi.org/10.1016/s0273-1177(02)80298-8
16. A. Racoviteanu, M. Williams, R. Barry, Optical remote sensing of glacier characteristics: a review with focus on the Himalaya. Sensors **8**(5), 3355–3383 (2008). https://doi.org/10.3390/s8053355
17. D. Alexander, Information technology in real-time for monitoring and managing natural disasters. Prog. Phys. Geogr.: Earth Environ. **15**(3), 238–260 (1991). https://doi.org/10.1177/030913339101500302
18. O. Montenbruck, E. Gill, *Satellite Orbits* (Springer Berlin Heidelberg, 2000). https://doi.org/10.1007/978-3-642-58351-3
19. A. Marshak, J. Herman, S. Adam, B. Karin, S. Carn, A. Cede, I. Geogdzhayev, D. Huang, L. Huang, Y. Knyazikhin, M. Kowalewski, N. Krotkov, A. Lyapustin, R. McPeters, K.G. Meyer, O. Torres, Y. Yang, Earth observations from DSCOVR EPIC instrument. Bull. Am. Meteorol. Soc. **99**(9), 1829–1850 (2018). https://doi.org/10.1175/bams-d-17-0223.1
20. E. Ustinov, Geophysical retrieval, forward models in remote sensing, in *Encyclopedia of Remote Sensing* (Springer, New York, 2014), pp. 241–247. https://doi.org/10.1007/978-0-387-36699-9_53
21. R. Richter, Atmospheric correction of DAIS hyperspectral image data. Comput. Geosci. **22**(7), 785–793 (1996). https://doi.org/10.1016/0098-3004(96)00016-7
22. C. Toth, G. Jóźków, Remote sensing platforms and sensors: a survey. ISPRS J. Photogramm. Remote Sens. **115**, 22–36 (2016). https://doi.org/10.1016/j.isprsjprs.2015.10.004
23. J. Veefkind, I. Aben, K. McMullan, H. Forster, J. de Vries, G. Otter, J. Claas, H. Eskes, J. de Haan, Q. Kleipool et al., TROPOMI on the ESA Sentinel-5 Precursor: a GMES mission for global observations of the atmospheric composition for climate, air quality and ozone layer applications. Remote Sens. Environ. **120**, 70–83 (2012)
24. Y. Ma, H. Wu, L. Wang, B. Huang, R. Ranjan, A. Zomaya, W. Jie, Remote sensing big data computing: challenges and opportunities. Futur. Gener. Comput. Syst. **51**, 47–60 (2015)
25. P. Liu, A survey of remote-sensing big data. Front. Environ. Sci. **3**, 45 (2015)
26. D. Laney, 3D data management: controlling data volume, velocity and variety. Technical Report (META Group, 2001)

27. M. Chi, A. Plaza, J.A. Benediktsson, Z. Sun, J. Shen, Y. Zhu, Big data for remote sensing: challenges and opportunities. Proc. IEEE **104**(11), 2207–2219 (2016). https://doi.org/10.1109/jproc.2016.2598228

28. Sentinel-5P Pre-Operationals Data Hub, https://s5phub.copernicus.eu/dhus. Accessed 8 Feb 2020

29. N.R. Ikhsanov, Particle acceleration and main parameters of ultra-high energy gamma-ray binaries. Astrophys. Space Sci. **184**(2), 297–311 (1991). https://doi.org/10.1007/bf00642978

30. E.M. Hume, N.S. Lucas, H.H. Smith, On the absorption of vitamin D from the skin. Biochem. J. **21**(2), 362–367 (1927). https://doi.org/10.1042/bj0210362

31. R. Voors, I.S. Bhatti, T. Wood, I. Aben, P. Veefkind, J. de Vries, D. Lobb, N. van der Valk, TROPOMI, the Sentinel 5 precursor instrument for air quality and climate observations: status of the current design, in *International Conference on Space Optics — ICSO 2012*, ed. by E. Armandillo, N. Karafolas, B. Cugny (SPIE, 2017). https://doi.org/10.1117/12.2309017

32. P.B. Fellgett, On the ultimate sensitivity and practical performance of radiation detectors. J. Opt. Soc. Am. **39**(11), 970 (1949). https://doi.org/10.1364/josa.39.000970

33. S. Beirle, J. Lampel, C. Lerot, H. Sihler, T. Wagner, Parameterizing the instrumental spectral response function and its changes by a super-gaussian and its derivatives. Atmos. Meas. Tech. **10**(2), 581–598 (2017). https://doi.org/10.5194/amt-10-581-2017

34. S. Liang, X. Li, J. Wang, A systematic view of remote sensing, in *Advanced Remote Sensing*, ed. by S. Liang, X. Li, J. Wang (Academic, Boston, 2012), Chap. 1, pp. 1–31. https://doi.org/10.1016/B978-0-12-385954-9.00001-0

35. R. Gray, L. Davisson, *An Introduction to Statistical Signal Processing* (Cambridge University Press, 2004)

36. R. Pu, *Hyperspectral Remote Sensing* (CRC Press, 2017). https://doi.org/10.1201/9781315120607

37. G. Hunt, Electromagnetic radiation: the communication link in remote sensing, in *Remote Sensing in Geology*, ed. by B. Siegal, A. Gillespia (Wiley, NewYork, 1980), pp. 5–45

38. H. Bovensmann, A. Doicu, P. Stammes, M.V. Roozendael, C. von Savigny, M.P. de Vries, S. Beirle, T. Wagner, K. Chance, M. Buchwitz, A. Kokhanovsky, A. Richter, A.V. Rozanov, V.V. Rozanov, From radiation fields to atmospheric concentrations- retrieval of geophysical parameters, in *SCIAMACHY - Exploring the Changing Earth's Atmosphere* (Springer Netherlands, 2010), pp. 99–127. https://doi.org/10.1007/978-90-481-9896-2_7

39. USGS Science for a changing world, https://www.usgs.gov. Accessed 8 Feb 2020

40. J.P. Burrows, M. Weber, M. Buchwitz, V. Rozanov, A. Ladstätter-Weißenmayer, A. Richter, R. DeBeek, R. Hoogen, K. Bramstedt, K. Eichmann, M. Eisinger, D. Perner, J. The global ozone monitoring experiment (GOME): mission concept and first scientific results. Atmos. Sci. **56**(2), 151–175 (1999). https://doi.org/10.1175/1520-0469(1999)056<0151:tgomeg>2.0.co;2

41. R. Munro, R. Lang, D. Klaes, G. Poli, C. Retscher, R. Lindstrot, R. Huckle, A. Lacan, M. Grzegorski, A. Holdak, A. Kokhanovsky, J. Livschitz, M. Eisinger, Atmos. The GOME-2 instrument on the Metop series of satellites: instrument design, calibration, and level 1 data processing – an overview. Atmos. Meas. Tech. **9**(3), 1279–1301 (2016). https://doi.org/10.5194/amt-9-1279-2016

42. H. Bovensmann, J. Burrows, M. Buchwitz, J. Frerick, S. Noël, V.V. Rozanov, K.V. Chance, A.P.H. Goede, SCIAMACHY: mission objectives and measurement modes. J. Atmos. Sci. **56**(2), 127–150 (1999). https://doi.org/10.1175/1520-0469(1999)056<0127:smoamm>2.0.co;2

43. R. Zurita-Milla, G. Kaiser, J. Clevers, W. Schneider, M. Schaepman, Downscaling time series of MERIS full resolution data to monitor vegetation seasonal dynamics. Remote Sens. Environ. **113**(9), 1874–1885 (2009). https://doi.org/10.1016/j.rse.2009.04.011

44. MODIS Web, https://modis.gsfc.nasa.gov/sci_team/pubs/. Accessed 29 May 2019

45. C. Cao, F.J. de Luccia, X. Xiong, R. Wolfe, F. Weng, Early on-orbit performance of the visible infrared imaging radiometer suite onboard the Suomi national polar-orbiting partnership (s-NPP) satellite. IEEE Trans. Geosci. Remote Sens. **52**(2), 1142–1156 (2014). https://doi.org/10.1109/tgrs.2013.2247768

46. S. Nag, C.K. Gatebe, T. Hilker, Simulation of multiangular remote sensing products using small satellite formations. IEEE J. Sel. Top. Appl. Earth Obs. Remote Sens. **10**(2), 638–653 (2017). https://doi.org/10.1109/jstars.2016.2570683

47. Y. Andre, J.M. Laherrere, T. Bret-Dibat, M. Jouret, J.M. Martinuzzi, J.L. Perbos, Instrumental concept and performances of the POLDER instrument, in *Remote Sensing and Reconstruction for Three-Dimensional Objects and Scenes*, ed. by T. Schenk (SPIE, 1995). https://doi.org/10.1117/12.216932

48. J.L. Bézy, R. Meynart, M. Porciani, M. Loiselet, G. Mason, U. Bruno, R.D. Vidi, I. Manolis, D. Labate, The 3MI instrument on the Metop second generation, in *International Conference on Space Optics — ICSO 2014*, ed. by B. Cugny, Z. Sodnik, N. Karafolas (SPIE, 2017). https://doi.org/10.1117/12.2304182

49. E. Sadeh (ed.), *Space Politics and Policy* (Springer Netherlands, 2004). https://doi.org/10.1007/0-306-48413-7

50. S. Jones, K. Reinke (eds.), *Innovations in Remote Sensing and Photogrammetry* (Springer Berlin Heidelberg, 2009). https://doi.org/10.1007/978-3-540-93962-7

51. Group on Earth Observations, *Global Earth Observation System of Systems GEOSS: 10-Year Implementation Plan Reference Document* (ESA Publication Division, 2005)

52. K. Anderson, B. Ryan, W. Sonntag, A. Kavvada, L. Friedl, Earth observation in service of the 2030 agenda for sustainable development. Geo-Spat Inf Sci **20**(2), 77–96 (2017). https://doi.org/10.1080/10095020.2017.1333230

53. G. Thomas, J. Lester, W. Sadeh, International cooperation in remote sensing for global change research: political and economic considerations. Space Policy **11**(2), 131–141 (1995). https://doi.org/10.1016/0265-9646(95)00008-z

54. K. Kondratyev, *Multidimensional Global Change*, Wiley - Praxis Series in Remote Sensing (Wiley-Praxis, Chichester, 1998)

55. L. Gunderson, C. Holling, *Panarchy Synopsys: Understanding Transformations in Human and Natural Systems* (Island Press, Washington, 2002)

Chapter 2
Physical Properties of the Terrestrial Atmosphere

2.1 Physical Properties of the Atmosphere at the Sea Level

In the upcoming sections we analyze the basic physical properties of the atmosphere. Although the atmospheric pressure and temperature are time-dependent and spatial-dependent, it is convenient to use the so-called static atmospheric models, i.e., models, in which the time dependency is excluded. The U.S. Standard Atmosphere [1] is one of the most common models. It provides the values for air temperature and pressure as a function of altitude above the sea level.

Although the atmosphere consists of a number of different gases, it is convenient to treat it as a homogeneous medium comprising air molecules when we consider the atmosphere as a thermodynamic system. The average molar mass of dry air is equal to the weighted average of the composite molecules and has a value of $M_{air} = 0.02896 \, \text{kg/mol}$ (mostly due to nitrogen). Its adiabatic index $\gamma = 1.4$ corresponds to that of a diatomic gas (due to nitrogen and oxygen).

It is easy to estimate the mass of atmosphere m_a by noting that the pressure p_0 at the surface should balance the force of gravity:

$$m_a g = p_0 S, \tag{2.1}$$

where S is the area of Earth's surface. We take the air pressure at sea level $p_0 = 101.3$ kPa and the standard acceleration due to gravity is $g \approx 9.807 \, \text{m/s}^2$. Considering Earth as a sphere, S can be estimated as

$$S = 4\pi R_E^2, \tag{2.2}$$

© Springer Nature Switzerland AG 2021
D. Efremenko and A. Kokhanovsky, *Foundations of Atmospheric Remote Sensing*,
https://doi.org/10.1007/978-3-030-66745-0_2

where R_E is the Earth radius. Taking $R_E = 6400$ km (as estimated already in the third century B.C. [2]),[1] the mass m_a of the atmosphere can be expressed from Eq. (2.1) and Eq. (2.2) as follows:

$$m_a = \frac{4\pi R_E^2 p_0}{g} \approx 5.3 \times 10^{18} \text{ kg}. \tag{2.3}$$

Then, the total number of molecules in the atmosphere is

$$N_{air} = \frac{m_a}{M_{air}} N_A \approx 1.1 \times 10^{44}, \tag{2.4}$$

where N_A is the Avogadro constant ($\approx 6.022 \times 10^{23} \text{ mol}^{-1}$).

Assuming the atmospheric air behaves as an ideal gas (which is a very good approximation[2]), i.e.,

$$p_0 = n_0 k_B T_0, \tag{2.5}$$

where n_0 is the molecular number density at the sea level and $k_B \approx 1.3806 \times 10^{-23}$ J/K is the Boltzmann constant, from Eq. (2.5) we can find $n_0 \approx 2.547 \times 10^{19} \text{ cm}^{-3}$ assuming that the air temperature at sea level $T_0 = 288.2$ K. The air volumetric mass density can be found from the ideal gas law:

$$p_0 V = \frac{m_a}{M_{air}} R T_0 \tag{2.6}$$

with $R = 8.3144598 \, J/(\text{mol} \cdot K)$ being the ideal gas constant. From Eq. (2.6) we have

$$\rho_0 = \frac{m_{air}}{V} = \frac{p_0 M_{air}}{R T_0} \tag{2.7}$$

or by using Eq. (2.5) and given that $R = k_B N_A$

$$\rho_0 = \frac{n_0 M_{air}}{N_A} \approx 1.225 \text{ kg/m}^3. \tag{2.8}$$

The speed of sound in the air at the sea level is

---

[1]The Earth's radius was calculated by Eratosthenes in the third century B.C. [2]. He noticed that on the summer solstice at local noon in Syene the Sun was directly overhead, while the angle of a shadow in Alexandria is $\varphi = 7.2° \approx 0.1256$ rad. Given the distance between Syene and Alexandria of $l = 787$ km, it is straight-forward to estimate the Earth's radius as $R_E = l/\varphi \approx 6266$ km—what an amazing accuracy!

[2]There is one component which is non-ideal, and that is water vapor. However, it is a trace gas (its concentration is below a percent), and the errors that result from assuming ideal behavior are generally negligible for our purposes.

$$u_0 = \sqrt{\gamma \frac{p_0}{\rho_0}} \approx 340 \, \text{m/s}. \tag{2.9}$$

Then, we assume that the system of air molecules has reached thermodynamic equilibrium. That means the velocities of these molecules follow Maxwell–Boltzmann statistics with the following probability density function for velocities v:

$$f(v) = \frac{4}{\sqrt{\pi}} \left(\frac{\bar{m}}{2\pi T} \right)^{3/2} v^2 e^{-\frac{mv^2}{2k_B T}}, \tag{2.10}$$

where $\bar{m} = M_{\text{air}}/N_A \approx 4.8 \times 10^{-26} \, \text{kg}$ is the mass of one air molecule. The mean velocity $\langle v \rangle$ is computed as follows:

$$\langle v \rangle = \int_0^\infty v f(v) \, dv = \sqrt{\frac{8 k_B T}{\pi \bar{m}}} \approx 460 \, \text{m/s}. \tag{2.11}$$

Given the air molecule collision cross section $\sigma_c \sim 10^{-19} \, \text{m}^2$ [3], the mean free path is given by

$$l = \frac{1}{n \sigma_c} \sim 4 \times 10^{-7} \, \text{m}, \tag{2.12}$$

while the typical collision time is

$$\tau = \frac{l}{\langle v \rangle} \sim 10^{-9} \, \text{s}. \tag{2.13}$$

2.2 Atmospheric Retention

The escape velocity v_{esc}, i.e., the minimum speed needed for a molecule with mass m to escape from the gravitational influence of the Earth, can be found from the energy conservation law:

$$\frac{m v_{\text{esc}}^2}{2} - G \frac{m m_E}{R_E} = 0, \tag{2.14}$$

where the first term is the kinetic energy while the second term is the potential energy in the gravitational field; the mass of Earth is $m_E = 5.97 \cdot 10^{24} \, \text{kg}$ while $G \approx 6.67 \cdot 10^{-11} \, \text{m}^3\text{kg}^{-1}\text{s}^{-2}$ is the gravitational constant. Thus,

$$v_{\text{esc}} = \sqrt{\frac{2 G m_E}{R_E}},$$

which is around 11 km/s. From kinetic molecular theory we estimate the velocity of chaotic movement (v_{th}) as follows:

$$\frac{m v_{\text{th}}^2}{2} = \frac{3}{2} k_{\text{B}} T, \tag{2.15}$$

which yields

$$v_{\text{th}} = \sqrt{\frac{3 k_{\text{B}} T}{m}}. \tag{2.16}$$

Formally, the molecule can escape from the planet if $v_{\text{th}} > v_{\text{esc}}$, i.e.,

$$\sqrt{\frac{3 k_{\text{B}} T}{m}} > \sqrt{\frac{2 G m_{\text{E}}}{R_{\text{E}}}} \tag{2.17}$$

or

$$T > \frac{2 G m_{\text{E}} m}{3 k_{\text{B}} R_{\text{E}}}. \tag{2.18}$$

From condition (2.17) it is clear that light molecules such as hydrogen and helium have more chances to escape from the Earth (lower value of T is required), so their concentrations are expected to be lower than those for heavier molecules. From Eq. (2.17) it also follows that small planets cannot retain an atmosphere. Indeed, assuming $m_{\text{E}} = \frac{4}{3} \pi R_{\text{E}}^3 \rho$ (here m_{E}, R_{E} and ρ refer to the mass, radius and mass density of a certain planet, respectively) we get

$$\sqrt{\frac{3 k_{\text{B}} T}{m}} > \sqrt{\frac{8 \pi \rho G R_{\text{E}}^2}{3}}. \tag{2.19}$$

Thus, for smaller planets the molecules require lower T to escape. In the same manner we can show that massive planets like Jupiter can retain hydrogen making it a giant gas.

We should note that our analysis here is qualitative. Even if condition (2.17) is not fulfilled, some molecules can escape due to the high-speed tail of the Maxwell distribution. Interested readers are encouraged to refer to [4–6], where the problem is considered in terms of the Jeans escape fluxes. However, for Earth the losses due to the Maxwell tail are negligibly small on a time scale of 10^5 years.

2.3   Pressure Profile

The atmospheric pressure changes with an altitude. This dependence can be found by dividing the atmosphere into layers between heights h and $h + dh$ and assuming the hydrostatic equilibrium for each of them. For simplicity, we assume that the temperature is constant along the vertical coordinate and is equal to T_0 (this is the isothermal atmosphere).

At any level, pressure should compensate for the weight of the overlying air column per unit area. We choose an element of air with mass $dm = \rho dV = \rho Sdh$, where S is the surface area. The gravitational force dmg should be balanced by the force produced by the pressure differences at the lower and upper level of element dm: $dp(h)S = (p(h + dh) - p(h))S$. Thus, the hydrostatic equilibrium condition reads as

$$dp(h) = -\rho gdh \tag{2.20}$$

or substituting Eq. (2.7) into Eq. (2.20):

$$dp(h) = -p(h)\frac{M_{air}}{RT_0}gdh. \tag{2.21}$$

Here the "minus" sign accounts for the fact that the pressure decreases as the height increases. Integrating Eq. (2.21) we obtain

$$p(h) = p_0 e^{-\frac{M_{air}g}{RT_0}h}. \tag{2.22}$$

Equation (2.22) is called the barometric formula. Using Eq. (2.5) in Eq. (2.22) we obtain a similar expression for the number density:

$$n(h) = n_0 e^{-\frac{M_{air}g}{RT_0}h}. \tag{2.23}$$

Taking into account that the temperature T depends on h, the pressure profile can be computed as

$$p(h) = p_0 e^{-\int_0^z \frac{Mg}{RT(h)}dh}. \tag{2.24}$$

We should note, however, that the hydrostatic equilibrium assumption fails for ozone and water since ozone is produced and destroyed by chemical reactions, while water experiences phase transitions.

Figure 2.1 shows the comparison between the profile computed with the barometric formula (2.22) and that from the climatological model "U.S. Standard atmosphere" [1], which is an average atmosphere with variations caused by weather, longitude, seasons- and time removed. For altitudes lower than 8 km, the barometric formula correctly predicts the pressure decrease rate with the increasing altitude. At higher altitudes it overestimates the pressure since the assumption about the isothermal atmosphere is not valid anymore.

Introducing the so-called atmospheric scale with respect to the atmospheric pressure as [7]

$$H(h) = -\left(\frac{d\ln p(h)}{dh}\right)^{-1}, \tag{2.25}$$

the variation of the pressure near the altitude h_0 can be expressed as follows:

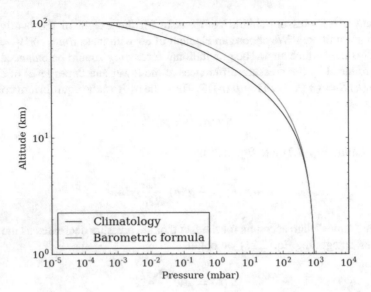

Fig. 2.1 Pressure profiles: comparison between the profile computed with the barometric formula and that for the climatological model "U.S. standard"

$$p(h) = p(h_0) e^{-\frac{h-h_0}{H(h_0)}}. \tag{2.26}$$

The atmospheric scale gives the distance at which the pressure decays by e times. Substituting Eq. (2.22) into Eq. (2.25), we find H for the isothermal atmosphere:

$$H = \frac{RT}{M_{\text{air}}g}. \tag{2.27}$$

At the sea level, we have $H \sim 8$ km.

As we can see from Eq. (2.23), the atmospheric number density progressively decreases with altitude. Therefore, a precise demarcation line between Earth's atmosphere and the outer space cannot be defined from the physical point of view. Moreover, there is no unique international definition of the top of the atmosphere level. Theodore von Kàrmàn proposed the following criterion: the atmosphere ends at the level where the aircraft velocity required for the lift force exceeds the orbital velocity. At this height, "aerodynamics ends and astronautics begins" [8]. This level is at about 80 km. Sometimes a rounded value of 100 km is used as the top of the atmosphere definition.

2.4 Temperature Profile

The atmospheric temperature is a non-monotonic function of the altitude. Considering the temperature profile in Fig. 2.2, the Earth's atmosphere can be divided into five layers: the troposphere, the stratosphere, the mesosphere, the thermosphere and the exosphere.

The **troposphere** is the atmospheric layer closest to the Earth's surface. It is about 6 to 18 km thick depending upon the latitude and the season (18 km in the tropics and 6 km in the polar regions). It encapsulates almost 80% of the mass of the entire atmosphere and 99% of the total water vapor and aerosols. The default temperature at the sea level in the standard atmosphere is 15 °C. The temperature decreases almost linearly with the increasing height at the rate of about 6.5 °C per km reaching a minimum value of about −60°C at the top of the troposphere. We should note that the tropospheric temperature has a very high variability depending on the latitude, season, weather and time of day. It is maintained by convective and turbulent transfer of heat due to absorption of solar radiation by the terrestrial surface. The troposphere ends with the tropopause, i.e., a region where the temperature is constant.

The next atmospheric layer is the **stratosphere**. It starts above the tropopause and ends at about 50 km. Here the temperature increases with the altitude reaching a maximum of about 0 °C at the top of the stratosphere. This increase is explained by the presence of ozone which absorbs the ultraviolet region of the solar spectrum. Since ozone determines the temperature behavior in this layer, the stratosphere is often referred to as the ozonosphere. The top of the stratosphere is called the stratopause.

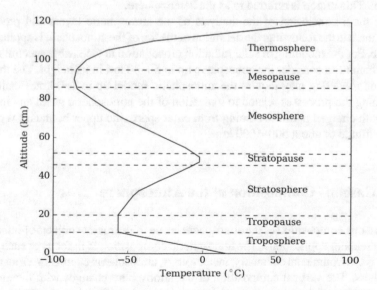

Fig. 2.2 Temperature profile according to the standard atmosphere model

The **mesosphere** starts at 50 km. The temperature again decreases with altitude reaching the lowest values in the Earth's atmosphere (about −90 °C). Meteors burn up in this layer. The mesosphere ends at a height of about 80 km. This level is called the mesopause.

The **thermosphere** starts from 80 to 90 km and ends between 500 and 1000 km. This layer is characterized by the temperature increase. Formally, the thermosphere is a part of the Earth's atmosphere. However, the air density here is so low that it can be thought as the outer space.[3] The atoms and molecules of oxygen and nitrogen present in this layer absorb the short-wave radiation harmful to life and get ionized. In addition, collisions between charged particles from space and atoms and molecules in the thermosphere lead to photon emission, which we see as the auroras. The layer of ions makes the ionosphere (another word for thermosphere), which reflects radio waves and allows their propagation at large distances. The temperature is determined by the solar activity and can reach the value of about 2000 K at 500 km altitude.

The **exosphere** is the highest layer. It is where the atmosphere merges into outer space. This layer mostly consists of widely dispersed molecules of hydrogen and helium. The molecules flight according to ballistic trajectories since the mean free path can be up to several kilometers. Their velocity is sufficient to overcome the gravitational field of the Earth and escape into outer space according to Eq. (2.17).

The atmosphere can be also divided into *homosphere and heterosphere* [10]. In the homosphere, the air molar mass changes slightly with the altitude and is about 28.96 g/mol. Roughly above 100 km, the processes of dissociation of O_2 become significant, thereby increasing the concentrations of atomic oxygen O. This leads to a decrease in the air molar mass reaching the value of about 11 g/mol at the height 600 km. This region is referred to as the heterosphere.

The main conclusion of the analysis of the atmospheric layers and processes within them is the following: the definition of the top of the atmosphere is application-specific. For the modeling of solar radiation propagation in the visible spectral range, it is sufficient to consider the atmosphere up to 50–80 km in its height. The thermal radiation modeling may require the atmospheric model up to 150 km. Finally, for examining the processes related to ionization of the atmospheric gases and interactions with charged particles coming from outer space, the upper boundary is placed at the altitude of about 500–1000 km.

2.5   Gaseous Composition of the Atmosphere

The gaseous composition of the atmosphere is one of the major subjects of interest of remote sensing applications. The atmospheric composition is in focus of such disciplines as environmental chemistry, meteorology, atmospheric chemistry, climatology and others. The vertical composition of the atmosphere changes with the altitude. Three-quarters of the whole mass of the atmosphere is within about 11 km above the

---

[3]In fact, the International Space Station orbits the Earth in the thermosphere [9].

Table 2.1 Permanent atmospheric gases

Constituent gas	By mass (%)	By volume (%)	Parts per million (ppm)	Molecular weight (g/mol)
Nitrogen (N_2)	75.5	78.09	780840	28.02
Oxygen (O_2)	23.1	20.92	209460	32.00
Argon (Ar)	1.3	0.93	9340	39.94

sea level. The three major constituents of the atmosphere are nitrogen (N_2), oxygen (O_2) and Argon (Ar), as shown in Table 2.1. Regarding water vapor, they account for 99.95% of the total volume and 99.9% of the total mass of the atmosphere. As we already noted, the mean molecular weight of air is $M_{air} = 28.97$ g/mol, which is mostly due to nitrogen ($M_{N_2} = 28.014$ g/mol). N_2, O_2, and Ar concentrations compositions vary little with time, so they are considered as permanent gases.

Other atmospheric gases (some of which are shown in Table 2.2) are referred to as trace gases, which in combination make up less than 0.01% of the total mass and the total volume (sometimes, Argon is also treated as a trace gas). Despite this fact, trace gases play very important roles in the Earth climate system because of their radiative and thermodynamic properties.

Several trace gases intensively absorb the Earth's thermal radiation while weakly absorbing (or even not absorbing) the solar radiation, as it passes through the atmosphere, and so cause the greenhouse effect. The primary greenhouse gases are water vapor (H_2O), carbon dioxide (CO_2), methane (CH_4), nitrous oxide (N_2O) and ozone (O_3). The concentrations of greenhouse gases are of special interest due to their impact on the climate and the environmental equilibrium. A remarkable research published by Keeling indicated a steady increase in CO_2 since 1958. This increase described by the Keeling curve [11] (shown in Fig. 2.3) was a subject of many political debates and scientific discussions. However, there is a consensus that the greenhouse concentrations should be monitored globally.

Some trace gases are the so-called volatile organic compounds (VOCs), which have a high vapor pressure at ordinary room temperature. That means, they may exist in the liquid or solid phase and they can easily evaporate. The main VOC produced by plants and animals is isoprene (C_5H_8). Some VOCs can be toxic or have long-term health effects (for instance, formaldehydes are classified as known human carcinogen [13]). The other VOCs are responsible for pleasant odors of flowers and perfumes.

In addition to gases, there are free radicals, such as hydroxyl radical (OH) and atomic oxygen (O). They exist due to dissociation processes induced by hard ultraviolet radiation. For instance, the process of stratospheric ozone formation can be described by two reactions:

$$O_2 \xrightarrow{\text{UV}} O + O \qquad (2.28)$$

Table 2.2 Atmospheric trace gases

Constituent gas	By volume	Parts per million (ppm)	Molecular weight
Carbon dioxide (CO_2)	0.04	400*	44.0
Neon (Ne)	$1.8 \cdot 10^{-3}$	18.18	20.18
Helium (He)	$5.2 \cdot 10^{-4}$	5.24	4.0
Methane (CH_4)	$1.7 \cdot 10^{-4}$	1.7	16.04
Krypton (Kr)	$1.1 \cdot 10^{-4}$	1.14	83.7
Hydrogen (H_2)	$5.5 \cdot 10^{-5}$	0.55	2.02
Nitrous oxide (N_2O)	$5.0 \cdot 10^{-5}$	0.5	44.0
Xenon (X)	$9.0 \cdot 10^{-6}$	0.09	131.3
Nitrogen dioxide (NO_2)	$2.0 \cdot 10^{-6}$	0.02	46.0
Ozone (O_3)	$0.1 \cdot 10^{-6}$	0.1	48.0
Radon (Rn)		$6.0 \cdot 10^{-18}$	222.0

*In 2013 this number surpassed 400. See Fig. 2.3

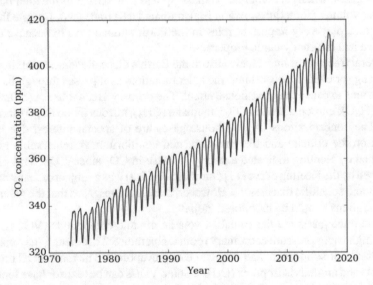

Fig. 2.3 Monthly CO_2 concentrations measured at Barrow, Alaska. Data is taken from [12]

and

$$O + O_2 + N_2 \longrightarrow O_3 + N_2. \tag{2.29}$$

Note that N_2 just stabilizes the O_3 molecule by taking some of the energy from it. The production of hydroxyl is given by two reactions which occur across the stratosphere and the troposphere:

$$O_3 \xrightarrow{\text{UV}} O_2 + O^* \tag{2.30}$$

and

$$O^* + H_2O \longrightarrow 2OH. \tag{2.31}$$

Here O^* is excited-state oxygen.

In the view of the strong influence of trace gases on the terrestrial climate system, it is crucial to monitor trace gas concentrations and to analyze their changes in time and space. Two types of trace gas sources and sinks can be distinguished—natural (i.e., caused by processes that occur in nature) and anthropogenic (i.e., caused by human activity). The balance between them determines trace gas amounts in the atmosphere. The concentrations of some gases (like water vapor, O_3, CO_2) are highly variable and have a seasonal behavior.

The amount of a certain gas is characterized by its mole fraction x, i.e., the number of moles ν_i of a specific gas per mole of air ν_{air}:

$$x_i = \frac{\nu_i}{\nu_{air}}. \tag{2.32}$$

Mole fractions for trace gases either unitless (mol/mol) or can be expressed in "parts per million by volume" units (ppmv): 1 ppmv $= 1 \cdot 10^{-6}$ mol/mol. Given that the atmospheric pressure is sufficiently low, the ideal gas law can be used:

$$pV = \nu RT, \tag{2.33}$$

where p, V and T are the pressure, volume and absolute temperature, respectively, ν is the number of moles of gas, and R is the ideal gas constant. Then the mole fraction equals to the volume mixing ratio (VMR), and

$$V_i = V_m \nu_i, \tag{2.34}$$

where V_m is the molar volume of an ideal gas $\left(\approx 0.022 \, \text{m}^3/\text{mol}\right)$. For each gas, the partial pressure can be introduced. According to Dalton's law, the total pressure p_{total} of a mixture of non-reacting gases is equal to the sum of their partial intensities p_i :

$$p_{total} = \sum_{i=1}^{n} p_i. \tag{2.35}$$

Using the definition of partial intensities,

$$p_{total} V_i = p_i V_{air}, \tag{2.36}$$

we conclude that

$$x_i = \frac{\nu_i}{\nu_{air}} = \frac{V_i}{V_{air}} = \frac{p_i}{p_{air}}. \tag{2.37}$$

Another important characteristic is the number density n_i, i.e., the number of molecules of an ith gas per unit volume of air. It is expressed in units of number of molecules per cm^3. The vertically integrated concentration over the atmospheric column gives the vertical column density (VCD):

$$\text{VCD}\left[\text{cm}^{-2}\right] = \int_0^{h_{\text{TOA}}} n_i(h)dh, \tag{2.38}$$

where h_{TOA} is the altitude of the top of the atmosphere. The VCD is expressed as number of molecules per unit area (e.g., cm^{-2}). The partial VCD of the ith gas between altitudes h_1 and h_2 is defined as

$$X_i\,(h_1, h_2)\left[\text{cm}^{-2}\right] = \int_{h_1}^{h_2} n_i(h)dh. \tag{2.39}$$

Thus, it can be regarded as an integral over the atmospheric layer. The slant column density (SCD) is the concentration integrated along the direction defined by the angle θ. In this case, for the plane-parallel atmosphere, the following integral should be evaluated:

$$\text{SCD} = \frac{1}{\cos\theta} \int_0^{h_{\text{TOA}}} n_i(h)dh. \tag{2.40}$$

When θ approaches $90°$, the sphericity of the atmosphere should be taken into account. If the integration in Eq. (2.40) is performed within a certain layer, then we obtain the partial SCD.

Dividing the partial column by Loschmidt's number n_L (i.e., the number of molecules of an ideal gas in a given volume) at standard temperature and pressure,

$$n_L = \frac{p_0}{kT_0} \approx 2.687 \cdot 10^{19}\,\text{cm}^{-3}, \tag{2.41}$$

gives the partial column in cm. Thus,

$$X_i\,(h_1, h_2)\,[\text{cm}] = \frac{kT_0}{p_0} \int_{h_1}^{h_2} n_i(h)dh. \tag{2.42}$$

Consequently, the partial column expressed in cm can be understood as the thickness of the layer of the ith gas which would be formed at standard conditions for temperature and pressure preserving the same amount of gas molecules. Atmospheric ozone is typically measured in Dobson units (DU):

$$X_i\,(h_1, h_2)\,[\text{DU}] = 0.01\frac{kT_0}{p_0} \int_{h_1}^{h_2} n_i(h)dh. \tag{2.43}$$

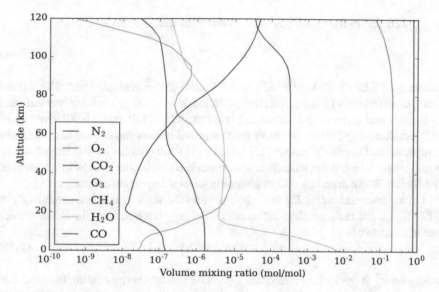

Fig. 2.4 Volume mixing ratios for various species versus height according to the standard atmosphere model [1]

One Dobson unit corresponds to 10 μm of the gas column at standard conditions or 2.687×10^{20} molecules per square meter. For instance, an ozone layer of 300 DU would form a layer of 3 mm at standard conditions for temperature and pressure.

Figure 2.4 shows the profiles of volume mixing ratios for various species versus height according to the standard atmosphere model [1]. Note that the profiles for major constituents of the atmosphere (e.g., O_2 and N_2) are constant, while the trace gas profiles change in time and thus are subject to continuous monitoring.

2.6 Radiation Transfer from the Sun to the Earth

Earth is a solar-powered system. The sunlight warms the Earth system, drives the climate system and makes our planet habitable. The solar energy defines not only temperature, but in indirect way, other characteristics such as humidity, windness, cloudiness etc. The important role of energy balance for the climate system was understood by early climatologists in the pre-satellite era. However, this topic appeared to be quite complex as proper modeling of the climatological system requires intensive computing. Moreover, only with satellites it was made possible to measure incoming and outgoing energy fluxes and analyze all factors which have an impact on the energy balance.

The basic qualitative analysis of the energy balance of the Sun–Earth system can be performed based on the Stefan–Boltzmann law [14], which states that the total radiant heat power E emitted from a surface of a black body across all wavelengths is proportional to the fourth power of its absolute temperature T:

$$E = \sigma T^4, \tag{2.44}$$

where $\sigma \approx 5.67 \cdot 10^{-8} \, \text{W} / \left(\text{m}^2 \text{K}^4 \right)$ is the Stefan–Boltzmann constant. The radiant heat power has units $[\text{W}/\text{m}^2]$. Note that m^2 in the denominator stands for "per unit area of the emitting surface". It is interesting that the Stefan–Boltzmann law derived in the nineteenth century and its discovery made it possible to estimate quite accurately the temperature of the Sun's surface [15]. This law is valid only for local thermodynamic equilibrium, which is not the case at atmospheric layers higher than 50 km. However, the Stefan–Boltzmann law allows obtaining several important estimations.

Giving the solar radius $R_\odot \approx 7 \cdot 10^5$ km and the solar surface temperature $T_\odot \approx 5777$ K, we can estimate the total amount of energy emitted from the Sun's surface per one second:

$$L_\odot = \sigma T_\odot^4 \times 4 \pi R_\odot^2 \approx 3.86 \cdot 10^{26} \, \text{W}, \tag{2.45}$$

where $4 \pi R_\odot^2$ is the surface area of the Sun. The radiant energy coming from the Sun crossing a square meter of the sphere at Earth's orbit each second is called the solar constant $S_\odot$. It can be found by noting that each second the same amount of energy $L_\odot$ must also cross the sphere of radius which is equal to the distance $d_\odot$ between the Sun and the Earth, $d_\odot \approx 150 \cdot 10^6$ km. Therefore,

$$S_\odot = \frac{L_\odot}{4 \pi d_\odot^2} \approx 1379 \, \text{W}/\text{m}^2. \tag{2.46}$$

Actually, the solar constant should not be considered as a physical constant (e.g., it is not like Planck's constant). Its changes are monitored using spaceborne observations directly outside the atmosphere (see, e.g., [16]), although its variations are not significant in most cases (excluding $\sim$6–7% (i.e., 90 W/m^2) variations attributed to the seasonal change in $d_\odot$) [17].

We can compute how much solar energy comes at the top of the atmosphere. Indeed, the amount of energy P_{in} striking the Earth is the product of the solar constant $S_\odot$ and the area of the equatorial cross-section $S = \pi R_E^2$, as shown in Fig. 2.5.[4] Thus,

$$P_{\text{in}} = S_\odot \times \pi R_E^2 \approx 1.8 \cdot 10^{17} \, \text{W}. \tag{2.47}$$

Let us estimate the amount of radiant energy W_{in} coming at the top of the atmosphere from the Sun in one year. For this, we multiply P_{in} by the time $t \approx 3.2 \cdot 10^7$ s:

$$W_{\text{in}} = P_{\text{in}} t \approx 5.7 \cdot 10^{24} \, \text{J} = 1.6 \cdot 10^9 \, \text{TWh}. \tag{2.48}$$

---

[4]In fact, here we should estimate a flux through a hemisphere. However, it is equal to the flux through the equatorial cross-section according to Gauss's theorem.

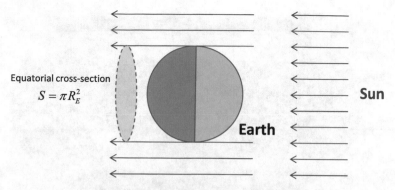

Fig. 2.5 The amount of energy striking the Earth equals the amount of energy striking the equatorial cross-section

For comparison, the humans consume $\sim 10^6$ TWh [18], i.e., 0.1% of that amount coming from the Sun.

The Sun doesn't heat the Earth evenly in time and space. The equatorial regions receive more energy than polar regions due to Earth's spherical shape. Figure 2.6 shows the map of monthly averaged incoming solar irradiance measured by the Cloud and the Earth's Radiant Energy System (CERES) on board the Suomi National Polar-orbiting Partnership (SNPP) and Joint Polar Satellite System (JSPP)—that is an instrument specifically designed to monitor Earth's energy budget [19]. The incident rays hit Earth's surface at larger angles with the surface normal as we move from equator to the pole. Therefore, the flux decreases approximately as the cosine of latitude. In addition, the 23.5° Earth's axis tilt causes seasonal variations. At each moment of time only one side of the planet gets to the Earth orbit the sunlight. Solar heating imbalance produces the climate's heat engine which acts as a driving force of the changes at the surface of the planet (seasonal cycles), and also atmosphere and hydrosphere circulations involving evaporation, convection, rainfall and winds. Interestingly, the uneven distribution of solar heating across the Earth is responsible for the so-called trade winds, i.e., east-to-west prevailing winds that flow mainly in the equatorial region, which the sailors were well aware of already in the fifteenth century. Together with Coriolis forces due to Earth's rotation, the atmospheric circulation has a complex multizonal structure consisting of Hadley cells, Ferrel cells and polar cells [20–22].

2.7 Earth's Energy Balance Based on the Stefan–Boltzmann Law

About 30% of the radiant energy falling on the planet is reflected back into space and therefore must be subtracted from the energy balance. To take it into account, it is convenient to introduce the albedo A as the fraction of incident solar radiation

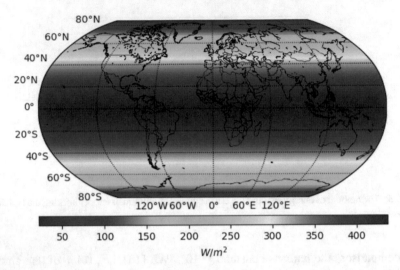

Fig. 2.6 Monthly mean incoming solar irradiance as provided by CERES for March 2012. Data is taken from [19]

reflected by the Earth.[5] The term "albedo" originates from the Latin word "albus", which means "white". Actually, the albedo is more complex physical parameter than assuming a single constant value for a surface. In fact, it is spectrally and angularly dependent. However, for the sake of simplicity, we model the reflection from the surface just by a single scalar value. Thus, the total energy absorbed by the Earth per second is

$$P_{abs} = S_{\odot} \times \pi R_E^2 \times (1 - A).$$ (2.49)

The albedo can be estimated from spacecraft measurements, for instance, by CERES. Figure 2.7 shows the map of Earth's albedo provided by CERES for March 2012 [19]. The values of albedo strongly depend on the type of surfaces, as shown in Table 2.3. High values of albedo are due to clouds and snow cover, while the albedo of Earth's surface is usually lower than 0.3. As we can see from Fig. 2.7, the average albedo for the Earth is about 0.3 [23]. Thus,

$$P_{abs} \sim 10^{17} \, \text{W}.$$ (2.50)

The total energy emitted by Earth per second is

$$P_{emit} = 4\pi R_E^2 \times \sigma T_e^4.$$ (2.51)

---

[5]Actually, the albedo depends on the geometry of observation and wavelength. Here we consider angular and spectrally averaged values.

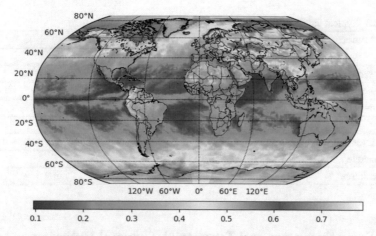

Fig. 2.7 The map of Earth's albedo for March 2012 provided by CERES. The albedo is computed as the ratio between reflected shortwave irradiance and incident solar irradiance. Data is taken from [19]

Here T should be understood as the effective temperature of the Earth, i.e., the temperature of the Earth as it appears from space according to its thermal emission. The effective temperature is also referred to as equilibrium temperature.

Then, we assume that the energy that absorbed from the sun's radiation is balanced by an equal energy that Earth radiates back into outer space. Were this not the case, the Earth would heat up or cool off.[6] Thus, we set

$$P_{abs} = P_{emit}. \tag{2.52}$$

Substituting Eqs. (2.49) and (2.51) into Eq. (2.52) one derives the following equation:

$$S_{\odot}(1 - A) = 4\sigma T_{e}^{4}, \tag{2.53}$$

from which we get the equilibrium temperature:

$$T_{e} = \sqrt[4]{\frac{S_{\odot}(1 - A)}{4\sigma}}. \tag{2.54}$$

Taking $A = 0.3$ for the Earth, we get $T_{e} = 253\,\text{K} = -20°C$. These values are much lower than the Earth's mean surface temperature $\bar{T} = 288\,\text{K} = 15°C$. As we will see later, such a discrepancy is a result of the greenhouse effect.

---

[6]It is interesting that there are exceptions from these assumptions. For instance, the planets Jupiter, Saturn and Neptune emit more energy than they absorb [24].

Table 2.3 Typical values of the surface albedo in the visible spectral range

Type of surface	Typical albedo values
Ocean	0.06
Desert sand	0.4
Fresh snow	0.75–1.0
Old snow, glaciers	0.4–0.7
Forest	0.05–0.2
Green grass	0.25
Dry concrete	0.2–0.3
Grain cultures	0.15–0.25

2.8 Solar Spectrum and Terrestrial Thermal Emission

So far, we have used only the Stefan–Boltzmann law to describe the solar incoming radiation, which provides the radiative flux throughout the whole spectrum. That was sufficient to estimate Earth's energy balance. However, more detailed consideration is required to take into account the spectral dependency of incoming radiation [25].

The examples of the solar spectrum which can be observed at the top of the atmosphere are shown in Fig. 2.8 and in Appendix 1. The blue line corresponds to the zero reference spectrum ASTM E-490 [26], which is a compilation of the data from satellites, space shuttle missions, rocket soundings and modeled data. The shape of the spectrum can be understood in terms of Planck's law, which describes the amount of electromagnetic radiation energy emitted by a black body in thermal equilibrium at a given temperature T in a unit solid angle by a unit surface at a certain wavelength

$$B\left(\lambda\right) = \frac{2hc^2}{\lambda^5} \frac{1}{e^{\frac{hc}{\lambda kT}} - 1}. \tag{2.55}$$

The unit for B is $W \cdot sr^{-1} \cdot m^{-2} \cdot nm^{-1}$. Note that here m^{-2} refers to "per unit area of the *emitting* surface". It can be shown that integrating Eq. (2.55) over all wavelengths and hemisphere gives the Stefan–Boltzmann law:

$$\int_0^\infty \int_{\text{hemisphere}} B\left(\lambda\right) \cos\theta d\Omega d\lambda = \sigma T^4, \tag{2.56}$$

where the Stefan–Boltzmann constant is expressed as follows:

$$\sigma = \frac{2k^4 \pi^5}{15c^2 h^3}. \tag{2.57}$$

Similar to Eqs. (2.45)-(2.46), the solar spectrum at the top of the Earth's atmosphere coming to a unit area surface is computed as

$$E_\odot (\lambda) = B (\lambda) \frac{R_\odot^2}{d_\odot^2}.$$ (2.58)

As it can be seen in Fig. 2.8, the radiation from the Sun can be modeled by Planck's law with T of about 5700 K—that is the effective temperature of the Sun. The radiation curve of the Sun is maximum in the visible spectral range. Naturally, the area below the spectrum gives the energy coming from the Sun per m^2 per second, i.e., the solar constant $S_\odot$, i.e.,

$$\int_0^\infty E_\odot (\lambda) \, d\lambda = S_\odot.$$ (2.59)

The energy is distributed over the spectrum in the following way: ~ 200 W/m^2 is due to the ultraviolet radiation, ~ 500 W/m^2 is due to the visible range, and ~ 600 W/m^2 is due to the infrared range.

Figure 2.9 shows the Planck's functions of the Earth and the Sun. One can see that the Earth emits similar to a 300K black body with a dominant wavelength around $10 \, \mu$m. The Earth's thermal radiation should be taken into account at wavelengths larger than $3 \, \mu$m. As predicted by the Wien's displacement law, the peak of the black body radiation curve shifts to shorter wavelengths as the temperature increases:

$$\lambda_p T = b,$$ (2.60)

where λ_p is the wavelength at which the radiation curve has a peak, while $b \approx 2898 \, \mu$m $\cdot$ K is the Wien's displacement constant. The solar Planck function with temperature ≈ 5700 K peaks at 500 nm while the Earth's Planck function with temperature ≈ 300 K peaks at 10000 nm. In this regard, when talking about the radiative balance, the solar radiation is often referred to as shortwave radiation, while the terrestrial thermal emission is referred to as longwave radiation. From this point of view, we can consider the Earth as a system which converts shortwave radiation into longwave radiation. Such transformation is behind all climatological processes occurring on our planet.

Again, referring to the CERES data, it is possible to obtain a map of net radiation, i.e., incoming shortwave sunlight irradiance minus reflected shortwave and outgoing longwave irradiances. Figure 2.10 reveals spatial energy imbalance: areas around the equator absorb about 150 W/m^2 more on average than they reflect and emit. Near to the poles the budget is inverted with more than 150 W/m^2 reflected and emitted over absorbed.

Coming back to the idea of climate's heat engine, we can estimate the velocity of trade winds v_w [27]. Indeed, the mechanical power is

$$P_m = \eta P_{in} (1 - A) \sim 10^{15} \text{ W},$$ (2.61)

where η is the thermodynamic efficiency of the climate's heat engine. Following [21, 28] it is equal to 2%. The absorbed energy is transferred into kinetic energy of air. Thus,

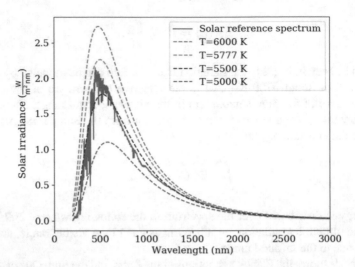

Fig. 2.8 Solar spectrum at the top of the atmosphere: the blue line shows the reference spectrum E-490, while the other curves are the spectra computed by using Planck's law

Fig. 2.9 Planck's functions for the Earth and the Sun

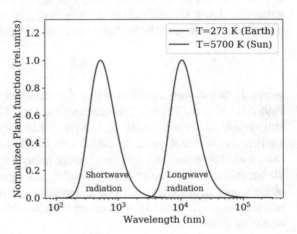

$$P_{\mathrm{m}}\tau_{\mathrm{t}} \sim \frac{m_{\mathrm{a}}v_{\mathrm{w}}^2}{2}, \qquad (2.62)$$

where τ_{t} is the characteristic time for which air moves a distance equal to the radius of the Earth. Estimating τ_{t} as

$$\tau_{\mathrm{t}} \sim \frac{R_{\mathrm{E}}}{v_{\mathrm{w}}} \qquad (2.63)$$

and substituting Eq. (2.63) into Eq. (2.62), we obtain

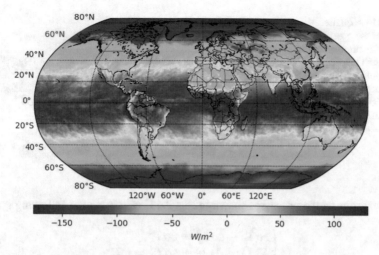

Fig. 2.10 Net radiation map for March 2012 as measured by CERES

$$v_{\mathrm{w}} \sim \sqrt[3]{\frac{\eta P_{\mathrm{in}} (1 - A) R_{\mathrm{E}}}{m_{\mathrm{a}}}}. \tag{2.64}$$

That gives $v_{\mathrm{w}} \sim 10\,\mathrm{m/s}$ and $\tau_{\mathrm{t}} \sim 1$ week. The latter is the characteristic time for weather change caused by large-scale air movement.

2.9 Qualitative Description of the Greenhouse Effect

The greenhouse effect was discussed already in the nineteenth century by Fourier and Arrhenius [29]. Today, the term "greenhouse effect" sounds somewhat misleading, since in the atmosphere the heat is prevented from escaping by using radiative mechanisms, while in real greenhouses, the elevated temperature is caused by the glass cover which suppresses the convective heat losses. The atmospheric greenhouse effect is based on the fact that the optical properties of the atmosphere are different for the shortwave and longwave radiation. The atmosphere is essentially transparent for the shortwave radiation, but mostly opaque to the longwave part of the spectrum. The gases that cause this absorption are referred to as greenhouse gases. The primary greenhouse gases are water vapor (H_2O), carbon dioxide (CO_2), ozone (O_3) and methane (CH_4).

A simplistic but very intuitive treatment can be found in [6, 30]. We model the atmosphere as a single-layer gray body with the atmospheric temperature T_{a} and the emissivity $\varepsilon < 1$. Thus, the atmosphere emits the radiation of the power $\varepsilon \sigma T_{\mathrm{a}}^4$ per unit surface area. The Earth's surface temperature is T_{s}. According to Kirchhoff's law of thermal radiation, the absorptivity (i.e., the ratio of the absorbed energy to the incident energy) equals the emissivity as long as the body is in thermodynamic equilibrium.

Fig. 2.11 A qualitative model of the greenhouse effect. The atmosphere is assumed to be totally opaque for the solar radiation. The emissivity (and absorptivity) of the Earth's thermal radiation for the atmosphere is ε

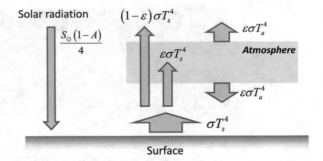

Thus, examining the processes shown in Fig. 2.11, we obtain the following system:

$$
\begin{cases}
\varepsilon\sigma T_a^4 + \varepsilon\sigma T_a^4 = \varepsilon\sigma T_s^4, \\
S_\odot\,(1-A)\,/4 + \varepsilon\sigma T_a^4 = \sigma T_s^4.
\end{cases}
\tag{2.65}
$$

Solving the system (2.65) we find that

$$
\begin{cases}
T_s = \sqrt[4]{\dfrac{S_\odot\,(1-A)}{4\sigma\,(1-\varepsilon/2)}}, \\
T_a = T_s/\sqrt[4]{2}.
\end{cases}
\tag{2.66}
$$

This solution shows that the surface temperature is higher than that predicted by Eq. (2.54), while the temperature of the surface T_s is higher than that of the atmosphere T_a. Taking $\varepsilon = 0.8$, we get $T_s \approx 290$ K, which is almost $40°$ higher than that follows from Eq. (2.54). On its turn, the temperature of the atmosphere is lower than the surface temperature.

2.10 Radiative Forcing and Climate Sensitivity

If condition (2.52) is not fulfilled, then the total amount of heat energy W is changing with time:

$$
\frac{\partial W}{\partial t} = P_{abs} - P_{emit} \equiv \text{RF}.
\tag{2.67}
$$

The difference (RF) on the right side of Eq. (2.67) is referred to as radiative forcing (or climate forcing) [31]. Here the word "forcing" can be understood as a force which pushes the Earth's climate system in the direction of warming or cooling. Radiative forcing is given in units of $W \cdot m^{-2}$. Usually, RF is evaluated at the tropopause.

Radiative forcing can be regarded as a measure of how the energy balance of the Earth responds when factors that affect the climate are altered [32]. It is used to assess the influence of anthropogenic and natural agents, which are responsible for the climate change. If $P_{abs} > P_{emit}$, then the Earth heats up. The surface temperature and corresponding P_{emit} increase until a new equilibrium temperature is reached. Thus, whenever RF $\neq 0$, there is a corresponding change in the surface temperature to balance the energy budget. Assuming linear dependence between RF and a corresponding temperature change ΔT we obtain [33]

$$RF + \frac{\partial RF}{\partial T} \Delta T = 0, \tag{2.68}$$

or rewriting Eq. (2.68)

$$\Delta T = \lambda RF, \tag{2.69}$$

with

$$\lambda = -\left(\frac{\partial RF}{\partial T}\right)^{-1}. \tag{2.70}$$

Here λ is the climate sensitivity parameter [34] and given in units of $\left(K \times m^2\right)/W$. Its reciprocal, $\alpha = 1/\lambda$, is called the climate feedback parameter [35].

2.11 Equilibrium Time

When a force is changed, a climate system requires some time to adjust to it. To illustrate this concept, we analyze what happens when the incoming radiation is changed. We rewrite Eq. (2.53) for the incoming solar constant perturbed by $\Delta S_\odot$:

$$4\sigma \left(T_e + \Delta T_e\right)^4 = \left(S_\odot + \Delta S_\odot\right)\left(1 - A\right), \tag{2.71}$$

where $T_e + \Delta T_e$ is the new equilibrium temperature, with ΔT_e being the corresponding temperature perturbation. Subtracting Eq. (2.53) from Eq. (2.71) and providing that $\Delta T_e/T_e \ll 1$, we get

$$\Delta T_e = \frac{\Delta S_\odot \left(1 - A\right)}{16\sigma T_e^3}. \tag{2.72}$$

Solution (2.72) shows how sensitive the temperature fluctuations are to the changes in the solar flux.

Now we estimate how fast the corresponding changes occur, i.e., we want to find the equilibrium temperature perturbation $\Delta T(t)$ in time providing the solar constant perturbation $\Delta S_\odot$. For that, we substitute Eqs. (2.49) and (2.51) into Eq. (2.67):

$$\frac{\partial W}{\partial t} = S_\odot \times \pi R_E^2 \times (1 - A) - 4\pi R_E^2 \times \sigma T_e^4 = 0. \tag{2.73}$$

Note, Eq. (2.73) is equal to zero as we consider equilibrium. Following [6] we consider the solar constant perturbation $\Delta S_\odot$, which leads us to the following equation:

$$\frac{\partial (W + \Delta W (t))}{\partial t} = (S_\odot + \Delta S_\odot) \times \pi R_E^2 \times (1 - A) - 4\pi R_E^2 \times \sigma (T_e + \Delta T (t))^4,$$

$$(2.74)$$

where ΔW is the perturbations in the amount of heat energy of the atmosphere. Subtracting Eq. (2.73) from Eq. (2.74) and providing

$$\Delta W = m_a c_p \Delta T, \tag{2.75}$$

where $c_p \approx 1005 \, \text{J}/(\text{kg} \times \text{C})$ is the specific heat capacity of air at constant pressure, we obtain the following first-order linear non-homogeneous differential equation:

$$\frac{m_a c_p}{4\pi R_E^2} \frac{\partial (\Delta T (t))}{\partial t} = \frac{\Delta S_\odot}{4}(1 - A) - 4\sigma T_e^3 \Delta T (t). \tag{2.76}$$

Its solution is given as

$$\Delta T (t) = \Delta T_e \left(1 - e^{-t/\tau}\right), \tag{2.77}$$

where τ is referred to as the radiation relaxation time constant given by

$$\tau = \frac{m_a c_p}{16\pi R_E^2 \sigma T_e^3}. \tag{2.78}$$

Examining Eq. (2.77) we see that $\Delta T (t)$ goes to Eq. (2.72) as t increases. The value of τ gives a time scale during which the change in temperature due to radiation emission becomes significant. For the Earth, we have $\tau \approx 30$ days. To make our consideration a bit more realistic, we substitute T_e in Eq. (2.78) with $\bar{T}$, in which case $\tau \approx 20$ days. Equation (2.77) can be simplified for $t \ll \tau$:

$$\Delta T (t) = \Delta T_e \frac{t}{\tau} \tag{2.79}$$

or taking into account Eqs. (2.72) and (2.78) we get a simplified expression:

$$\Delta T (t) = \frac{\Delta S_\odot (1 - A) \pi R_E^2}{m_a c_p}t, \tag{2.80}$$

which allows us to estimate a diurnal variation of the temperature associated with day/night cycle. Taking $\Delta S_\odot \sim S_\odot$, we get $\Delta T \sim 1$ K. Here we just assumed that the Sun does not illuminate the atmosphere-underlying surface system during the night at all (which is of course not true). Essentially, we obtained the upper boundary for temperature fluctuations. However, we know that the day/night variations of temperature usually exceed 5 K. Thus can conclude that the daily variations are mostly due to the interaction between the atmospheric boundary layer and the surface, rather than radiation emission to outer space. Table 2.4 provides a comparison between Venus, Earth and Mars. As we can see, Venus has the strongest greenhouse effect, while the daily variations of the temperature are pronounced only for Mars and negligible for the others.

Table 2.4 Characteristics of three planets (data is taken from [6, 37])

Planet	Venus	Earth	Mars
Atmospheric mass (kg)	$4.8 \cdot 10^{20}$	$5.2 \cdot 10^{18}$	$2.5 \cdot 10^{16}$
Solar flux (W/m^2)	2600	1400	600
Equilibrium temperature (K)	227	253	216
Surface temperature (K)	750	288	240
Radiative time constant, terrestrial days	130	20	1
Day length, terrestrial days	233	1	1
Albedo	0.77	0.3	0.15

Table 2.5 Equilibration times for Earth's subsystems (data is taken from [36])

Earth's subsystems	Order of magnitude for the equilibration time
Boundary atmospheric layer	1 day
Atmosphere	10 days
Lakes and rivers	10 days
Deep ocean	100–1000 years
Snow and surface ice layer	1 day
Mountain glaciers	100 years
Ice sheets	1000s of years
Soil/vegetation	10 days—100s of years
Lithosphere	Millions of years

Actually, in Eq. (2.75) we should have taken into account the heat energy in the ocean and Earth's surface. However, the equilibrium times for hydrosphere and lithosphere are several orders of magnitude larger than for the atmosphere [36] (as shown in Table 2.5) and therefore, in this illustrative example, these processes are neglected. So large difference in equilibration times of the Earth's subsystems creates a difficulty in understanding the climate. In particular, it becomes a challenge to assess time-dependent responses from the ocean, as well as to estimate the climate sensitivity parameters.

2.12 Feedbacks

Broadly speaking, feedback occurs when the result of a certain action modifies a reason which has caused that action. The result of such a looped interaction can be either an amplification (i.e., a positive feedback) of the action or an attenuation (i.e., a negative feedback). The climate system incorporates many radiative forcing agents, where each of them produces certain feedback, thereby making the climate modeling a very challenging task. Below we consider several important feedbacks.

(a) **Water vapor feedback**: When the global surface temperature increases, the evaporation of water is more intensive and the amount of water vapor in the atmosphere increases. Water vapor is a greenhouse gas as it absorbs the longwave radiation emitted by the Earth's surface. Thus, the greenhouse effect becomes stronger and the temperature further increases. That is a positive feedback that amplifies the original warming.

(b) **Surface feedback**: When the global surface temperature increases, then snow and ice melt. From Table 2.3 it follows that it leads to the decrease in albedo. Thus, more portions of incoming radiation are absorbed by the surface, thereby further increasing the global surface temperature. That is a positive feedback.

(c) **Biogenic feedback**: When CO_2 level increases, the Earth becomes warmer and it intensifies the growth of plants, which tend to absorb CO_2 due to photosynthesis. That is an example of a negative feedback.

2.13   Atmospheric Aerosols

Atmospheric aerosol is a global phenomenon present worldwide. Essentially, atmospheric aerosol is composed of microscopic solid and liquid particles (except water droplets and ice crystals, which form the clouds) suspended in the terrestrial atmosphere. Aerosols are often referred to as PM (stands for "particulate matter") equipped with a low index which characterizes the particle diameter. For instance, $PM_{2.5}$ corresponds to particles with a diameter of $d_p = 2.5$ µm or less. Such particles are considered to be fine, while those with a diameter of about $d_p = 10$ µm are called coarse particles. Sources of particles can be natural or anthropogenic. Some examples of natural sources are the oceans, volcanoes, forest fires, gas-to-particle conversion, living vegetation, dust storms and aerosol of cosmic origin. Anthropogenic sources are due to traffic based on fossil fuels, power plants, coal combustion etc. However, sometimes it is not straight-forward to distinguish anthropogenic from natural aerosols. Emission fluxes for some aerosols are shown in Table 2.6. Precise description of aerosol composition requires *in situ* chemical measurements, which of course impose time and location restrictions. Therefore, to estimate the anthropogenic part of aerosols, remote sensing measurements equipped with aerosol models and information on urban and agricultural activities are involved. In addition to observations from space, ground-based photometers are used, for instance, the AERONET (Aerosol Robotic NETwork) consisting of more than 500 automated instruments [38].

These tiny yet ubiquitous aerosol particles have various impacts on global environmental change, climate, and chemical heterogeneous reactions in the atmosphere [39]. The main climate effect consists in changing the amount of incoming solar radiation and outgoing thermal radiation due to the presence of aerosols. There are several mechanisms that cause such a change. Depending on their chemical composition, they can contribute to the warming or the cooling of the atmosphere. The aerosols can affect the net energy balance directly by absorbing or scattering light. In the case of absorption, for instance, by black carbon (soot) in aerosol particles, the atmosphere is heated. Scattering leads to an increase in albedo values and cooling the atmosphere. Also, the deposition of absorbing aerosol on snow causes more solar radiation to be absorbed at the terrestrial surface [40]. Here aerosol acts as a radiative forcing agent. As opposed to direct impact, the indirect effect of aerosols consists of a modification of cloud properties, which results in the changes of the Earth's radiative budget and modulates precipitations [41]. For instance, aerosols may lead to smaller droplets and higher albedo of the cloud, thereby cooling the Earth.

The effect of aerosols on the climate is quantified in terms of aerosol radiative forcing. Solar radiation absorption caused by aerosols is comparable in magnitude to that caused by trace gases. That means poor knowledge of aerosols may cause the uncertainty in the climate forcing models as large as the current estimation of the radiative forcing itself. Complex studies of the impact of aerosols on the climate were started in the 1960s [42]. The direct radiative effect of anthropogenic aerosols is estimated at around -1.5 W/m^2 [43]. And yet aerosols remain the major scientific objective for many research missions and one of the toughest questions about the Earth's energy balance. Long-term satellite records on aerosol content are required to understand the aerosol impact on climate of our planet [44].

Aerosols have several health effects. For instance, long-term exposure to ambient air pollution causes lung cancer [45], while inhalation of particulate matter with the diameter of particles below 2.5 microns (PM$_{2.5}$) can cause low birth weight and high blood pressure in children [46]. The other health problems caused by aerosols include asthma, respiratory diseases, premature delivery and others. In this regard, strict regulations are adopted in several countries for hourly, daily and annual aerosol concentrations in the atmosphere [47] (around $30\,\text{mg} \times \text{m}^{-3}$ and $15\,\mu\text{g} \times \text{m}^{-3}$ for daily and annual PM$_{2.5}$, respectively).

There are several factors which complicate the study of aerosols. Aerosol particles come in all sizes, shapes and chemical compositions, as well as they can have internal structure (e.g., inhomogeneous particles, water/ice shells) [48]. Accurate simulations of interactions between light and aerosol particles involve in situ measurements. There are different types of aerosols making the determination of the aerosol properties highly complicated. The amount of aerosol particles changes over time due to several processes such as diffusion, gravitational setting, evaporation, coagulation, chemical reaction etc. Moreover, the aerosol particles can be transported at distances of thousands of kilometers from their sources, as shown in Table 2.7. The corresponding transport models are validated by ground-based measurements as well as by observations from space that can provide an estimate on the global distribution of aerosols in time and space [49]. However, retrieval of aerosol distribution from reflectance measurements is a challenging task and often involves several a priori assumptions about particle sphericity, aerosol physical properties, loads etc [50].

Primary aerosols are emitted directly from the source, while secondary aerosols originate from gaseous emissions that are later converted into aerosols particles as a result of chemical reactions (gas-to-particle conversion). Typically, particles made by gas-to-particle conversion have a size smaller than $d_p = 0.1$ μm (those are so-called nucleation particles), while coarse particles (e.g., PM_{10}) are primary aerosols. Nucleation particles coagulate (mostly due to reactions with OH) and make larger particles, which can travel or hang in the atmosphere for several weeks. Particles with diameters between $d_p = 0.1$ μm and $d_p = 1$ μm have the longest atmospheric lifetime. In addition, aerosol particles are essential for cloud formation.

The optical properties of spherical aerosol particles are determined by their complex refractive index ($\tilde{n} = n - i\chi$) (which is defined by their composition) and the so-called size parameter x is defined as

$$x = \frac{2\pi r_p}{\lambda}, \tag{2.81}$$

where r_p is the particle's radius. Consequently, a complete description of an ensemble of particles would require describing the composition and geometry of each particle, which is impossible in view of a large amount of particles. An alternative approach is statistical. It is based on the particle size distribution.

An important parameter of aerosols composed of homogeneous spherical particles is the particle size distribution $f(r_p)$ such that the fraction of particles that have the diameter between r_1 and r_2 is given by

$$\int_{r_1}^{r_2} f(r_p) \, dr_p, \tag{2.82}$$

while the normalization condition reads as

$$\int_0^\infty f(r_p) \, dr_p = 1. \tag{2.83}$$

The average radius of particles is then

$$\langle r \rangle = \int_0^\infty r f(r) \, dr. \tag{2.84}$$

It is convenient to approximate a particle size distribution by a certain mathematical function. The log-normal distribution is a widely used model which tends to be the best fit for single-source aerosols (i.e., $\log r_p$ rather than r_p follows the normal distribution) [51][7]:

$$f(r) = \frac{1}{\sqrt{2\pi} r \ln \sigma} e^{-\frac{(\ln r - \ln r_{mod})^2}{2(\ln \sigma)^2}}, \tag{2.85}$$

---

[7]Theoretical justification why coarse aerosol size distributions should be log-normal was elaborated by Kolmogorov [52].

Table 2.6 Emission of aerosol particles (data is taken from [39]

Particles	Global emission (10^6 ton per year)
Sea salt	
$d_p < 1\,\mu m$	54
$d_p = 1 - 16\,\mu m$	3290
Mineral dust	
$d_p < 1\,\mu m$	110
$d_p = 1 - 2\,\mu m$	290
$d_p = 2 - 20\,\mu m$	1750
Industrial dust ($d_p > 1\,\mu m$)	100
Black carbons ($d_p = 0 - 2\,\mu m$)	
Biomass burning	5.7
Fossil fuel	6.6
Organic matter ($d_p = 0 - 2\,\mu m$)	
Biomass burning	54
Fossil fuel	28
Biogenic ($d_p > 1\,\mu m$)	56

where $\log \sigma$ is the standard deviation of the size distribution, while r_{mod} corresponds to a peak of the distribution (note that $\log r_{mod}$ gives the mean value of $\log r$). The actual size distribution can be modeled as a superposition of two size distributions, representing the fine and coarse fractions of aerosols—that is the bimodal size distribution. However, in nature the aerosol size distributions are not necessarily log-normal (or bimodal), so other than log-normal distribution can be used (e.g., the modified Gamma distribution, outlined in the next section) [53].

In the troposphere, the properties of aerosols are strongly dependent on location and the sources near the surface. To describe the wide range of possible aerosols, the aerosol particles are modeled as mixtures of certain components [54]. Each component corresponds to a certain origin. Parameters of the distributions for some aerosol components included in the aerosol database OPAC [55] are shown in Table 2.8, while the refraction indices for different types of aerosol components can be found in [56].

If $r_p \gg \lambda$, the energy removed from a light beam by a particle is proportional to the particle's area. In this regard, it is useful to introduce the area-weighted mean radius referred to as the effective radius:

$$r_e = \frac{\int_0^\infty r f(r) \pi r^2 dr}{\int_0^\infty f(r) \pi r^2 dr}. \tag{2.86}$$

The optical properties of aerosol particles with different particle size distributions but the same r_e are similar.

Table 2.7 Lifetime and transfer scales for aerosol particles depending on their radius

Radius (mm)	Lifetime (days)	Horizontal transfer scale (km)	Vertical transfer scale (km)
10^{-3}	10^{-2}	10	10^{-2}
10^{-2}	1	10^3	1
0.1	10	10^4	10
1	10	10^4	10
10	1	10^3	1
100	10^{-2}	10	10^{-2}

Table 2.8 Parameters of the size distribution for different aerosol components (data is taken from [55])

Aerosol component	r_{mod}	σ	$\ln \sigma$
Sea salt (accumulation mode)	0.209	2.03	0.71
Sea salt (coarse mode)	1.75	2.03	0.71
Desert dust (nucleation mode)	0.07	1.95	0.67
Desert dust (accumulation mode)	0.39	2.0	0.69
Desert dust (coarse mode)	1.9	2.15	0.77
Water-insoluble aerosol	0.471	2.51	0.92
Water-soluble aerosol	0.0212	2.24	0.81
Soot aerosol	0.0118	2.0	0.69

2.14   Clouds

A cloud is a suspension of microscopic water droplets and/or frozen ice crystals. Clouds are an important component of the global hydrological cycle. They play a prominent role in the Earth's climate system through their strong impact on radiation processes, atmospheric dynamics, chemistry and thermodynamics.

The formation of cloud droplets is a complicated physical process. It was described by H. Köhler [57]. According to his theory, the water vapor is condensed onto the so-called cloud condensation nuclei (CCN) when the supersaturation of air exceeds a certain critical value. Essentially, CCN is a particle made of hydrophilic substances (i.e., to which water can attach). The relative humidity (also referred to as the saturation ratio) is

$$s = \frac{p}{p_s}, \tag{2.87}$$

where p is the water vapor pressure, and p_s is the saturation water vapor pressure. The value $s - 1$ is called supersaturation. Clearly, $s - 1 = 0$ at saturation, negative for a subsaturated cases and positive for supersaturated cases. For supersaturation, the relative humidity is greater than 100%. If there are CCNs in the system, then s does not rise much above 1. Instead, the water vapor is condensed into water droplets. The Köhler theory takes into account two factors. The first factor is the curvature effect (the so-called Kelvin effect), which consists in the fact that surface tension at the water droplet "skin" increases the equilibrium water vapor pressure. This increases the supersaturation required for condensation to take place. The second factor is the solute effect (the so-called Raoult's law), which takes into account that the atmosphere always contains contamination. The solute molecules prevent water molecules from evaporating by physically blocking them from the surface and by electrostatic interactions between them. Thus, the equilibrium vapor pressure of the solution is always less than that for the case without solutes. Through these effects, aerosols have an indirect impact on climate by changing the cloud formation processes (see a review in [58]).

Clouds can be characterized by the parameter called the liquid water content (LWC) defined as the mass of liquid water in the unit air volume. LWC is usually in the range 0.1–3 g/m^3 depending on the atmospheric conditions.

The physics behind the influence of clouds on the net radiation balance in general, and on spaceborne measurements in particular including trace gas total columns and profiles come down to three main contributions:

1. The albedo effect associated with the enhancement of reflectivity for cloudy scenes compared to cloud-free sky scenes;
2. The so-called shielding effect for which that part of the trace gas column below the cloud is hidden by the clouds themselves; and
3. The increase in absorption related to multiple scattering inside clouds which leads to enhancement of the optical path length.

In essence, the effect of clouds as radiative forcing agent depends on the cloud height, coverage and thickness. Lower clouds tend to reflect solar radiation, thereby cooling the atmosphere. On the other hand, high clouds absorb the longwave radiation which would escape to space otherwise. This process has a heating effect. The albedo and in-cloud absorption effects can increase the retrieved total column values of trace gases at and above the cloud-top, while the shielding effect normally results in an underestimation of the trace gas column.

One of the important cloud parameters is the cloud cover fraction, i.e., the fractional area covered by clouds as observed from above by satellites. If the cloud fraction is 1, then the pixel is fully covered by the cloud, while the zero cloud fraction corresponds to the pixel without any cloud. Note that the global cloud cover is about 0.7, i.e., almost 70% of ground pixels are "contaminated" with clouds (see Fig. 2.12), which hide the information about the surface and trace gas properties. Considering the fact that almost two-thirds of Earth are covered by water, there are

Fig. 2.12 Clouds as they are seen from space (the picture is taken from [59])

not so many pixels left with reliable information about the properties of underlying ground. Therefore the influence of clouds on the radiance measured from space should be taken into account in retrieval algorithms.

The cloud-top height parameter represents the location of the top of the clouds. In the case of a very tenuous upper portion, this parameter may be lower than the altitude of the first cloud particles. Cloud-top height can be converted into cloud-top pressure by using the barometric formula (see Eqs. (2.22), (2.24) and (2.26)).

Based on the cloud-top height (pressure) and cloud thickness values, several cloud types can be distinguished:

1. High-level clouds (5–13 km), which include cirrocumulus, cirrus and cirrostratus;
2. Mid-level clouds (2–7 km), which can be altocumulus, altostratus or nimbostratus; and
3. Low-level clouds (0–2 km), which are either stratus, cumulus, cumulonimbus or stratocumulus.

Note that such a distinction is not unique, and different cloud databases can adopt other classification schemes.

The optical properties of clouds are due to the properties of particles they are made of. The interaction between light and the particles is defined by the size parameter and the complex refractive index. The refractive indices for water and ice are shown in Fig. 2.13 and in Table 2.9. The retrieval of ice refractive index in the visible and UV is a subject of large uncertainties due to difficulties in the preparation of homogeneous ice samples. Recent updates of the imaginary part of the ice refractive index have been provided by Picard et al. [60]. The water clouds are usually modeled as a collection of spherical water particles, whose size distributions follow a certain mathematical law. For instance, the modified Gamma distribution [61] can be used:

$$f(r) = ar^{\alpha}e^{-br^{\gamma}}, \qquad (2.88)$$

where a, α, b and γ are real constants. The constants a and α describe the slope of the size distribution. Cirrus clouds are made of ice crystals. The presence of non-spherical particles [62] introduces computational challenges for simulating the interaction between the radiation and the cirrus clouds (as will be discussed in Chap. 3).

Table 2.9 Real and imaginery parts of refractive indices (data is taken from [63] and [64] for water and ice, respectively)

Wavelength (nm)	Water		Ice	
	Real ($\tilde{n}$)	Im ($\tilde{n}$)	Real ($\tilde{n}$)	Im ($\tilde{n}$)
300	1.349	1.6E-8	1.334	< 2E-11
325	1.346	1.08E-8	1.329	< 2E-11
350	1.343	6.5E-9	1.325	< 2E-11
375	1.341	3.5E-9	1.322	< 2E-11
400	1.339	1.86E-9	1.319	2.365E-11
425	1.338	1.3E-9	1.317	3.637E-11
450	1.337	1.02E-9	1.316	9.239E-11
475	1.336	9.35E-10	1.314	2.405E-10
500	1.335	1.00E-9	1.313	5.890E-10
525	1.334	1.32E-9	1.312	1.242E-9
550	1.333	1.96E-9	1.311	2.289E-9
575	1.333	3.60E-9	1.310	3.810E-9
600	1.332	1.09E-8	1.309	5.730E-9
625	1.332	1.39E-8	1.309	9.490E-9
650	1.331	1.64E-8	1.308	1.430E-8
675	1.331	2.23E-8	1.308	1.990E-8
700	1.331	3.35E-8	1.307	2.900E-8
725	1.330	9.15E-8	1.306	4.165E-8
750	1.330	1.56E-7	1.306	5.870E-8
775	1.330	1.48E-7	1.305	9.390E-8
800	1.329	1.25E-7	1.305	1.340E-7
825	1.329	1.82E-7	1.305	1.440E-7
850	1.329	2.93E-7	1.304	1.830E-7
875	1.328	3.91E-7	1.304	3.000E-7
900	1.328	4.86E-7	1.303	4.200E-7
925	1.328	1.06E-6	1.303	4.925E-7
950	1.327	2.93E-6	1.302	6.020E-7
975	1.327	3.48E-6	1.302	1.023E-6
1000	1.327	2.89E-6	1.302	1.620E-6
1200	1.324	9.89E-6	1.298	6.710E-6
1400	1.321	1.38E-4	1.294	1.980E-5
1600	1.317	8.55E-5	1.289	2.882E-4
1800	1.312	1.15E-4	1.283	1.411E-4
2000	1.306	1.10E-3	1.274	1.640E-3
2200	1.296	2.89E-4	1.263	2.547E-4
2400	1.279	9.56E-4	1.242	5.781E-4

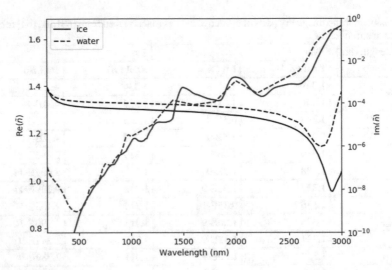

Fig. 2.13 Real and imagenery parts of refractive indices of water and ice, according to [63] and [64], respectively

Appendix 1: Solar Spectrum

Figure 2.14 shows the solar spectra measured at the top and at the bottom of the atmosphere. One can see the impact of atmospheric gases on the spectrum of radiation transmitted through the atmosphere. Numerical values are provided in Table 2.10.

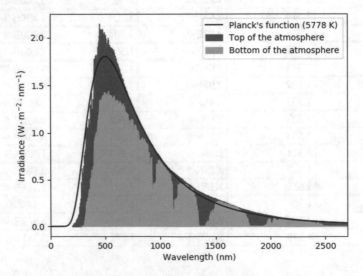

Fig. 2.14 Solar spectra at the top of the atmosphere and at the bottom of the atmosphere

Table 2.10 Example of the spectrum of the incoming solar radiation at the top of the atmosphere

Wavelength (nm)	Irradiance $(Wm^{-2}nm^{-1})$	Wavelength (nm)	Irradiance $(Wm^{-2}nm^{-1})$
3.000E+02	4.579E-01	1.400E+03	3.390E-01
3.250E+02	8.292E-01	1.425E+03	3.326E-01
3.500E+02	1.012E+00	1.450E+03	3.177E-01
3.750E+02	9.850E-01	1.475E+03	3.035E-01
4.000E+02	1.688E+00	1.500E+03	3.008E-01
4.250E+02	1.755E+00	1.525E+03	2.785E-01
4.500E+02	2.069E+00	1.550E+03	2.696E-01
4.750E+02	2.080E+00	1.575E+03	2.474E-01
5.000E+02	1.916E+00	1.600E+03	2.526E-01
5.250E+02	1.928E+00	1.625E+03	2.375E-01
5.500E+02	1.863E+00	1.650E+03	2.284E-01
5.750E+02	1.834E+00	1.675E+03	2.162E-01
6.000E+02	1.770E+00	1.700E+03	2.054E-01
6.250E+02	1.644E+00	1.725E+03	1.969E-01
6.500E+02	1.526E+00	1.750E+03	1.852E-01
6.750E+02	1.499E+00	1.775E+03	1.760E-01
7.000E+02	1.422E+00	1.800E+03	1.680E-01
7.250E+02	1.347E+00	1.825E+03	1.626E-01
7.500E+02	1.274E+00	1.850E+03	1.534E-01
7.750E+02	1.208E+00	1.875E+03	1.323E-01
8.000E+02	1.125E+00	1.900E+03	1.404E-01
8.250E+02	1.070E+00	1.925E+03	1.345E-01
8.500E+02	9.100E-01	1.950E+03	1.263E-01
8.750E+02	9.356E-01	1.975E+03	1.215E-01
9.000E+02	9.138E-01	2.000E+03	1.167E-01
9.250E+02	8.764E-01	2.025E+03	1.118E-01
9.500E+02	8.287E-01	2.050E+03	1.059E-01
9.750E+02	7.735E-01	2.075E+03	1.009E-01
1.000E+03	7.426E-01	2.100E+03	9.624E-02
1.025E+03	6.992E-01	2.125E+03	9.218E-02
1.050E+03	6.612E-01	2.150E+03	8.971E-02
1.075E+03	6.177E-01	2.175E+03	8.576E-02
1.100E+03	6.000E-01	2.200E+03	8.279E-02
1.125E+03	5.701E-01	2.225E+03	7.884E-02
1.150E+03	5.462E-01	2.250E+03	7.537E-02
1.175E+03	5.185E-01	2.275E+03	7.263E-02
1.200E+03	5.000E-01	2.300E+03	6.964E-02
1.225E+03	4.793E-01	2.325E+03	6.556E-02

(continued)

Table 2.10 (continued)

Wavelength (nm)	Irradiance $(Wm^{-2}nm^{-1})$	Wavelength (nm)	Irradiance $(Wm^{-2}nm^{-1})$
1.250E+03	4.586E-01	2.350E+03	6.434E-02
1.275E+03	4.407E-01	2.375E+03	6.141E-02
1.300E+03	4.154E-01	2.400E+03	5.974E-02
1.325E+03	3.937E-01	2.425E+03	5.733E-02
1.350E+03	3.708E-01	2.450E+03	5.459E-02
1.375E+03	3.575E-01	2.475E+03	5.337E-02

References

1. *U.S. Standard atmosphere* (U.S. Government Printing Office, Washington, 1976). https://ntrs.nasa.gov/archive/nasa/casi.ntrs.nasa.gov/19770009539.pdf
2. N. Nicastro, *Circumference: Eratosthenes and the Ancient Quest to Measure the Globe* (St. Martin's Press, New York 2008). https://www.xarg.org/ref/a/0312372477/
3. D. Benyoucef, T. Tahri, Air molecule collision cross sections: calculation and validation. Can. J. Phys. **95**(4), 346–352 (2017). https://doi.org/10.1139/cjp-2016-0406
4. F.H. Shu, *The Physical Universe: An Introduction to Astronomy* (Series of Books in Astronomy) (University Science Books, 1981)
5. I. de Pater, *Planetary Sciences* (Cambridge University Press, 2010)
6. G. Visconti, *Fundamentals of Physics and Chemistry of the Atmosphere* (Springer International Publishing, 2016). https://doi.org/10.1007/978-3-319-29449-0
7. B. Smirnov, *Microphysics of Atmospheric Phenomena* (Springer International Publishing, 2017). https://doi.org/10.1007/978-3-319-30813-5
8. T. von Karman, L. Edson, *The Wind and Beyond: Theodore von Karman, Pioneer in Aviation and Pathfinder in Space* (Little, Brown and Company, 1967)
9. Reference guide to the International Space Station. – Utilization Edition. np-2015-05-022-jsc. NASA. https://www.nasa.gov/sites/default/files/atoms/files/np-2015-05-022-jsc-iss-guide-2015-update-111015-508c.pdf (2015). Accessed 29 May 2019
10. Y. Timofeyev, A. Vasil'ev, *Theoretical fundamentals of Atmospheric Optics* (Cambridge, 2008)
11. C. Keeling, S. Piper, R. Bacastow, M. Wahlen, T. Whorf, M. Heimann, H. Meijer, Atmospheric CO_2 and $^{13}CO_2$ exchange with the terrestrial biosphere and oceans from 1978 to 2000: observations and carbon cycle implications, in *A History of Atmospheric CO_2 and Its Effects on Plants, Animals, and Ecosystems* (Springer-Verlag), pp. 83–113. https://doi.org/10.1007/0-387-27048-5_5
12. K. Thoning, P. Tans, A. Crotwell, D. Kitzis, NOAA Atmospheric Carbon Dioxide Mole Fractions from Mauna Loa Observatory, Hawaii, United States (2014). Accessed 15 June 2018 . https://doi.org/10.7289/v54x55rg. https://data.nodc.noaa.gov/cgi-bin/iso?id=gov.noaa.ncdc:C00890
13. L. Zhang, L. Beane Freeman, J. Nakamura, S. Hecht, J. Vandenberg, M. Smith, B. Sonawane, Formaldehyde and leukemia: Epidemiology, potential mechanisms, and implications for risk assessment, in *Environmental and Molecular Mutagenesis*, pp. NA–NA (2009). https://doi.org/10.1002/em.20534
14. L. Boltzmann, Ableitung des Stefan'schen Gesetzes, betreffend die Abhängigkeit der Wärmestrahlung von der Temperatur aus der electromagnetischen Lichttheorie. Annalen der Physik **258**(6), 291–294 (1884). https://doi.org/10.1002/andp.18842580616. [in German]

15. J. Stefan, Ueber die beziehung zwischen der waermestrahlung und der temperatur. Sitzungs-berichte der Kaiserlichen Akademie der Wissenschaften in Wien **79**, 391–428 (1879). Aus der k.k. Hof-und Staatsdruckerei
16. D. Crommelynck, A. Fichot, V. Domingo, L. Lee, SOLCON solar constant observations from the ATLAS missions. Geophys. Res. Lett. **23**(17), 2293–2295 (1996). https://doi.org/10.1029/96gl01878
17. C. Fröhlich, J. Lean, Solar radiative output and its variability: evidence and mechanisms. Astron. Astrophys. Rev. **12**(4), 273–320 (2004). https://doi.org/10.1007/s00159-004-0024-1
18. Key world energy statistics 2019 (2019). International Energy Agency
19. D. Doelling. CERES energy balanced and filled (EBAF) TOA monthly means data in netCDF Edition4.1 (2019). https://doi.org/10.5067/TERRA-AQUA/CERES/EBAF-TOA_L3B004.1. https://eosweb.larc.nasa.gov/project/ceres/ebaf_ed4.1
20. D. Frierson, J. Lu, G. Chen, Width of the Hadley cell in simple and comprehensive general circulation models. Geophys. Res. Lett. **34**(18) (2007). https://doi.org/10.1029/2007gl031115
21. J. Huang, M. McElroy, Contributions of the Hadley and Ferrel circulations to the energetics of the atmosphere over the past 32 years. J. Clim. **27**(7), 2656–2666 (2014). https://doi.org/10.1175/jcli-d-13-00538.1
22. G. Vallis, *Atmospheric and Oceanic Fluid Dynamics* (Cambridge University Press, 2017). https://doi.org/10.1017/9781107588417
23. P.R. Goode, J. Qiu, V. Yurchyshyn, J. Hickey, M.C. Chu, E. Kolbe, C.T. Brown, S.E. Koonin, Earthshine observations of the earth's reflectance. Geophys. Res. Lett. **28**(9), 1671–1674 (2001). https://doi.org/10.1029/2000gl012580
24. I. de Pater, J. Lissauer, *Planetary Sciences* (Cambridge University Press, 2015). https://doi.org/10.1017/cbo9781316165270
25. J. Marshall, *Atmosphere Ocean Climate Dynamics* (Academic Press, 2019)
26. 2000 ASTM Standard Extraterrestrial Spectrum Reference E-490-00. https://www.nrel.gov/grid/solar-resource/spectra-astm-e490.html. Accessed 08 Feb 2018]
27. A. Byalko, *Our planet – Earth* (Kvant, 1989). [in russian]
28. B. Smirnov, Earth's heat balance. Kvant **1**, 10–13 (1973). [in russian]
29. S. Arrhenius, XXXI. On the influence of carbonic acid in the air upon the temperature of the ground. Lond., Edinb., Dublin Philos. Mag. J. Sci. **41**(251), 237–276 (1896). https://doi.org/10.1080/14786449608620846
30. G. Petty, *A First Course in Atmospheric Radiation* (Sundog Publishing, Madison, WI, 2006)
31. V. Ramaswamy, M.L. Chanin, J. Angell, J. Barnett, D. Gaffen, M. Gelman, P. Keckhut, Y. Koshelkov, K. Labitzke, J.J.R. Lin, A. O'Neill, J. Nash, W. Randel, R. Rood, K. Shine, M. Shiotani, R. Swinbank, Stratospheric temperature trends: observations and model simulations. Rev. Geophys. **39**(1), 71–122 (2001). https://doi.org/10.1029/1999rg000065
32. P. Forster, V. Ramaswamy, P. Artaxo, P.e.a, Changes in atmospheric constituents and in radiative forcing, in *Climate Change 2007: The Physical Science Basis*. Contribution of Working Group I to the Fourth Assessment Report of the Intergovernmental Panel on Climate, ed. by S. Solomon, D. Qin, M.e.a Manning, (Cambridge University Press, Cambridge, United Kingdom), Chap.2
33. K. Stamnes, G. Thomas, J. Stamnes, *Radiative Transfer in the Atmosphere and Ocean* (Cambridge University Press, 2017). https://doi.org/10.1017/9781316148549
34. S.H. Schneider, R. Dickinson, Climate modeling. Rev. Geophys. **12**(3), 447 (1974). https://doi.org/10.1029/rg012i003p00447
35. J.M. Gregory, T. Andrews, Variation in climate sensitivity and feedback parameters during the historical period. Geophys. Res. Lett. **43**(8), 3911–3920 (2016). https://doi.org/10.1002/2016gl068406
36. K. McGuffie, A. Henderson-Sellers, Forty years of numerical climate modelling. Int. J. Climatol. **21**(9), 1067–1109 (2001). https://doi.org/10.1002/joc.632
37. R. Goody, J. Walker, *Atmospheres*. Foundations of Earth Science Series (Prentice-Hall, 1972)
38. G. Shaw, B. Holben, L. Remer, Passive shortwave remote sensing from the ground, in *Aerosol Remote Sensing* (Springer Berlin Heidelberg, 2013), pp. 137–157. https://doi.org/10.1007/978-3-642-17725-5_6

39. J.T. Houghton, Y. Ding, D.J. Griggs, M. Noguer, P.J. van der Linden, X. Dai, K. Maskell, C.A. Johnson, *Climate Change 2001: The Scientific Basis*. Contribution of Working Group I to the Third Assessment Report of the Intergovernmental Panel on Climate Change (Cambridge University Press, Cambridge, UK, 2001)
40. C. Jiao, M.G. Flanner, Y. Balkanski, S.E. Bauer, N. Bellouin, T.K. Berntsen, H. Bian, K.S. Carslaw, M. Chin, N.D. Luca, T. Diehl, S.J. Ghan, T. Iversen, A. Kirkevåg, D. Koch, X. Liu, G.W. Mann, J.E. Penner, G. Pitari, M. Schulz, Ø. Seland, R.B. Skeie, S.D. Steenrod, P. Stier, T. Takemura, K. Tsigaridis, T. van Noije, Y. Yun, K. Zhang, AeroCom assessment of black carbon in arctic snow and sea ice. Atmos. Chem. Phys. **14**(5), 2399–2417 (2014). https://doi.org/10.5194/acp-14-2399-2014
41. Y. Kaufman, D. Tanré, O. Boucher, A satellite view of aerosols in the climate system. Nature **419**(6903), 215–223 (2002). https://doi.org/10.1038/nature01091
42. K. Kondratyev, L. Ivlev, V. Krapivin, C. Varostos, *Atmospheric Aerosol Properties* (Springer Berlin Heidelberg, 2006). https://doi.org/10.1007/3-540-37698-4
43. B. Grandey, D. Rothenberg, A. Avramov, Q. Jin, H. Lee, X. Liu, Z. Lu, S. Albani, C. Wang, Effective radiative forcing in the aerosol – climate model CAM5.3-MARC-ARG. Atmos. Chem. Phys. **18**(21), 15783–15810 (2018). https://doi.org/10.5194/acp-18-15783-2018
44. M.I. Mishchenko, I.V. Geogdzhayev, W.B. Rossow, B. Cairns, B.E. Carlson, A.A. Lacis, L. Liu, L.D. Travis, Long-term satellite record reveals likely recent aerosol trend. Science **315**(5818), 1543–1543 (2007). https://doi.org/10.1126/science.1136709
45. O. Raaschou-Nielsen, Z. Andersen, R. Beelen, E. Samoli, M. Stafoggia, G. Weinmayr, B. Hoffmann, P. Fischer, M. Nieuwenhuijsen, B. Brunekreef, W. Xun, K. Katsouyanni, K. Dimakopoulou, J. Sommar, B. Forsberg, L. Modig, A. Oudin, B. Oftedal, P.E. Schwarze, P. Nafstad, U.D. Faire, N.L. Pedersen, C.G. Östenson, L. Fratiglioni, J. Penell, M. Korek, G. Pershagen, K.T. Eriksen, M. Sørensen, A. Tjønneland, T. Ellermann, M. Eeftens, P.H. Peeters, K. Meliefste, M. Wang, B.B. de Mesquita, T.J. Key, K. de Hoogh, H. Concin, G. Nagel, A. Vilier, S. Grioni, V. Krogh, M.Y. Tsai, F. Ricceri, C. Sacerdote, C. Galassi, E. Migliore, A. Ranzi, G. Cesaroni, C. Badaloni, F. Forastiere, I. Tamayo, P. Amiano, M. Dorronsoro, A. Trichopoulou, C. Bamia, P. Vineis, G. Hoek, Air pollution and lung cancer incidence in 17 European cohorts: prospective analyses from the European study of cohorts for air pollution effects (ESCAPE). Lancet Oncol. **14**(9), 813–822 (2013). https://doi.org/10.1016/s1470-2045(13)70279-1
46. M. Zhang, N. Mueller, H. Wang, X. Hong, L. Appel, X. Wang, Maternal exposure to ambient particulate matter ≤2.5⁻m during pregnancy and the risk for high blood pressure in childhood. Hypertension **72**(1), 194–201 (2018). https://doi.org/10.1161/hypertensionaha.117.10944
47. M. Shiraiwa, K. Ueda, A. Pozzer, G. Lammel, C. Kampf, A. Fushimi, S. Enami, A. Arangio, J. Fröhlich-Nowoisky, Y. Fujitani, A. Furuyama, P.S.J. Lakey, J. Lelieveld, K. Lucas, Y. Morino, U. Pöschl, S. Takahama, A. Takami, H. Tong, B. Weber, A. Yoshino, K. Sato, Aerosol health effects from molecular to global scales. Environ. Sci. Technol. **51**(23), 13545–13567 (2017). https://doi.org/10.1021/acs.est.7b04417
48. V. Babenko, *Electromagnetic Scattering in Disperse Media: Inhomogeneous and Anisotropic Particles (Springer Praxis Books)* (Springer, 2003)
49. I. Tegen, P. Hollrig, M. Chin, I. Fung, D. Jacob, J. Penner, Contribution of different aerosol species to the global aerosol extinction optical thickness: estimates from model results. J. Geophys. Res.: Atmos. **102**(D20), 23895–23915 (1997). https://doi.org/10.1029/97jd01864
50. A. Kokhanovsky, *Cloud Optics* (Springer Netherlands, 2006). https://doi.org/10.1007/1-4020-4020-2
51. J. Heintzenberg, Properties of the log-normal particle size distribution. Aerosol Sci. Technol. **21**(1), 46–48 (1994). https://doi.org/10.1080/02786829408959695
52. A. Kolmogorov, On the log-normal law of particle size distribution during crushing.Proc. USSR Acad. Sci. **31**(2), 99–101 (1941). [in Russian]
53. J. Lenoble, L. Remer, D. Tanré, Introduction, in *Aerosol Remote Sensing* (Springer Berlin Heidelberg, 2013), pp. 1–11. https://doi.org/10.1007/978-3-642-17725-5_1
54. A. Deepak, H.E. Gerber, *Report on Expert Meeting on Aerosols and Their Climatic Effects* (WMO publication WCP-55, 1983)

55. M. Hess, P. Koepke, I. Schult, Optical properties of aerosols and clouds: The software package OPAC. Bull. Am. Meteorol. Soc. **79**(5), 831–844 (1998). https://doi.org/10.1175/1520-0477(1998)079<0831:opoaac>2.0.co;2

56. The database and associated software package OPAC. http://cds-espri.ipsl.fr/etherTypo/?id=989. Accessed 20 Apr 2020

57. H. Köhler, The nucleus in and the growth of hygroscopic droplets. Trans. Faraday Soc. **32**(0), 1152–1161, (1936). https://doi.org/10.1039/tf9363201152

58. J. Fan, Y. Wang, D. Rosenfeld, X. Liu, J. Review of aerosol–cloud interactions: mechanisms, significance, and challenges. Atmos. Sci. **73**(11), 4221–4252 (2016). https://doi.org/10.1175/jas-d-16-0037.1

59. NASA Worldview. https://worldview.earthdata.nasa.gov. Accessed 3 Jan 2020

60. G. Picard, Q. Libois, L. Arnaud, Refinement of the ice absorption spectrum in the visible using radiance profile measurements in Antarctic snow. The Cryosphere **10**(6), 2655–2672 (2016). https://doi.org/10.5194/tc-10-2655-2016, https://tc.copernicus.org/articles/10/2655/2016/

61. D. Deirmendjian, *Electromagnetic Scattering on Spherical Polydispersions* (Elsevier, 1969)

62. K.N. Liou, P. Yang, *Light Scattering by Ice Crystals* (Cambridge University Press, 2016). https://doi.org/10.1017/cbo9781139030052

63. G. Hale, M. Querry, Optical constants of water in the 200-nm to 200-μm wavelength region.Appl. Opt. **12**(3), 555 (1973). https://doi.org/10.1364/ao.12.000555

64. S. Warren, R. Brandt, Optical constants of ice from the ultraviolet to the microwave: a revised compilation. J. Geophys. Res. **113**(D14) (2008). https://doi.org/10.1029/2007jd009744

Chapter 3
Light Scattering, Absorption, Extinction, and Propagation in the Terrestrial Atmosphere

3.1 Radiative Transfer Theory: Phenomenological Approach

Radiative transfer is the physical phenomenon of energy transfer in the form of electromagnetic radiation, while the radiative transfer equation describes this phenomenon mathematically. The radiative transfer theory found its applications in astrophysics, atmospheric physics and engineering sciences. In atmospheric remote sensing, it is the basis for the so-called forward models which are used to simulate the measurements for a given set of atmospheric parameters. Forward models reflect our understanding of physical processes in the atmosphere and serve as a precursor for retrieval methods.

The geometrical optics laws and the extinction law of Bouguer were one of the first attempts to understand and quantitatively characterize the interaction between light and the medium. The radiative transfer theory, as we know it now, originated from the classical works of Chwolson [1], Lommel [2], Shuster [3, 4], Schwarzschild [5], Milne [6] and others. Let us consider a slab of clear glass and a slab of milk glass, as shown in Fig. 3.1. In the first case, the light beam striking the layer after two refractions continues its monodirectional travel in space. However, in the case of a turbid medium, the escaping light is not monodirectional anymore but rather turns into a diffuse beam.

To describe this phenomenon mathematically, the radiative transfer equation is used, which is formulated as a balanced equation for the amount of radiant energy (or energy flow) passing through a given imaginary surface. This idea was further exploited by Ambartsumian [7], Sobolev [8], Chandrasekhar [9] and Rozenberg [10]. Such treatment is often referred to as a phenomenological approach since it is not based on the rigorous Maxwell theory describing the interaction between the electromagnetic waves and the medium. The basic assumptions used in the phenomenological approach can be summarized as follows [11]:

© Springer Nature Switzerland AG 2021
D. Efremenko and A. Kokhanovsky, *Foundations of Atmospheric Remote Sensing*,
https://doi.org/10.1007/978-3-030-66745-0_3

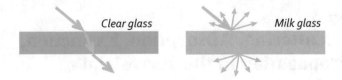

Fig. 3.1 Light going through the clear glass and the milk glass

1. The radiation propagates in the form of rays (that is the ray approximation); this approximation is valid in the geometrical optics approximation limit (when the wavelength λ of the light is much smaller than the size of the object it interacts with);

2. Rays are assumed to arrive at a given point from different directions being totally incoherent; that means, rays can be algebraically summed up disregarding the phases and interference terms;

3. The measured parameters are assumed to be neither local nor current values, but rather some time and space averaged, squared field characteristics; and the radiative transfer theory operates with these characteristics;

4. Radiation has a property of ergodicity, i.e., averaging over time for one specific realization of a random scattering process is equivalent to ensemble averaging;

5. The medium is assumed to be consisting of randomly distributed scatterers, while the interaction between the ray and the medium is statistically independent from the outcome of subsequent scattering events (that is the so-called Markov process).

These assumptions allow one to model the radiation propagation as transport of localized particles (i.e., photons). In this regard, the resulting radiative transfer model becomes similar to those formulated for other types of particles [12] (e.g., electrons and neutrons).

The main advantage of the phenomenological approach is its simplicity. It leads to efficient numerical models for simulating the scattered light characteristics, although it gives wrong results when the wave properties of light are of importance. One of such cases is the so-called coherent backscattering [13]. This phenomenon is intimately connected to the weak localization of the radiation [14] and consists of radiance enhancement in the backscattering direction. This effect can be observed in an extremely sharp cone around the backscattering direction with the cone angle given as a ratio of the wavelength of the light to the photon mean free path in the medium.

3.2 Radiance

In the framework of phenomenological approach, we can start directly by defining radiance as the most fundamental quantity. In particular, the monochromatic radiance L_λ (sometimes referred to as the specific intensity) is defined as the amount of radiant energy dW in the wavelength interval $[\lambda, \lambda + d\lambda]$ transported across an element of

Fig. 3.2 To the definition of
the radiance

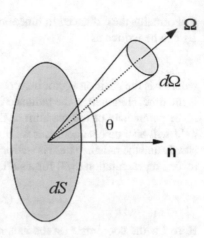

area dS into a solid angle $d\Omega$ oriented with an angle θ relative to the normal **n**
of the area, within a time interval dt as shown in Fig. 3.2 (strictly speaking, in the
following definition, we should write d^4W instead of dW since the differential in
the numerator should be balanced by four differentials in the denominator, as in the
classical textbook [15]):

$$L_\lambda = \frac{dW}{\cos\theta \, d\Omega \, d\lambda \, dS \, dt}. \tag{3.1}$$

The radiance has units of $[\text{W} \cdot \text{m}^{-2} \cdot \text{sr}^{-1} \cdot \text{nm}^{-1}]$. Analogously, considering the
wavenumber space $[\nu, \nu + d\nu]$, we introduce L_ν as

$$L_\nu = \frac{dW}{\cos\theta \, d\Omega \, d\nu \, dS \, dt}. \tag{3.2}$$

The amount of transferred energy should not be changed after conversion from L_λ
to L_ν. Hence,

$$L_\lambda \, |d\lambda| = L_\nu \, |d\nu|. \tag{3.3}$$

Since

$$\nu = \frac{1}{\lambda}, \tag{3.4}$$

the corresponding differentials can be expressed as follows:

$$d\nu = -\frac{d\lambda}{\lambda^2} = -\nu^2 d\lambda, \tag{3.5}$$

where the sign "minus" tells that ν increases when λ decreases. Substituting Eq.
(3.5) in Eq. (3.3) leads to the following result:

$$L_\lambda = L_\nu \nu^2 = L_\nu / \lambda^2. \tag{3.6}$$

Formally, the radiance is a function of coordinates, direction and time. Therefore, L_λ can be written as

$$L_\lambda = L_\lambda (\mathbf{r}, \boldsymbol{\Omega}, t), \tag{3.7}$$

where $\mathbf{r}$ is the position vector incorporating the coordinates, $\boldsymbol{\Omega}$ is the direction and t is the time. Hereinafter, the radiance is assumed not to change with time. Of course, if the parameters of the medium or the radiance source are changing, the radiance field will also be changed. But we assume that the medium and the source change slowly and the radiance field is stationary. For the medium stratified in plane-parallel layers, representation (3.7) for a stationary radiance field is reduced to

$$L_\lambda = L_\lambda (z, \boldsymbol{\Omega}) = L_\lambda (z, \mu, \varphi). \tag{3.8}$$

Here z is the coordinate on the axis normal to the plane of stratification, while the direction $\boldsymbol{\Omega} = \{\mu, \varphi\}$ is defined in spherical coordinates through the cosine μ of the zenith angle θ and azimuthal angle φ as shown in Fig. 3.3, $\mu > 0$ for downwelling radiation and $\mu < 0$ for the upwelling radiation [8]. Note that

$$d\Omega = d\mu d\varphi. \tag{3.9}$$

If the radiance field is isotropic, then it is independent of direction, that is

$$L_\lambda = L_\lambda (z). \tag{3.10}$$

If the radiance is integrated over the all wavelengths, it is called the integrated radiance and is given by

$$L = \int_0^\infty L_\lambda d\lambda. \tag{3.11}$$

3.3 Irradiance

The amount of radiant energy transferred through a unit area is called the irradiance and reads as

$$E_\lambda = \frac{dW}{d\lambda dS dt}. \tag{3.12}$$

It has units of [$W \cdot m^{-2} \cdot nm^{-1}$]. Analogously to Eq. (3.2), we define

$$E_\nu = \frac{dW}{d\nu dS dt}, \tag{3.13}$$

and in view of Eq. (3.5) we obtain the following transformation rule:

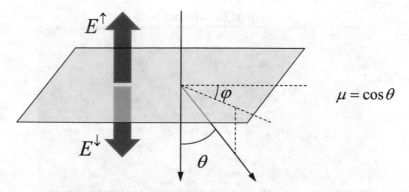

Fig. 3.3 To the definition of the irradiance

$$E_\lambda = E_\nu v^2 = E_\nu / \lambda^2. \tag{3.14}$$

By comparison Eq. (3.1) with Eq. (3.12), we have

$$dE_\lambda(r) = L_\lambda(r, \Omega) \mu d\Omega. \tag{3.15}$$

For the stratified medium we define (according to Fig. 3.3)

$$E_\lambda^\downarrow(z) = \int_0^{2\pi} \int_0^1 L_\lambda(z, \mu, \varphi) \mu d\mu d\varphi \tag{3.16}$$

and

$$E_\lambda^\uparrow(z) = \int_0^{2\pi} \int_{-1}^0 L_\lambda(z, \mu, \varphi) \mu d\mu d\varphi \tag{3.17}$$

as irradiances of the upward radiation and downward radiation, respectively. Note that in some textbooks (e.g., [16]) the irradiance is referred to as flux as it might look similar to the current density coming from the particle transport theory. The net irradiance is defined as

$$E_\lambda(z) = E_\lambda^\uparrow(z) - E_\lambda^\downarrow(z). \tag{3.18}$$

Let us consider an atmospheric layer bounded at altitudes z and $z+dz$, as shown in Fig. 3.4. The energy balance for the layer can be written as follows:

$$mc_p dT = dW, \tag{3.19}$$

where m is the layer mass, c_p is the isobaric heat capacity, dT is the temperature change, while dW is the net energy given as follows:

$$dW = (E^\downarrow(z + dz) - E^\downarrow(z) + E^\uparrow(z) - E^\uparrow(z + dz))Sdt, \tag{3.20}$$

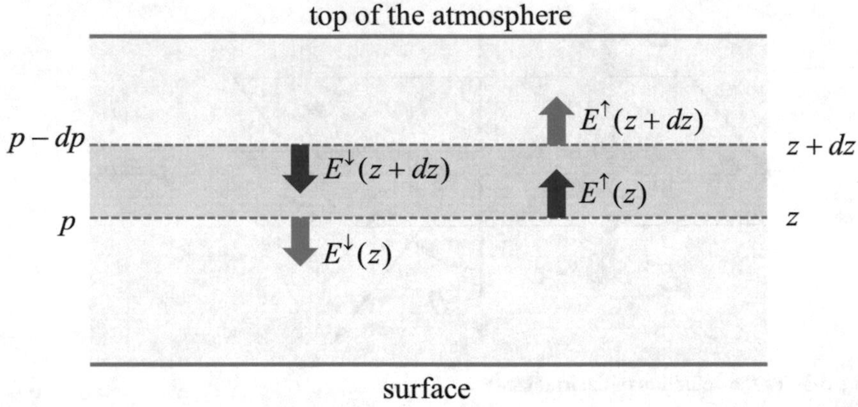

top of the atmosphere

Fig. 3.4 To the definition of the heating rate. Red arrows correspond to the incoming radiation, while blue arrows show the outgoing radiation

where S is the area of the layer, t is the time. Regrouping terms in Eq. (3.20) and taking into account Eq. (3.18), we obtain:

$$dW = (E(z) - E(z + dz))Sdt = -dE(z)Sdt. \qquad (3.21)$$

The mass of the layer can be expressed as

$$m = \rho S dz. \qquad (3.22)$$

Substituting Eq. (3.22) and Eq. (3.21) into Eq. (3.19) yields

$$\left(\frac{\partial T}{\partial t}\right)_p = -\frac{1}{\rho c_p}\frac{dE(z)}{dz}. \qquad (3.23)$$

The ratio $dE(z)/dz$ is called the radiative heating rate.

For a collimated incident plane-parallel light beam (unidirectional illumination) we have

$$L_\lambda(z, \mu, \varphi) = \frac{E_\lambda^\downarrow}{\mu_0}\delta(\mu_0 - \mu)\delta(\varphi), \quad \mu < 0. \qquad (3.24)$$

Here δ is the Dirac delta-function. Indeed, substituting Eq. (3.24) into Eq. (3.16) gives identity.

The integrated irradiance across all wavelengths reads as

$$E = \int_0^\infty E_\lambda d\lambda. \qquad (3.25)$$

Fig. 3.5 To the definition of
the radiation density

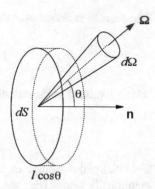

3.4 Radiation Density and Moments of the Radiation Field

The amount of energy per unit volume per unit wavelength interval is called the
radiation density. Formally it is defined as follows:

$$dU_\lambda = \frac{dW}{dVd\lambda}, \tag{3.26}$$

where U is the radiation density, while V is the volume. Consider a small cylindrical
volume of area dS in Fig. 3.5. The height of the cylinder is $l\cos\theta$, where l is the path
of the light inside the cylindrical volume. Since

$$l = cdt, \tag{3.27}$$

where c is the light velocity and dt is the time interval, the volume of this element is

$$dV = dSl\cos\theta = dS\cos\theta cdt. \tag{3.28}$$

Then, substituting Eq. (3.1) and Eq. (3.28) in Eq. (3.26) gives:

$$dU_\lambda = \frac{L_\lambda}{c}d\Omega \tag{3.29}$$

and

$$U_\lambda(z) = \frac{1}{c}\int_{4\pi} L_\lambda(z, \Omega)\,d\Omega, \tag{3.30}$$

where the integration is performed over the entire solid angle. Integrating both parts
of Eq. (3.30) over λ, we obtain [17]:

$$U(z) = \frac{1}{c}\int_{4\pi} L(z, \Omega)\,d\Omega. \tag{3.31}$$

The average radiance over the entire solid angle can be defined as

$$J\left(z\right) = \frac{1}{4\pi} \oint_{4\pi} L\left(z, \Omega\right) d\Omega, \tag{3.32}$$

which is related to the radiation density as follows:

$$U\left(z\right) = \frac{4\pi}{c} J\left(z\right). \tag{3.33}$$

Essentially, the average radiance is the zero moment of the radiation field. For the azimuthally symmetric case, we have

$$J\left(z\right) = \frac{1}{2} \int_{-1}^{1} L\left(z, \mu\right) d\mu. \tag{3.34}$$

Similar to the average radiance, higher-order moments can be defined, namely, the Eddington flux (i.e., the first moment):

$$H\left(z\right) = \frac{1}{2} \int_{-1}^{1} L\left(z, \mu\right) \mu d\mu, \tag{3.35}$$

and the so-called K-integral (i.e., the second moment):

$$K\left(z\right) = \frac{1}{2} \int_{-1}^{1} L\left(z, \mu\right) \mu^2 d\mu. \tag{3.36}$$

3.5 Radiance and Photon Flux

Radiation can be considered as a "photon gas" [18] and described as a statistical ensemble of photons [19] by using the corresponding distribution function $f\left(v, \boldsymbol{r}, \boldsymbol{\Omega}\right)$ so that $f\left(v, \boldsymbol{r}, \boldsymbol{\Omega}\right) dv dV d\Omega$ gives the number of photons with frequencies from ν to $\nu + d\nu$ in a volume dV centered at a point $\mathbf{r}$ and propagating within a solid angle $d\Omega$ around direction $\boldsymbol{\Omega}$ (Fig. 3.5). The amount of energy dW transferred by photons in the volume $dV = dS \cos \theta c dt$ (see Eq. (3.28)) through the area dS during a time interval from t to $t + dt$ is

$$dW = h\nu \times f\left(\nu, \mathbf{r}, \boldsymbol{\Omega}\right) d\nu \times dS \cos \theta c dt \times d\Omega, \tag{3.37}$$

where $h\nu$ is the energy of one photon. Comparing Eq. (3.2) and Eq. (3.37) we have:

$$L_\nu = ch\nu \times f\left(\nu, \mathbf{r}, \boldsymbol{\Omega}\right). \tag{3.38}$$

You might note that such a description of radiation is somewhat similar to the ray approximation. Indeed, the photon, just as the ray, is localized both in direction and space.

3.6 Radiance and Poynting Vector

In the electromagnetic theory, the energy transport is associated with the Poynting vector defined as follows [20]:

$$\mathbf{S} = \mathbf{E} \times \mathbf{H}, \tag{3.39}$$

where $\mathbf{E}$ is the electric field vector, while $\mathbf{H}$ is the magnetic field strength vector. The absolute value of the Poynting vector is measured in W/m^2. For a harmonic, linearly polarized plane wave traveling along direction z we have:

$$\mathbf{E}(z, t) = E_0 \cos(\omega t - kz) \, \mathbf{e}_x, \tag{3.40}$$

$$\mathbf{H}(z, t) = H_0 \cos(\omega t - kz) \, \mathbf{e}_y, \tag{3.41}$$

where E_0 and H_0 are the amplitudes of electric and magnetic fields, respectively, ω is the circular frequency, k is the wavenumber, while $\mathbf{e}_x$ and $\mathbf{e}_y$ are unit vectors along the x and y axes, respectively. Substitution of Eqs. (3.40) and (3.41) into Eq. (3.39) gives:

$$\mathbf{S}(z, t) = E_0 H_0 \cos^2(\omega t - kz) \, \mathbf{e}_z. \tag{3.42}$$

where $\mathbf{e}_z = \mathbf{e}_x \times \mathbf{e}_y$. Note that the Poynting vector $\mathbf{S}$ oscillates in time due to the term $\cos^2(\omega t - kz)$ with a period of π/ω, which is $\sim 10^{15}$ seconds at optical frequencies. Optical detectors are characterized by a time constant, which is much larger than such a period. Therefore, the instantaneous value of the Poynting vector is impossible to measure. Instead, a time-averaged value of the Poynting vector is measured:

$$\langle S \rangle = \frac{1}{T} \int_0^T E_0 H_0 \cos^2(\omega t - kz) \, dt = \frac{E_0 H_0}{2}, \tag{3.43}$$

where $T \gg \pi/\omega$. The quantity $\langle S \rangle$ is often referred to as the irradiance.

To compute the time-averaged Poynting vector, it is convenient to consider $\mathbf{E}$ and $\mathbf{H}$ as complex fields. In this case, the time-averaged Poynting vector can be represented as

$$\langle S \rangle = \frac{1}{2} \text{Re} \left\{ \mathbf{E} \times \mathbf{H}^* \right\}, \tag{3.44}$$

where "*" denotes complex conjugate (see Appendix 1 for details).

We can imagine that the energy "flows" along the lines generated by the Poynting vector; such a model is called "hydrodynamic". However, only under special conditions there is a bridge between the "hydrodynamic" model and the electromagnetic theory [21]. The first problem is that the radiance is a function of directions (i.e., it gives the angular distribution of the radiation flow), while the Poynting vector is unique at each point. Secondly, the physical energetic meaning of the Poynting vector is ambiguous. To illustrate it, we consider the Poynting theorem [20] in the differential form:

$$\frac{dU}{dt} = -\nabla \cdot \mathbf{S}\,(\mathbf{r}, t)\,, \qquad (3.45)$$

and the corresponding integral form:

$$\frac{dW}{dt} = -\oint_s \mathbf{S}\,(\mathbf{r}, t)\,\mathbf{n}\,(\mathbf{r})\,d^2\mathbf{r}, \qquad (3.46)$$

where $W = VU$, V is the volume surrounded by a closed surface s, and $\mathbf{n}\,(\mathbf{r})$ is the unit vector directed along the local outward normal to the boundary. Let us add a curl of an arbitrary vector field $\mathbf{Q}$ to $\mathbf{S}$, namely,

$$\mathbf{S}' = \mathbf{S} + \nabla \times \mathbf{Q}. \qquad (3.47)$$

Since $\nabla \cdot \nabla \times \mathbf{Q} = 0$ for any field, the augmented Poynting vector $\mathbf{S}'$ also solves Eq. (3.45) and, consequently, Eq. (3.46). Therefore, $\mathbf{S}$ and $\mathbf{S}'$ contribute equally to the change in energy density. In this regard, as pointed out in [21], it is not the Poynting vector itself, but rather the integral in Eq. (3.46) that has a strictly defined energetic meaning.

3.7   Radiative Transfer and the First Principle Derivations

An alternative to the phenomenological approach is the microphysical approach. It considers radiative transfer as propagation of electromagnetic waves in the discrete medium consisting of a set of scatterers. The starting point in deriving the radiative transfer equation is the Maxwell equations [22]. A comprehensive review of discussions on microphysical and phenomenological approaches can be found in [23]. The rigorous derivations, which can be found in [24–26], are out of the scope of our book. We only briefly outline the main ideas behind those derivations.

The medium is assumed to consist of discrete particles; each particle scatters the electromagnetic wave, and in the case of set of particles, the radiation is scattered multiple times before it is measured by the detector at the observation point $\mathbf{r}'$. The total electric and magnetic fields at $\mathbf{r}'$ are superpositions of the respective incident field and fields scattered by N particles

$$\mathbf{E}\left(\mathbf{r}',t\right) = \mathbf{E}^{\mathrm{inc}}\left(\mathbf{r}',t\right) + \sum_{i=1}^{N}\mathbf{E}_{i}^{\mathrm{sca}}\left(\mathbf{r}',t\right),\qquad(3.48)$$

where $\mathbf{E}_{i}^{\mathrm{sca}}$ corresponds to the wave, scattered by the ith particle (for $\mathbf{H}\left(\mathbf{r}',t\right)$ the equation is similar to Eq. (3.48)). Furthermore, it is assumed that the particles are distributed randomly and uniformly, while the positions of particles are uncorrelated. Naturally, it would be required to consider an ensemble average over all position realizations. However, assuming the property of ergodicity, the ensemble averaging is substituted with time averaging.

The scattered wave becomes spherical at a sufficiently large distance from the scattering particle. The far-field approximation implies that all particles are in the far-field of each other. In this sense the medium is said to be sparse.

The main complexity of the problem consists in the fact that the ith particle scatters waves that might have been scattered by other particles earlier. That means, the field which strikes the ith particle is given by

$$\mathbf{E}\left(\mathbf{r}_{i},t\right) = \mathbf{E}^{\mathrm{inc}}\left(\mathbf{r}_{i},t\right) + \sum_{\substack{j=1\\i\neq j}}^{N}\mathbf{E}_{j}^{\mathrm{sca}}\left(\mathbf{r}_{i},t\right).\qquad(3.49)$$

In each turn, $\mathbf{E}_{j}^{\mathrm{sca}}$ is a result of scattering of the striking wave $\mathbf{E}\left(\mathbf{r}_{j},t\right)$ (i.e., the jth particle scatters the field, which is also described by Eq. (3.49)). By using the so-called Foldy equations [27, 28], the resulting electric and magnetic fields at the observation point are expressed as order-of-scattering series (the so-called Neumann series). There the resulting field is represented as a sum of fields being scattered by all possible sequences of particles. As noted in [29], such formulation should be treated as a mathematical abstraction rather than a real physical phenomenon, since all mutual excitations occur simultaneously and are not temporally discrete and ordered events.

To evaluate $\mathbf{E}$ and $\mathbf{H}$ at the observation point, contributions from all sequences of scattering have to be summed up. In order to be able to perform analytical summation, it is assumed that the radiation follows the paths, in which particles appear only once (these are the so-called self-avoiding scattering paths), as shown in Fig. 3.6. Such an assumption is referred to as the Twersky approximation [30]. By using the Twersky approximation, we neglect the so-called coherent component, making it possible to sum up the contributions from different scattering sequences disregarding the phases of waves.

Furthermore, to compute the Poynting vector, it is required to compute the product of $\mathbf{E}$ and $\mathbf{H}$. Note that $\mathbf{H}$ is also expanded in series similar to Eqs. (3.48) and (3.49). When considering the product of $\mathbf{E}$ and $\mathbf{H}$, it may happen that there are common particles in the $\mathbf{E}$ and $\mathbf{H}$ scattering series. In the so-called ladder approximation, it is assumed that the orders of common particles are the same in both sequences (in this case the Feynman diagrams look like ladders, as shown in Fig. 3.7). Note that the

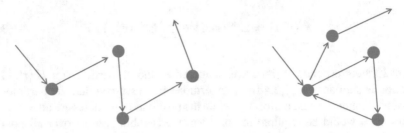

Fig. 3.6 Scattering sequences: (left) self-avoiding scattering path, which is allowed by the Twersky approximation; (right) non-self-avoiding scattering path, which is not allowed by the Twersky approximation

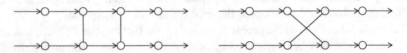

Fig. 3.7 Feynman diagrams: (left) ladder-diagram, (right) diagram with crossing connectors

Twersky approximation, as well as the ladder approximation, is valid provided that the number of particles is sufficiently large. The aforementioned assumptions allow to perform analytical summation of scattering sequences, and the radiative transfer equation can be derived.

Speaking about the first-principle derivations of the radiative transfer equation, we must at least mention the statistical-wave foundation of the radiative transfer theory. As a matter of fact, this field deserves to be a topic of a separate book. Early theoretical works (see, e.g., [31]) revealed a direct connection between the radiance and the coherence function of the wave field. The radiation transfer equation was derived from stochastic wave field equations. These developments resulted in formulating the so-called non-classical radiometry [32]. Unlike classical radiometry, it accounts for diffraction effects. An interesting feature of this theory is that the radiance can have both positive and negative values. Interested readers are encouraged to refer to [11, 21, 33–35].

3.8 Polarization: Historical Outlook

So far, we did not take into account the polarization of the light. Such models are referred to as scalar models since the most fundamental parameter to describe the energy flow—the radiance—is a scalar quantity. As opposite to scalar models, vector models deal with the Stokes vector and take into account the polarization state of the electromagnetic radiation.

Humanity has comprehended the phenomenon of polarization over several centuries. There are several chronological surveys on this topic [36]. A Danish archeologist, Thorkild Ramskou [37], suggested that about 1000 years ago the Vikings were already aware of the polarization phenomenon. Perhaps, they could observe the polarization of the blue sky through cordierite and thus were able to navigate in the absence of sun (although this assumption should be treated with caution [38]). The interest in polarization in the new time began with the discovery of double-refraction of calcite crystals by Erasmus Bartolinus [39]. The double-refraction property of this crystal was important for understanding the nature of light as a wave, as it was considered in the theoretical works of Christian Huygens. Etinne Louis Malus studied the polarization of the reflected light [40] by looking through the crystal of Iceland spar while rotating it. In particular, he formulated the law which gives the dependency of the quantity of transmitted light on a position of the polarizing filter given that the light is linearly polarized. The polarization of the blue sky was studied by Arago and later by J. Tindall [41]. David Brewster [42] discovered that at a certain incident angle (known as Brester's angle), the unpolarized light is totally converted into linearly polarized light. Brewster's angle θ_B was found to be

$$\theta_B = \arctan \frac{n_2}{n_1}, \tag{3.50}$$

where n_1 is the refractive index of the medium in which the incident light and the reflected light travel, while n_2 is the refractive index of the medium which reflects the incident light. A theoretical explanation of polarization was formulated by Augustin-Jean Fresnel [43].

A mathematically convenient formalism describing polarization was proposed by Sir George Stokes [44]. He discovered that any state of polarized light can be completely described by using four measurable quantities. They are known as the Stokes polarization parameters. Chandrasekhar, Rozenberg and Sobolev formulated the radiative transfer theory for partially polarized light using the Stokes parameters [10]. After that they became an integral part of the scientific literature on polarized light.

3.9 Stokes Parameters, Polarization Ellipse for Monochromatic Light

Now we consider how to describe quantitatively the polarized radiation. Consider a parallel monochromatic polarized wave of circular frequency ϖ traveling along the z-axis. Here z-axis is perpendicular to the plane of the paper and goes from us into the paper. The projections of the electric field onto two mutually perpendicular directions (corresponding to unit vectors $\mathbf{x}$ and $\mathbf{y}$) are given as

$$E_x(t) = E_{x0} \cos(\varpi t - \varepsilon_x) \tag{3.51}$$

and

$$E_y(t) = E_{y0} \cos(\varpi t - \varepsilon_y), \tag{3.52}$$

where t is the time, $\varepsilon_x, \varepsilon_y$ are initial phases, and E_{x0} and E_{y0} are the amplitudes. Thus, a completely polarized electromagnetic wave is characterized by four real-value numbers: $E_{0x}, E_{0y}, \varepsilon_x$ and ε_y. Excluding ϖt dependency from Eqs. (3.51) and (3.52) by applying cosine of sum formula we get

$$\left(\frac{E_x(t)}{E_{x0}}\right)^2 + \left(\frac{E_y(t)}{E_{y0}}\right)^2 - 2\frac{E_x(t) E_y(t)}{E_{x0} E_{y0}} \cos\delta = \sin^2\delta, \qquad (3.53)$$

where $\delta = \varepsilon_x - \varepsilon_y$. Equation (3.53) is the ellipse equation. That means the endpoint of the electric vector $E_x(t)\mathbf{x} + E_y(t)\mathbf{y}$ describes the so-called vibration ellipse, shown in Fig. 3.8. We introduce an angle χ between the major axis of the ellipse and the positive $\mathbf{x}$-axis so that $0 \leq \chi < \pi$. Also, we mark an angle β so that $\tan\beta$ gives the ellipticity, i.e., the ratio of the semi-minor and the semi-major axes of the ellipse, and $-\pi/4 \leq \beta \leq \pi/4$. The sign of β is positive if polarization is right-handed, i.e., the electric vector rotates clockwise as viewed by an observer looking in the direction of propagation, and negative for left-handed polarization (sometimes the opposite convention is used, e.g., as in [45]).

The Stokes parameters are defined as follows:

$$I = E_{x0}^2 + E_{y0}^2, \qquad (3.54)$$

$$Q = E_{x0}^2 - E_{y0}^2, \qquad (3.55)$$

$$U = 2E_{x0} E_{y0} \cos\delta \qquad (3.56)$$

and

$$V = 2E_{x0} E_{y0} \sin\delta. \qquad (3.57)$$

Here $I \geq 0$ is referred to as the total intensity and is equivalent to the radiance defined through Eq. (3.1); Q and U describe linearly polarized light, while V characterizes circularly polarized radiation. Thus, four quantities fully characterize the polarized radiation. Moreover, any other observable quantity is a linear combination of the four Stokes parameters.

It is easy to see that for totally polarized light Eqs. (3.54)–(3.57) imply

$$I^2 = Q^2 + U^2 + V^2. \qquad (3.58)$$

The angle of polarization ψ defines the polarization ellipse orientation and is given by

$$\chi = \frac{1}{2}\arctan\frac{U}{Q}. \qquad (3.59)$$

For the angle β describing ellipticity we have

Fig. 3.8 The vibration ellipse for the electric vector. The direction of propagation of the wave is into the paper and perpendicular to **x** and **y**. Right-handed polarization is considered

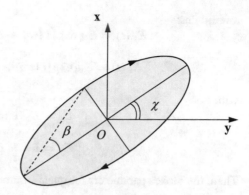

$$\beta = \frac{1}{2} \arcsin \frac{V}{I}. \tag{3.60}$$

3.10 Polarization of Quasi-monochromatic Light

Strictly monochromatic light is a rare case. For quasi-monochromatic light, i.e., light with circular frequency variations $\Delta\varpi \ll \varpi$, the amplitudes of the electric field as well as the phases are time-dependent:

$$
\begin{aligned}
E_x(t) &= E_{x0}(t) \cos(\varpi t - \varepsilon_x(t)), \\
E_y(t) &= E_{y0}(t) \cos(\varpi t - \varepsilon_y(t)).
\end{aligned}
\tag{3.61}
$$

For long time intervals compared to $2\pi/\varpi$, the amplitudes and phase differences are small and fluctuate independently of each other or with some correlation. If there are no correlations at all, the light is said to be completely unpolarized. In this case, we should take the time-average of the polarization ellipse. In particular, the Stokes parameters for quasi-monochromatic light are given as follows:

$$I = \left\langle (E_{x0}(t))^2 + (E_{y0}(t))^2 \right\rangle, \tag{3.62}$$

$$Q = \left\langle (E_{y0}(t))^2 - (E_{x0}(t))^2 \right\rangle, \tag{3.63}$$

$$U = 2 \left\langle E_{x0}(t) E_{y0}(t) \cos(\varepsilon_x(t) - \varepsilon_y(t)) \right\rangle \tag{3.64}$$

and

$$V = 2 \left\langle E_{x0}(t) E_{y0}(t) \sin(\varepsilon_x(t) - \varepsilon_y(t)) \right\rangle. \tag{3.65}$$

Following the same reasoning as in Appendix 1, we compute time-average quantities by representing the real optical amplitudes of the electric field in terms of complex amplitudes:

$$E_y(t) = E_{y0} \exp\left[i\left(\varpi t + \delta_y\right)\right] = E_y \exp(i\varpi t), \tag{3.66}$$

$$E_x(t) = E_{x0} \exp\left[i\left(\varpi t + \delta_x\right)\right] = E_x \exp(i\varpi t) \tag{3.67}$$

with

$$E_x = E_{x0} \exp(i\delta_x), \tag{3.68}$$

$$E_y = E_{y0} \exp(i\delta_y). \tag{3.69}$$

Then, the Stokes parameters for quasi-monochromatic light are given as follows:

$$I = \left\langle E_x E_x^* + E_y E_y^* \right\rangle, \tag{3.70}$$

$$Q = \left\langle E_x E_x^* - E_y E_y^* \right\rangle, \tag{3.71}$$

$$U = \left\langle E_x E_y^* + E_y E_x^* \right\rangle \tag{3.72}$$

and

$$V = i\left\langle E_x E_y^* - E_y E_x^* \right\rangle. \tag{3.73}$$

That is the equivalent representation to that obtained formally from the polarization ellipse.

Note that all Stokes parameters are expressed as sums of time-averaged products of the type $E_\alpha E_\beta$, where E_α and E_β are the linear components (projections) of the electric field. They can be arranged in the form of the so-called coherence matrix [46] (also referred to as the density matrix):

$$\Psi = \begin{bmatrix} \langle E_x E_x^* \rangle & \langle E_x E_y^* \rangle \\ \langle E_y E_x^* \rangle & \langle E_y E_y^* \rangle \end{bmatrix} = \frac{1}{2} \begin{bmatrix} I+Q & U-iV \\ U+iV & I-Q \end{bmatrix}, \tag{3.74}$$

which also fully describes the polarization state. The radiative transfer in terms of density matrices was considered in [47]. Although the algebraic rules for Stokes parameters and the coherence matrix are different, these two formalisms are completely equivalent. Nevertheless, in papers devoted to optical remote sensing, the Stokes parameter formalism is more common.

One can see that the Stokes vector is defined with respect to the certain coordinate system, which is attached to the direction of light propagation. Obviously, this direction is changed many times during the process of light scattering in a random medium. Consequently, the Stokes parameters have to be redefined whenever the propagation is changed. As we will see further, it can be done by using special rotation matrices.

This problem can be avoided by using the so-called coordinate-free approach, which is based on the light beam tensor. For a given Cartesian coordinate system (x, y, z), let us consider the electric field $E = E_x \mathbf{x} + E_y \mathbf{y} + E_z \mathbf{z}$. Then, the corresponding light beam tensor is given by

$$F = \begin{pmatrix} E_x E_x^* & E_x E_y^* & E_x E_z^* \\ E_y E_x^* & E_y E_y^* & E_y E_z^* \\ E_z E_x^* & E_z E_y^* & E_z E_z^* \end{pmatrix}. \tag{3.75}$$

If z is the direction of light propagation so that $E_z = 0$, the tensor reduces to the density matrix (3.74). The properties of the light tensor were studied by Fedorov [48] in particular for optics of anisotropic and gyrotropic media (see an excellent review [49] for details). The radiative transfer equation for the light tensor was derived in [50].

3.11 Stokes Vector and Degree of Polarization

The Stokes parameters can be arranged into a four-element Stokes vector:

$$L = \begin{bmatrix} I & Q & U & V \end{bmatrix}^T. \tag{3.76}$$

Note that each of these parameters can be measured by optical instruments. So, I is measured by using a standard photomultiplier. The components Q, U and V are measured using retarders and polarizers.

The degree of polarization P is given as

$$P = \frac{\sqrt{Q^2 + U^2 + V^2}}{I}. \tag{3.77}$$

For completely polarized light, $P = 1$, it is sufficient to measure only three Stokes parameters. The degree of linear polarization is defined as follows:

$$P_l = \frac{\sqrt{Q^2 + U^2}}{I}. \tag{3.78}$$

As the dimensions of the Stokes parameters are the same, it is convenient to represent particular cases of polarized light:

1. Unpolarized light

$$L = I \begin{bmatrix} 1 & 0 & 0 & 0 \end{bmatrix}^T; \tag{3.79}$$

2. Linearly horizontally polarized light

$$L = I \begin{bmatrix} 1 & 1 & 0 & 0 \end{bmatrix}^T; \tag{3.80}$$

3. Linearly vertically polarized light

$$L = I \begin{bmatrix} 1 & -1 & 0 & 0 \end{bmatrix}^T ; \tag{3.81}$$

4. Linearly polarized light at 45°

$$L = I \begin{bmatrix} 1 & 0 & 1 & 0 \end{bmatrix}^T ; \tag{3.82}$$

5. Linearly polarized light at −45°

$$L = I \begin{bmatrix} 1 & 0 & -1 & 0 \end{bmatrix}^T ; \tag{3.83}$$

6. Right circularly polarized light

$$L = I \begin{bmatrix} 1 & 0 & 0 & 1 \end{bmatrix}^T ; \tag{3.84}$$

7. Left circularly polarized light

$$L = I \begin{bmatrix} 1 & 0 & 0 & -1 \end{bmatrix}^T . \tag{3.85}$$

Using Stokes parameters, it is straight-forward to show that an arbitrary beam of quasi-monochromatic light can be regarded as a mixture of a beam of natural light and a beam of completely polarized light. Indeed, let us consider the following representation:

$$L = L_1 + L_2, \tag{3.86}$$

where

$$L_1 = \begin{bmatrix} I - \sqrt{Q^2 + U^2 + V^2} & 0 & 0 & 0 \end{bmatrix}^T \tag{3.87}$$

describes the unpolarized component, while

$$L_2 = \begin{bmatrix} \sqrt{Q^2 + U^2 + V^2} & Q & U & V \end{bmatrix}^T \tag{3.88}$$

corresponds to the completely polarized component.

The set of parameters (I, P, ψ, β) also completely describes polarized light. One might say that they are more intuitive than the Stokes parameters. However, the Stokes parameters can be directly measured, as mentioned earlier, and have the same units. Therefore, they can be used to formulate the vector radiative transfer equation. The conversion from one formalism to another one can be done by using Eqs. (3.59), (3.60) and (3.77).

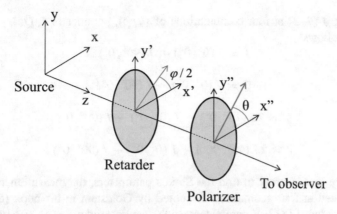

Fig. 3.9 The polarization analysis system

3.12 Measurement Principles of Stokes Polarization Parameters

Measurements of Stokes polarization parameters can be performed by using a retarder-polarizer system, shown in Fig. 3.9.

A retarder is a phase-shifting optical element, which has the property that the phase of the x-component of the electric field is advanced by $\varphi/2$ and the phase of the y-component is retarded by $\varphi/2$. A polarizer has the property that the optical field can only pass along an axis known as the transmission axis. If one places a polarizer after a retarder, the component of E_x along the transmission axis becomes $E_x \cos \theta$, where θ is the angle between the axis specified by the vector $\mathbf{x}$ and the transmission axis of a polarizer. Similarly, the component of E_y becomes $E_y \sin \theta$.

Thus, the total field emerging from the retarder-polarizer system is

$$E = E_x e^{i\varphi/2} \cos \theta + E_y e^{-i\varphi/2} \sin \theta. \tag{3.89}$$

The intensity of the beam (associated with the Poynting vector) is defined as

$$J = E \cdot E^*. \tag{3.90}$$

Then, taking the complex conjugate of Eq. (3.89) and forming the product following Eq. (3.90) we obtain:

$$\begin{aligned} J(\theta, \varphi) &= E_x E_x^* \cos^2 \theta + E_y E_y^* \sin^2 \theta \\ &+ 0.5 E_x^* E_y \sin 2\theta \exp(-i\varphi) + 0.5 E_y^* E_x \sin 2\theta \exp(i\varphi). \end{aligned} \tag{3.91}$$

Applying Eqs (3.70)-(3.73) for Eq. (3.91) we obtain

$$J(\theta, \varphi) = 0.5 (I + Q \cos 2\theta + U \cos \varphi \sin 2\theta + V \sin \varphi \sin 2\theta). \tag{3.92}$$

Measuring $J(\theta, \varphi)$ at four combinations of (θ, φ), parameters I, Q, U, V can be found as follows:

$$I = J(0°, 0°) + J(90°, 0°), \tag{3.93}$$

$$Q = J(0°, 0°) - J(90°, 0°), \tag{3.94}$$

$$U = 2J(45°, 0°) - J(0°, 0°) - J(90°, 0°), \tag{3.95}$$

$$V = 2J(45°, 90°) - J(0°, 0°) - J(90°, 0°). \tag{3.96}$$

We can see that in order to find the Stokes parameters, the measurements should be performed at four geometries. As noted by Goldstein in his book [51], when Stokes introduced I, Q, U, V parameters, such detectors did not exist, and hence, Eqs. (3.93)–(3.96) could not be used to measure the Stokes parameters.

3.13 Rotation Transformation Rule for the Stokes Parameters

Note that Q and U depend on χ, i.e., if we rotate the coordinate system (x,y) in Fig. 3.8, Q and U will be changed. The Stokes parameters are said to be defined with respect to a given plane of reference (e.g., the plane through x and the direction of propagation). If the coordinate system (or the plane of reference) is rotated, then the Stokes parameters are changed. The rotation matrix is given by

$$\boldsymbol{R}(\eta) = \begin{bmatrix} 1 & 0 & 0 & 0 \\ 0 & \cos 2\eta & -\sin 2\eta & 0 \\ 0 & \sin 2\eta & \cos 2\eta & 0 \\ 0 & 0 & 0 & 1 \end{bmatrix}, \tag{3.97}$$

where η is the angle of rotation. The Stokes parameters in the system with the plane of reference rotated are expressed as follows:

$$\boldsymbol{L}' = \boldsymbol{R}(\eta)\boldsymbol{L}. \tag{3.98}$$

Note that the Stokes parameters I and V are invariant with respect to such transformation, while Q and U will be changed:

$$I' = I, \tag{3.99}$$

$$Q' = Q\cos 2\eta + U\sin 2\eta, \tag{3.100}$$

$$U' = -Q\sin 2\eta + U\cos 2\eta, \tag{3.101}$$

$$V' = V, \tag{3.102}$$

where I', Q', U' and V' are the Stokes parameters in the new coordinate system. From Eqs. (3.100) and (3.101) it follows that the degree of linear polarization is also unchanged for transformation given by Eq. (3.98).

3.14 Circular Basis Representation of the Stokes Parameters

The Stokes parameters can be represented in the so-called circular polarization (CP) basis [52]:

$$\mathbf{L}_c = \frac{1}{2} \begin{pmatrix} Q + iU \\ I + V \\ I - V \\ Q - iU \end{pmatrix}. \tag{3.103}$$

The advantage of comparing it with the conventional (sometimes referred to as energetic) representation consists in the fact that the rotation matrix in the CP-basis $\mathbf{R}_c$ such that

$$\mathbf{L}'_c = \mathbf{R}_c(\eta)\,\mathbf{L}_c \tag{3.104}$$

has a diagonal form:

$$\mathbf{R}_c(\eta) = \begin{pmatrix} e^{-2i\eta} & 0 & 0 & 0 \\ 0 & 1 & 0 & 0 \\ 0 & 0 & 1 & 0 \\ 0 & 0 & 0 & e^{2i\eta} \end{pmatrix}. \tag{3.105}$$

The transformation into the CP-basis can be performed by using matrix multiplication [53]:

$$\mathbf{L}_c = \mathbf{A}\mathbf{L}, \tag{3.106}$$

where

$$\mathbf{A} = \frac{1}{2} \begin{pmatrix} 0 & 1 & i & 0 \\ 1 & 0 & 0 & 1 \\ 1 & 0 & 0 & -1 \\ 0 & 1 & -i & 0 \end{pmatrix}. \tag{3.107}$$

For backward transformation we have

$$\mathbf{L} = \mathbf{A}^{-1}\mathbf{L}_c, \tag{3.108}$$

where

$$A^{-1} = \frac{1}{2} \begin{pmatrix} 0 & 1 & 1 & 0 \\ 1 & 0 & 0 & 1 \\ -i & 0 & 0 & i \\ 0 & 1 & -1 & 0 \end{pmatrix}. \tag{3.109}$$

As we will see, the CP-representation is extremely convenient when we consider scattering of polarized light.

3.15 Light Scattering, Absorption and Extinction by Atmospheric Particles

The terrestrial atmosphere absorbs and scatters the incoming solar light. Let us consider how we can describe such interaction.

The light is scattered by air density fluctuations (molecular scattering), aerosol particulate matter and cloud particles. Looking at clouds or at the clear sky, our eyes capture scattered sunlight. Scattering is a complicated physical process which can be explained in terms of electromagnetic theory [45]. The incident electromagnetic wave sets electric charges in the scattering particle into oscillatory motion, which in their turn, radiates the secondary electromagnetic wave–that is the scattered electromagnetic wave. If the frequencies of the incident radiation and scattered radiation are the same, then we may say about elastic scattering, although the word "elastic" is often omitted.

In addition, the part of the incident electromagnetic energy can be absorbed and then transformed into thermal energy. There is also a possibility for more complicated processes in which the incident radiation is absorbed in a certain wavelength and then reemitted at the other wavelength. Such processes that are accompanied by a change in the frequency of the electromagnetic wave (and the energy of the photon) are called inelastic scattering processes. A detailed consideration of inelastic scattering effects is out of the scope of this book.

Let E_0 be the irradiance of the incident radiation striking the particle, and W_{sca} and W_{abs} are the amounts of energy scattered and absorbed per unit time interval in the scattering process, respectively. The scattering cross-section can be defined as follows:

$$C_{sca} = \frac{W_{sca}}{E_0}, \tag{3.110}$$

while the absorption cross-section is given as

$$C_{abs} = \frac{W_{abs}}{E_0}. \tag{3.111}$$

The cross-sections have the units of area $[m^2]$. The extinction cross-section is the sum of the scattering and absorption cross-sections:

$$C_{ext} = C_{sca} + C_{abs}. \tag{3.112}$$

For non-absorbing particles, we have $C_{ext} = C_{sca}$. Such a case is called conservative scattering.

The cross-sections can be defined considering the radiation as a flux of photons in the same manner as they are defined for particles [54]. Let j_0 be the number of photons, crossing the unit area per unit time, N_{sca} and N_{abs} are the number of scattered and absorbed photons per unit time, respectively. Then the scattering cross-section is

$$C_{sca} = \frac{N_{sca}}{j_0}, \tag{3.113}$$

while the absorption cross-section is

$$C_{abs} = \frac{N_{abs}}{j_0}. \tag{3.114}$$

In addition to cross-sections C_{ext}, C_{sca} and C_{abs} we define extinction, scattering and absorption coefficients ($[m^{-1}]$), i.e., [54]

$$\sigma_{ext} = n_d \bar{C}_{ext}, \tag{3.115}$$

$$\sigma_{sca} = n_d \bar{C}_{sca}, \tag{3.116}$$

and

$$\sigma_{abs} = n_d \bar{C}_{abs}, \tag{3.117}$$

respectively, where n_d is the number density of scatterers, $\bar{C}_{ext}$, $\bar{C}_{sca}$ and $\bar{C}_{abs}$ are the average extinction, scattering and absorption cross-sections, respectively. The averaging must be performed with respect to the size and shape of scatterers. The extinction coefficient is inversely proportional to the photon mean free path l_{ext} in the medium, i.e., the mean path of a photon between two consecutive scattering events. In view of Eq. (3.112), we have

$$\sigma_{ext} = \sigma_{abs} + \sigma_{sca}. \tag{3.118}$$

For a homogeneous layer with the geometrical thickness Δz, the optical thickness τ is defined as follows:

$$\tau = \sigma_{ext} \Delta z = \frac{\Delta z}{l_{ext}}. \tag{3.119}$$

The optical thickness can be regarded as a unitless characteristic length that defines the scale of a scattering medium. If the layer is inhomogeneous (i.e., $\sigma_{ext} = \sigma_{ext}(z)$, where z is the vertical coordinate), it can be split by infinitely thin homogeneous sublayers with the geometrical thickness dz. The optical thickness of a sublayer is equal to $\sigma_{ext}dz$ according to Eq. (3.119). The optical thickness of the whole inhomogeneous layer is then given by

$$\tau = \int_{z_1}^{z_2} \sigma_{ext}(z)\,dz, \qquad (3.120)$$

where z_1 and z_2 are the altitudes of the lower and upper boundaries of the atmospheric layer, respectively.

The single-scattering albedo ω is defined as

$$\omega = \frac{\sigma_{sca}}{\sigma_{ext}}, \qquad (3.121)$$

and can be understood as the probability for a photon to survive in the single-scattering event. An albedo of zero means that the medium does not scatter light (it is only absorbed), while an albedo of one says that there is no absorption in the medium.

Scattering, absorption and extinction coefficients change with altitude. Discretizing a vertically inhomogeneous atmosphere into a discrete set of layers, for each layer corresponding optical thickness and the single-scattering albedo can be computed.

3.16 Single-Scattering Phase Functions

In the scattering process, the photon's direction is changed. That leads to the redirection of radiation. The direction of propagation is defined with respect to a certain axis. We choose an axis OZ, shown in Fig. 3.10. The so-called principal plane contains the axis OZ and the incident beam. The polar angle θ_0 of incidence is defined with respect to OZ. The scattering angle Θ is the angle between the direction of the incident beam and the scattered beam. The polar angle of the scattered beam is θ. Note that, in general, the scattered beam may lie off the principal plane. The azimuthal angle φ is the angle between the principal plane and the plane which contains OZ and the scattered beam.

According to the spherical law of cosines, the cosine of the scattering angle can be expressed as

$$\cos\Theta = \mu_0\mu + \sqrt{1-\mu^2}\sqrt{1-\mu_0^2}\cos\varphi. \qquad (3.122)$$

The differential scattering cross-section $\Xi = dC_{scat}\left(\boldsymbol{\Omega}, \boldsymbol{\Omega'}\right)/d\Omega$ is the function of the scattering angle Θ, i.e., the angle between the incident direction $\boldsymbol{\Omega} = \{\mu, \varphi\}$ and the direction of observation $\boldsymbol{\Omega'} = \{\mu', \varphi'\}$. Again, the angle Θ can be computed using the cosine law of spherical geometry

$$\cos\Theta = \mu\mu' + \sqrt{1-\mu^2}\sqrt{1-\mu'^2}\cos\left(\varphi' - \varphi\right). \qquad (3.123)$$

Using Eq. (3.110), the amount of energy going in the direction Ω scattered in the direction Ω' into the solid angle $d\Omega$ by a single scatterer can be expressed as

$$dW_{\text{sca}}(\cos\Theta) = E_0 \frac{dC_{\text{sca}}(\Omega, \Omega')}{d\Omega} d\Omega. \tag{3.124}$$

It is convenient to use a function that characterizes the angular distribution of the scattered radiation after a single-scattering event. That is the phase function $p_s(\Omega, \Omega')$ which has dimension $[\text{sr}^{-1}]$ and is defined as

$$p_s(\Omega, \Omega') = \frac{4\pi}{C_{\text{sca}}} \frac{dC_{\text{sca}}(\Omega, \Omega')}{d\Omega} \tag{3.125}$$

for a single particle. The differential scattering coefficient β of a polydispersed medium is defined as

$$\beta = n_d \bar{\Xi}, \tag{3.126}$$

where $\bar{\Xi}$ is the average differential scattering cross-section. The phase function for a polydispersed medium is defined as

$$p(\Omega, \Omega') = \frac{4\pi\beta}{\sigma_{\text{sca}}}$$

or

$$p(\Omega, \Omega') = \frac{4\pi\bar{\Xi}}{\bar{C}_{\text{sca}}}.$$

Physically, realistic phase functions obey the Helmholtz reciprocity rule, i.e.,

$$p(\Omega', \Omega) = p(\Omega, \Omega'). \tag{3.127}$$

Fig. 3.10 Scattering geometry definition

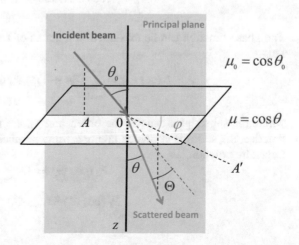

That means the value of the phase function remains unchanged if the locations of the light source and the viewer are swapped. In most atmospheric applications, scattering is assumed to be axisymmetric with respect to the incident direction. In this case, p is a function of the scattering angle, namely $p(\cos \Theta)$. In addition, the phase function can be considered as a function of spherical coordinates, namely $p(\mu, \varphi, \mu', \varphi')$. The normalization condition for p reads as follows:

$$\int_{\substack{\text{the whole} \\ \text{sphere}}} p(\cos \Theta)\, d\Omega = \int_0^{2\pi} \int_{-1}^1 p(\mu, \varphi, \mu', \varphi')\, d\mu'd\varphi = 4\pi, \quad (3.128)$$

so that the quantity $p(\cos \Theta)/4\pi$ can be thought of as a probability density function, expressing the chances of a photon of light being scattered in a particular direction. Since the scattering angle depends on the difference between the incident and the scattered azimuthal angles (see Eq. (3.123)), it is convenient to consider the phase function as a function of the relative azimuthal angle $(\varphi' - \varphi)$, namely $p(\mu, \mu', \varphi - \varphi')$ instead of $p(\mu, \mu', \varphi, \varphi')$. The azimuthally averaged phase function is given by

$$p(\mu, \mu') = \frac{1}{2\pi} \int_0^{2\pi} p(\mu, \mu', \varphi - \varphi')\, d(\varphi - \varphi'), \quad (3.129)$$

for which the normalization condition reads

$$\int_{-1}^1 p(\mu, \mu')\, d\mu' = 2. \quad (3.130)$$

Setting $\mu = 1$ and $\mu' = \cos \Theta$ in Eq. (3.130) and associating $p(1, \cos \Theta)$ with $p(\cos \Theta)$, we conclude that

$$\int_{-1}^1 p(\cos \Theta)\, d(\cos \Theta) = 2. \quad (3.131)$$

The phase function can be expanded in a series of Legendre polynomials P_n [55], i.e.,

$$p(\cos \Theta) = \sum_{n=0}^{\infty} (2n+1)\chi_n P_n(\cos \Theta), \quad (3.132)$$

where the subscript n denotes the order, and χ_n are the expansion coefficients. In practice, the series is evaluated taking a finite number $N_{\max}$ of terms. The first few Legendre polynomials are:

$$P_0(\cos \Theta) = 1, \quad (3.133)$$

$$P_1(\cos \Theta) = \cos \Theta, \quad (3.134)$$

$$P_2 (\cos \Theta) = \frac{1}{2} \left(3 \cos^2 \Theta - 1 \right), \tag{3.135}$$

and

$$P_3 (\cos \Theta) = \frac{1}{2} \left(5 \cos^3 \Theta - 3 \cos \Theta \right). \tag{3.136}$$

The system of Legendre polynomials is often used for the phase function expansion due to its important properties:

1. it is orthogonal,
2. it is complete, and
3. the addition theorem can be applied.

Let us have a closer look at these properties.

The *orthogonality* property means that

$$\int_{-1}^{1} P_m (\cos \Theta) P_n (\cos \Theta) d (\cos \Theta) = \frac{2}{2n+1} \delta_{mn}, \tag{3.137}$$

where δ_{mn} denotes the Kronecker delta, which is equal to 1 if $m = n$ and to 0 otherwise. Multiplying Eq. (3.132) by $P_m (\cos \Theta)$ and integrating over $\cos \Theta$ from -1 to 1 taking into account Eq. (3.137) we obtain the explicit formula for the expansion coefficient χ_n:

$$\chi_n = \frac{1}{2} \int_{-1}^{1} P_n (\cos \Theta) p (\cos \Theta) d (\cos \Theta). \tag{3.138}$$

From Eqs. (3.131), (3.133) and (3.138) it follows that

$$\chi_0 = 1. \tag{3.139}$$

For χ_1 we have

$$\chi_1 = \int_{-1}^{1} p (\cos \Theta) \cos \Theta d (\cos \Theta). \tag{3.140}$$

Since $p (\cos \Theta)$ can be regarded as a probability density function, χ_1 is the average cosine of the scattering angle and is often denoted as g. It is referred to as the asymmetry factor and shows how forward-peaked the phase function is. The positive values of g correspond to dominant forward scattering, and the negative values give dominant back-scattering. If $g = 1$, then the phase function describes complete forward scattering, while if $g = -1$, the phase function corresponds to complete backscattering. The case $g = 0$ corresponds to symmetric scattering about $\cos \Theta = 0$ and, in particular, to isotropic and molecular scattering.

The *completeness* of the system of Legendre polynomials means that

$$\sum_{n=0}^{\infty} \frac{2n+1}{2} P_n (\cos \Theta) P_n (\cos \Theta') = \delta (\cos \Theta - \cos \Theta'). \tag{3.141}$$

In fact, the completeness property makes expansion (3.132) possible. Indeed, consider

$$p\left(\cos\Theta\right) = \int_{-1}^{1} \delta\left(\cos\Theta - \cos\Theta'\right) p\left(\cos\Theta'\right) d\left(\cos\Theta'\right). \tag{3.142}$$

Substituting the delta-function in Eq. (3.142) by Eq. (3.141) and inverting the order of integration and summation we get

$$p\left(\cos\Theta\right) = \sum_{n=0}^{\infty} (2n+1) P_n\left(\cos\Theta\right) \left[\frac{1}{2}\int_{-1}^{1} P_n\left(\cos\Theta'\right) p\left(\cos\Theta'\right) d\left(\cos\Theta'\right)\right]. \tag{3.143}$$

Noting that the expression in square brackets is χ_n (according to Eq. (3.138)) we prove expansion (3.132).

Finally, the Legendre addition theorem states that for $\cos\Theta$ defined as in Eq. (3.123), the following representation for $P_n\left(\cos\Theta\right)$ is valid[1]:

$$P_n\left(\cos\Theta\right) = \sum_{m=0}^{n} (2 - \delta_{m0}) \frac{(n-m)!}{(n+m)!} P_n^m(\mu) P_n^m(\mu') \cos[m(\varphi - \varphi')]. \tag{3.144}$$

where P_n^m are the associated Legendre polynomials. Note that

$$P_n^0(\mu) \equiv P_n(\mu) \tag{3.145}$$

and

$$P_n^m(\mu) = 0, \quad \text{if } |m| > n. \tag{3.146}$$

Inserting Eq. (3.144) into Eq. (3.132) and taking into account Eq. (3.146), we get the phase function expansion in spherical coordinates:

$$p\left(\mu, \mu', \varphi - \varphi'\right) = \sum_{m=0}^{n} (2 - \delta_{m0}) \sum_{n=m}^{\infty} (2n+1) \frac{(n-m)!}{(n+m)!} \chi_n P_n^m(\mu) P_n^m(\mu') \cos[m(\varphi - \varphi')]. \tag{3.147}$$

Introducing the normalized associated Legendre polynomials as

$$\bar{P}_n^m(\mu) = \sqrt{(2n+1)\frac{(n-m)!}{(n+m)!}} P_n^m(\mu), \tag{3.148}$$

Eq. (3.147) can be written in a more compact form:

$$p\left(\mu, \mu', \varphi - \varphi'\right) = \sum_{m=0}^{n} (2 - \delta_{m0}) \sum_{n=m}^{\infty} \chi_n \bar{P}_n^m(\mu) \bar{P}_n^m(\mu') \cos[m(\varphi - \varphi')]. \tag{3.149}$$

---

[1] The addition theorem can be used in the following form: $P_n\left(\cos\Theta\right) = P_n(\mu) P_n(\mu') + 2\sum_{m=1}^{n} \frac{(l-m)!}{(l+m)!} P_n^m(\mu) P_n^m(\mu') \cos m\left(\varphi - \varphi'\right)$, which is equivalent to Eq. (3.144).

In practice, it is preferable to use the normalized polynomials since their computations are computationally more stable [56] (an algorithm for computing the Legendre polynomials is outlined in Appendix 2).

3.17 Modeled Phase Functions

Various analytical phase functions were proposed to mimic the scattering properties of particles [57]. The simplest phase function is isotropic:

$$p(\cos \Theta) = 1. \tag{3.150}$$

The expansion of the isotropic phase function is simple: $\chi_0 = 1$ and all other coefficients vanishing. This phase function is not realistic for atmospheric remote sensing, although it is convenient in analytical derivations due to its simplicity. Also in theoretical radiative transfer the linear approximation for the phase function is used:

$$p(\cos \Theta) = 1 + b \cos \Theta, \tag{3.151}$$

where b is a constant. The radiative transfer calculations for this phase function are simpler as compared to the cases with other phase functions and produce results accurate enough for a number of engineering applications [58].

The opposite to isotropic is the needle phase function given by

$$p(\cos \Theta) = 2\delta(1 - \cos \Theta), \tag{3.152}$$

which refers to scattering only in a forward direction.

One of the most commonly used non-isotropic phase functions is the Henyey–Greenstein phase function [59]:

$$p_{\text{HG}}(g, \cos \Theta) = \frac{1 - g^2}{\left(1 + g^2 - 2g \cos \Theta\right)^{3/2}} \tag{3.153}$$

with g being the asymmetry factor. This phase function was empirically derived to model the scattering of light by intergalactic dust. It is used to parametrize real phase functions. Note that p_{HG} has a simple expansion in terms of the Legendre polynomials

$$\chi_n = g^n.$$

The ratio of forward scattering to backward scattering is

$$\frac{p_{HG}(g, 1)}{p_{HG}(g, -1)} = \left(\frac{1 + g}{1 - g}\right)^3. \tag{3.154}$$

It is evident that for g close to 1, p_{HG} is forward-peaked, while for negative g backscattering dominates. The case $g = 0$ corresponds to the isotropic phase function. While the Henyey–Greenstein phase function can describe the forward peak reasonably well, it often fails to capture the features of realistic phase functions in the backward hemisphere. To improve the description of backward scattering, the double Henyey–Greenstein phase function can be used. It is defined as follows:

$$p(\alpha, g_1, g_2, \cos \Theta) = \alpha p_{HG}(g_1, \cos \Theta) + (1 - \alpha) p_{HG}(g_2, \cos \Theta) \tag{3.155}$$

with α being the weighting factor. Another modification of the Henyey–Greenstein phase function was proposed in [60]:

$$p(g, \cos \Theta) = \frac{3}{2} \frac{1 - g^2}{2 + g^2} \frac{1 + g^2}{\left(1 + g^2 - 2g \cos \Theta\right)^{3/2}}. \tag{3.156}$$

The expansion coefficients can be expressed analytically as follows:

$$\chi_l = \frac{3}{2} \frac{1}{2 + g^2} \left(\frac{l(l - 1)}{2l - 1} g^{l-2} + \left[\frac{5l^2 - 1}{2l - 1} + \frac{(l + 1)^2}{2l + 3}\right] g^l + \frac{(l + 1)(l + 2)}{2l + 3} g^{l+2}\right) \tag{3.157}$$

The main advance of this phase function is that it describes realistic phase functions for small particles (even it converges to Rayleigh scattering (see below) as the mean cosine of scattering angle given by $3g(4 + g^2)/[5(2 + g^2)]$ goes to 0) in the backward hemisphere. We also mention the Kagiwada-Kalaba phase function [61] given by

$$p(\cos \Theta) = \frac{2}{(b - \cos \Theta) \ln \frac{b+1}{b-1}} \tag{3.158}$$

with

$$b = \frac{r + 1}{r - 1}, \tag{3.159}$$

where $r = p(1)/p(-1)$ is the ratio of forward to backward scattering. To mimic scattering phase function for realistic clouds, r is chosen around 2000 [62].

3.18 Phase Matrix

The Stokes vector of incident light is changed after a single-scattering event. Obviously, this phenomenon corresponds to a 4-by-4 matrix which translates the Stokes parameter vector of the incident radiation to that of the scattered radiation. Thus, a single-scattering event of the polarized radiation is described by the 4-by-4 scattering matrix $\mathbf{F}$ [10]. The elements of $\mathbf{F}$ can be expressed in terms of the amplitudes of the scattered light [63]. For mirror symmetric media with randomly oriented non-spherical particles, there are at most six non-zero independent parameters and the scattering matrix has the following block-diagonal structure [54]:

$$\mathbf{F}(\Theta) = \begin{bmatrix} a_1(\Theta) & b_1(\Theta) & 0 & 0 \\ b_1(\Theta) & a_2(\Theta) & 0 & 0 \\ 0 & 0 & a_3(\Theta) & b_2(\Theta) \\ 0 & 0 & -b_2(\Theta) & a_4(\Theta) \end{bmatrix}, \tag{3.160}$$

where Θ is the scattering angle. The parameter $a_1(\Theta)$ is the phase function, considered previously, and describes the angular distribution of the scattered intensity provided that the incident light is unpolarized. For a_1 there is the following normalization condition:

$$\int_{4\pi} a_1(\Theta)\, d\Theta = 4\pi. \tag{3.161}$$

For the scattering matrix derived for a spherical particle, the following properties are valid [64]:

$$a_1(\Theta) = a_2(\Theta), \tag{3.162}$$

$$a_3(\Theta) = a_4(\Theta), \tag{3.163}$$

$$a_3(0) = a_1(0), \tag{3.164}$$

$$a_3(\pi) = -a_1(\pi), \tag{3.165}$$

$$b_1(0) = b_1(\pi) = 0 \tag{3.166}$$

and

$$b_2(0) = b_2(\pi) = 0. \tag{3.167}$$

These properties are also valid for the ensemble-averaged scattering matrices of a small group of spherically symmetric particles. The example of the phase matrix computed for an ensemble of spherical particles is shown in Fig. 3.11.

Just like the single-scattering phase functions can be expanded into Legendre series, the elements of the phase matrix can be expanded over the generalized spherical functions ($\mu \equiv \cos \Theta$):

$$a_1(\Theta) = \sum_{l=0}^{\infty} \beta_l P_l(\mu), \tag{3.168}$$

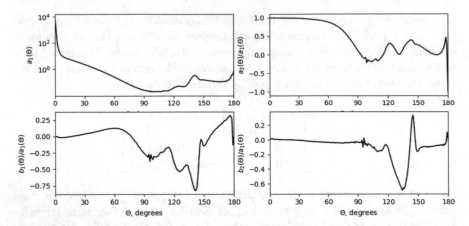

Fig. 3.11 Phase matrix elements for ensemble of spherical particles of water droplets. The particle size follows the modified gamma distribution (see Eq. 2.88) with $\alpha = 5$, $\gamma = 1.05$ and $b = 0.938$. The wavelength is 500 nm

$$a_2(\Theta) = \sum_{l=2}^{\infty} \sqrt{\frac{(l-2)!}{(l+2)!}} \left\{ \alpha_l R_l^2(\mu) + \xi_l T_l^2(\mu) \right\}, \tag{3.169}$$

$$a_3(\Theta) = \sum_{l=2}^{\infty} \sqrt{\frac{(l-2)!}{(l+2)!}} \left\{ \xi_l R_l^2(\mu) + \alpha_l T_l^2(\mu) \right\}, \tag{3.170}$$

$$a_4(\Theta) = \sum_{l=0}^{\infty} \delta_l P_l(\mu), \tag{3.171}$$

$$b_1(\Theta) = \sum_{l=2}^{\infty} \sqrt{\frac{(l-2)!}{(l+2)!}} \gamma_l P_l^2(\mu), \tag{3.172}$$

$$b_2(\Theta) = -\sum_{l=2}^{\infty} \sqrt{\frac{(l-2)!}{(l+2)!}} \varepsilon_l P_l^2(\mu), \tag{3.173}$$

where the expansion coefficients α_j, β_j, δ_j, γ_j, ε_j and ξ_j are called the Greek coefficients, while $R_l^m(\mu)$ and $T_l^m(\mu)$ are the generalized spherical functions. For convenience of the reader, the algorithm for computing $R_l^m(\mu)$ and $T_l^m(\mu)$ is summarized in Appendix 3. Note that the coefficients β_l correspond to the expansion coefficients of the single-scattering phase function x_l:

$$\beta_l = (2l+1) x_l. \tag{3.174}$$

The Henyey–Greenstein phase function can be generalized to the corresponding phase matrix [65]:

$$
\mathbf{F}_{HG}(\Theta) = p_{HG}(\cos\Theta, g) \times
$$
$$
\begin{bmatrix}
1 & P_m\left(1 - \cos^2\Theta\right) & 0 & 0 \\
P_m\left(1 - \cos^2\Theta\right) & 0.25\left(1 + \cos^2\Theta\right) & 0 & 0 \\
0 & 0 & 0.25\left(1 + \cos^2\Theta\right) & Q_m\left(1 - \cos^2\Theta\right) \\
0 & 0 & -Q_m\left(1 - \cos^2\Theta\right) & 1
\end{bmatrix}.
$$
$$(3.175)$$

where P_m and Q_m are the maximum degrees of linear polarization and ellipticity, respectively. The analytical representation given by this equation is useful in theoretical studies of light propagation in various media.

If the Stokes vector was defined in the scattering plane of reference, the scattered vector would be just $\mathbf{F}(\Theta)\mathbf{L}$. However, usually the Stokes parameters are defined with respect to the reference meridional plane. Referring to Fig. 3.12, let us consider single-scattering of polarized radiation by angle Θ. The incident Stokes parameters are defined with respect to the meridian plane OZP_1. The description of a single-scattering event for the Stokes vector includes three steps:

1. A rotation of the reference plane so that the Stokes vector refers to the direction parallel to the scattering plane OP_1P_2,
2. Applying the scattering matrix $\mathbf{F}(\Theta)$;
3. A rotation of the reference plane again so that the Stokes vector refers to the direction along the meridian plane OZP_2.

The rotations are performed by applying rotation matrices defined by Eq. (3.97). The scattering process for the polarized light is described by the phase matrix $\mathbf{Z}(\mu, \mu', \varphi - \varphi')$. It can be represented as follows:

$$
\mathbf{Z}(\mu, \mu', \varphi - \varphi') = \mathbf{R}(\pi - \sigma_2)\mathbf{F}(\Theta)\mathbf{R}(-\sigma_1). \tag{3.176}
$$

The phase matrix defined by Eq. (3.176) can be expanded in a Fourier series as [66]

$$
\mathbf{Z}(\mu, \mu', \varphi - \varphi') =
$$
$$
\sum_{m=0}^{\infty} \boldsymbol{\Phi}_1^m(\varphi - \varphi')\mathbf{W}^m(\mu, \mu')\mathbf{D}_1 + \boldsymbol{\Phi}_2^m(\varphi - \varphi')\mathbf{W}^m(\mu, \mu')\mathbf{D}_2, \tag{3.177}
$$

where

$$
\boldsymbol{\Phi}_1^m(\varphi) = (2 - \delta_{m0})\,\mathrm{diag}\left\{\cos\varphi, \cos\varphi, \sin\varphi, \sin\varphi\right\}, \tag{3.178}
$$

$$
\boldsymbol{\Phi}_2^m(\varphi) = (2 - \delta_{m0})\,\mathrm{diag}\left\{-\sin\varphi, -\sin\varphi, \cos\varphi, \cos\varphi\right\}, \tag{3.179}
$$

$$
\mathbf{D}_1 = \mathrm{diag}\{1, 1, 0, 0\}, \tag{3.180}
$$

and

$$
\mathbf{D}_2 = \mathrm{diag}\{0, 0, 1, 1\}. \tag{3.181}
$$

Fig. 3.12 Scattering of the polarized radiation at a point O. The direction of the incident light is OP_1 and that of the scattered light is OP_2. The incident angle with respect to OZ is θ, while the angle of scattered radiation is θ'. The scattering plane contains the points O, P_1 and P_2. The local meridian plane of incidence goes through the axis OZ and the point P_1, while the local meridian plane of scattering goes through the axis OZ and the point P_2

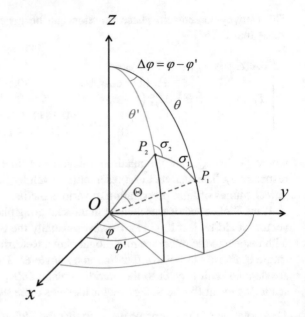

Matrices $\mathbf{W}^m$ can be expressed as follows [67]:

$$\mathbf{W}^m \left(\mu, \mu' \right) = \sum_{l=m}^{M} \mathbf{\Pi}_l^m \left(\mu \right) \mathbf{B}_l \mathbf{\Pi}_l^m \left(\mu' \right), \tag{3.182}$$

where

$$\mathbf{\Pi}_l^m \left(\mu \right) = \begin{bmatrix} P_l^m \left(\mu \right) & 0 & 0 & 0 \\ 0 & R_l^m \left(\mu \right) & -T_l^m \left(\mu \right) & 0 \\ 0 & -T_l^m \left(\mu \right) & R_l^m \left(\mu \right) & 0 \\ 0 & 0 & 0 & P_l^m \left(\mu \right) \end{bmatrix}, \tag{3.183}$$

while the matrix $\mathbf{B}$ comprises the Greek coefficients as follows:

$$\mathbf{B}_l = \begin{bmatrix} \beta_l & \gamma_l & 0 & 0 \\ \gamma_l & \alpha_l & 0 & 0 \\ 0 & 0 & \xi_l & -\varepsilon_l \\ 0 & 0 & \varepsilon_l & \delta_l \end{bmatrix}. \tag{3.184}$$

3.19 Techniques for Computing Phase Functions/Matrices

After we have considered the mathematical properties of the phase functions and phase matrices, let us outline how we can compute them based on the properties of the particles and Maxwell's theory.

The phase matrix and also absorption/scattering cross-sections for an isolated spherical particle of any size are described by using Mie theory [68]. It is based on the solution of a system of Maxwell equations in spherical coordinates. Two boundary conditions are imposed. First, the electric and magnetic fields are required to satisfy continuity conditions on the surface of the spherical particle. Second, at infinity the fields should be spherical waves (i.e., decay as $1/r$). If it were not the case, then the Poynting vector would decay slower as $1/r^2$ and the integral (3.46) would diverge.

Since the scattering wave is proportional to the incident wave, and the scattered wave behaves as a spherical wave in the far-field region, we may write

$$\begin{bmatrix} E_l^s \\ E_r^s \end{bmatrix} = \frac{\exp(-ikr + ikz)}{ikr} \begin{bmatrix} S_2 & S_3 \\ S_4 & S_1 \end{bmatrix} \begin{bmatrix} E_l^i \\ E_r^i \end{bmatrix}, \tag{3.185}$$

where upper indices s and i correspond to the scattered and incident waves, respectively, lower indices l and r refer to the electric fields parallel and perpendicular to the plane of scattering, respectively, while S_1, S_2, S_3 and S_4 are the angular-dependent amplitude functions. For spherical particles, $S_3 = 0$ and $S_4 = 0$. Thus,

$$\begin{bmatrix} E_l^s \\ E_r^s \end{bmatrix} = \frac{\exp(-ikr + ikz)}{ikr} \begin{bmatrix} S_2 & 0 \\ 0 & S_1 \end{bmatrix} \begin{bmatrix} E_l^i \\ E_r^i \end{bmatrix}. \tag{3.186}$$

The key step in deriving Mie solution is the expansion of the incident field, the scattered field and the field inside the particle into vector spherical wave function (VSWF) series. The VSWFs form a basis for the solutions to the Maxwell equations (i.e., any function expressed as a linear combination of VSWFs automatically satisfies Maxwell equations). Moreover, the incident field and the field inside the particle are expressed in terms of regular VSWFs which ensure the fields are bounded (finite). The scattered field is expressed in terms of radiating VSWFs, which automatically satisfy the boundary condition at infinity. The Mie solution is represented in the form of two infinite series:

$$S_1(\Theta) = \sum_{n=1}^{\infty} \frac{2n+1}{n(n+1)} \{a_n \pi_n(\cos\Theta) + b_n \tau_n(\cos\Theta)\} \tag{3.187}$$

and

$$S_2(\Theta) = \sum_{n=1}^{\infty} \frac{2n+1}{n(n+1)} \{b_n \pi_n(\cos\Theta) + a_n \tau_n(\cos\Theta)\}, \tag{3.188}$$

where π_n and τ_n are the angular functions, related to the associated Legendre functions by the following formulas:

$$\pi_n (\cos \Theta) = \frac{P_n^1 (\cos \Theta)}{\sin \Theta}, \tag{3.189}$$

$$\tau_n (\cos \Theta) = \frac{d P_n^1 (\cos \Theta)}{d\Theta}. \tag{3.190}$$

Equations (3.187)–(3.188) can be regarded as a multipole expansion of the scattered light, wherein the coefficients a_1, a_2 and a_3 correspond to an electric dipole, quadrupole and octupole radiation, respectively (the same is valid for b_n, but for magnetic multipole radiation) [68].

By considering the boundary condition at the particle surface and applying orthogonality properties of VSWFs, the unknown expansion coefficients a_n and b_n can be found (see full derivation in [45, 54]). The Mie solution depends on the complex refractive index m of the particle with respect to the surrounding medium and the so-called size parameter defined as

$$x = \frac{2\pi r}{\lambda}, \tag{3.191}$$

where r is the particle radius, while λ is the wavelength. The final expressions for a_n and b_n read as follows:

$$a_n = \frac{\psi_n' (y) \psi_n (x) - m\psi_n (y) \psi_n' (x)}{\psi_n' (y) \zeta_n (x) - m\psi_n (y) \zeta_n' (x)}, \tag{3.192}$$

$$b_n = \frac{m\psi_n' (y) \psi_n (x) - \psi_n (y) \psi_n' (x)}{m\psi_n' (y) \zeta_n (x) - \psi_n (y) \zeta_n' (x)}, \tag{3.193}$$

where $y = mx$, and prime denotes the derivatives and

$$\psi_n(z) = z j_n (z), \tag{3.194}$$

$$\zeta (z) = z h_n^{(2)} (z) \tag{3.195}$$

with j and $h^{(2)}$ being spherical first and third-kind Bessel functions (the latter is referred to as the Hankel function).

Next, it is convenient to define extinction Q_{ext}, scattering Q_{sca} and absorption Q_{abs} efficiencies as ratios between corresponding cross-sections (C_{ext}, C_{sca}, C_{abs}) and the geometrical cross-section πr^2 of a particle. Efficiencies Q_{ext} and Q_{sca} can be expressed in terms of expansion coefficients a_n and b_n :

$$Q_{ext} = \frac{2}{x^2} \sum_{n=1}^{\infty} (2n + 1) \operatorname{Re} (a_n + b_n), \tag{3.196}$$

$$Q_{\text{sca}} = \frac{2}{x^2} \sum_{n=1}^{\infty} (2n + 1) \, \text{Re} \left(|a_n|^2 + |b_n|^2 \right). \tag{3.197}$$

From Eqs. (3.196) and (3.197) the extinction and scattering cross-sections can be computed.

As noted by Hansen and Travis [69], slightly more than x terms have to be summed up in Eqs. (3.187)–(3.188), since a_n and b_n rapidly approach zero as n becomes larger than x. They explained it physically by referring to the concept of ray localization, in which the nth term in Eqs. (3.187)–(3.188) roughly corresponds to the contribution from the light ray passing the center of the sphere at a distance $n\lambda/(2\pi)$ for large values of x. Thus the contribution of terms $n > x$ is small, as higher terms correspond to light rays missing the sphere.

For $x < 0.1$ (small particles compared to the wavelength) the Rayleigh scattering approximation becomes valid, for which the phase function can be expressed analytically:

$$p_R \left(\cos \Theta \right) = \frac{3}{4} \left(1 + \cos^2 \Theta \right) \tag{3.198}$$

(here the atmospheric air depolarization effects are neglected). The corresponding phase matrix is given by the following equation:

$$\mathbf{F_R} \left(\Theta \right) = \frac{3}{4} \begin{bmatrix} 1 + \cos^2 \Theta & \cos^2 \Theta - 1 & 0 & 0 \\ \cos^2 \Theta - 1 & 1 + \cos^2 \Theta & 0 & 0 \\ 0 & 0 & 2 \cos \Theta & 0 \\ 0 & 0 & 0 & 2 \cos \Theta \end{bmatrix}. \tag{3.199}$$

The Rayleigh phase function is symmetric and smooth, and its asymmetry factor g is equal to 0. Then, as the size parameter increases, the asymmetry factor also increases (i.e., the so-called forward peak becomes dominant, as shown in Fig. 3.13). Note that this dependency is not monotonic, as shown in Fig. 3.14. For $x > 100$ (corresponding to large particles compared to the wavelength), the geometrical optics becomes valid. The intermediate case ($x \sim 1$) is sometimes referred to as Mie scattering.

Mie's theory gives an interesting result for the extinction efficiency Q_{ext}. Not only Q_{ext} depends non-monotonically on the size parameter, but also it goes to 2 (not to 1) as $x \to \infty$, as shown in Fig. 3.15. Such a phenomenon is called "the extinction paradox" (a particle scatters radiation outside its own geometrical boundaries). The rigorous explanation of the extinction paradox can be found in [45].

Note that the phase function computed for one particle might have several peaks, as shown in Fig. 3.13. If there is an ensemble of particles of different sizes, the size distribution is split into bins, the Mie computations are performed for each bin and then averaged across the size distribution. Most of these peaks are smoothed out after

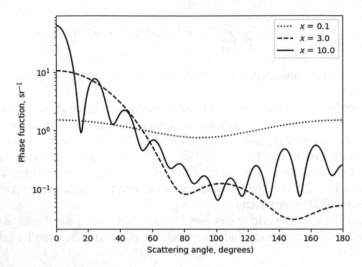

Fig. 3.13 Phase functions computed for different size parameters by using Mie theory. The refractive index is 1.33+0.001i

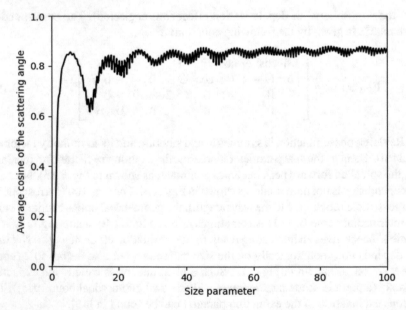

Fig. 3.14 Average cosine of the scattering angle as a function of the size parameter, as predicted by Mie theory. Computations are performed for the particle with the refractive index 1.33+0.00001i

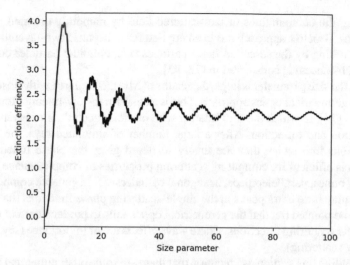

Fig. 3.15 Extinction efficiency as a function of the size parameter, as predicted by Mie theory. Computations are performed for the particle with refractive index 1.33+0.001i

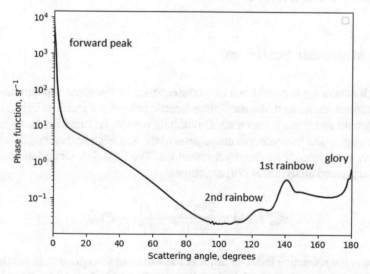

Fig. 3.16 Single-scattering phase function computed for water droplets. The particle size follows the modified gamma distribution, as in Fig. 3.11

the averaging, but some of them may remain. Figure 3.16 shows an example of the single-scattering phase function computed by using Mie theory for water droplets with the modified gamma size distribution. Besides the forward peak, we can see the first and second rainbows and the so-called glory in the backward direction.

For non-spherical particles, one can use the T-matrix method [70, 71]. There is an analytical technique to average the solution over all particles orientations, thus facilitating the computations of the scattered field by randomly oriented particles. In [72], the T-matrix approach has been applied for concentric two-layered spheres, while scattering by the so-called Janus particles (i.e., colloidal particles containing several substances) is considered in [72, 73].

When the size parameter is large, the results of Mie theory agree with those derived using the geometrical optics approach. This is a basis for the ray-tracing Monte-Carlo technique [74], in which the rays interact with particles according to Fresnel's laws of reflection and refraction. After a large number of simulated rays, the statistics of directions into which they are finally scattered gives the phase function. This approach is efficient for computing scattering properties of complex shape particles including cubes, parallelepipeds, hexagonal cylinders etc. In general, complex shape particles may have extra peaks in the single-scattering phase functions (halos) [75]. It should be pointed out that the geometrical optics fails to produce correct results in the specific scattering directions, where wave effects are of importance (say, rainbow and glory scattering).

Concluding this section, we mention that there are some other numerical methods, e.g., the finite-difference time-domain (FDTD) technique [76] and discrete dipole approximation (DDA) [77, 78].

3.20 Molecular Scattering

Rayleigh scattering is considered as a consequence of the electric polarizability of the air molecules. Indeed, the oscillating electric field of the incident light polarizes the molecule and turns it into a small radiating dipole. Its radiation is regarded as scattered light. The macroscopic description of the Rayleigh scattering phenomenon considers scattering by air density fluctuations. The Rayleigh scattering coefficient can be expressed analytically [79] as follows:

$$\sigma_{\text{sca}}^{\text{mol}} (z, \lambda) = \frac{24\pi^3 \left(n^2 - 1\right)^2}{N^2 \left(n^2 + 2\right)^2} \frac{1}{\lambda^4} F(\lambda), \qquad (3.200)$$

where n is the refractive index of air, λ is the wavelength expressed in centimeters, N is the molecular density, and $F(\lambda)$ is the so-called King factor. The latter is given by

$$F(\lambda) = \frac{6 + 3\rho(\lambda)}{6 - 7\rho(\lambda)}, \qquad (3.201)$$

where ρ is the depolarization ratio, which accounts for the anisotropy of air molecules.

The computations of $\sigma_{\text{sca}}^{\text{mol}}$ are complicated by the fact that the refractive index of air and the King factor depend on the wavelength. In this regard, there are several models for computing $\sigma_{\text{sca}}^{\text{mol}}$ (e.g., Bates approximation [80], Bucholtz model [81], Bodhaine model [79]) and, in particular, Rayleigh optical thickness τ_R. For instance, the empirical expression for τ_R was proposed in [82]:

Table 3.1 Rayleigh scattering coefficient at the Earth's surface, Rayleigh optical thickness of the atmosphere and the depolarization factor as a function of wavelength (data is taken from [81])

Wavelength (nm)	Scattering coefficient (km^{-1})	Rayleigh optical thickness	Depolarization factor
200	0.9202	7.788	0.04545
250	0.3208	2.714	0.03565
300	0.01437	1.216	0.03178
350	0.07450	0.6304	0.03010
400	0.04261	0.3606	0.02955
450	0.02616	0.2214	0.02899
500	0.01696	0.1435	0.02842
550	0.01149	0.09721	0.02842
600	0.00853	0.06815	0.02786
650	0.005819	0.04924	0.02786
700	0.004310	0.03647	0.02786
750	0.003261	0.02759	0.02786
800	0.002510	0.02124	0.02730
900	0.001561	0.01321	0.02730
1000	0.001022	0.008645	0.02730

$$\tau_R\,(\lambda) = \frac{p}{p_0} 0.00877 \lambda^{-4.05}, \tag{3.202}$$

where p is the site pressure, p_0 is 1013.25 mb, and λ is the wavelength in micrometers. The values of the Rayleigh scattering coefficient, the Rayleigh optical thickness of the atmosphere and the depolarization ratio are given in Table 3.1. The overview of approximate models for computing the Rayleigh scattering coefficient is given in [83].

Neglecting depolarization (setting ρ to zero in Eq. (3.201) leads to underestimation of the Rayleigh cross-section by $\sim 5\%$. The anisotropy of air molecules also affects the Rayleigh single-scattering phase function, which can be expressed as

$$p_{Ray}\,(\cos\Theta) = \frac{3}{4\,(1+2\gamma)}\left[(1+3\gamma) + (1-\gamma)\cos^2\Theta\right], \tag{3.203}$$

where γ is defined by

$$\gamma = \frac{\rho}{2-\rho}. \tag{3.204}$$

The expansion coefficients for the Rayleigh phase function are given as follows:

$$x_0 = 1, \quad x_1 = 0, \quad x_2 = \frac{1}{5}\frac{1-\rho}{2+\rho}. \tag{3.205}$$

Table 3.2 The Greek coefficients for Rayleigh scattering matrix

l	α_l	β_l	γ_l	δ_l	ε_l	ξ_l
0	0	1	0	0	0	0
1	0	0	0	$\frac{3(1-2\rho)}{2+\rho}$	0	0
2	$\frac{6(1-\rho)}{2+\rho}$	$\frac{1-\rho}{2+\rho}$	$-\frac{\sqrt{6}(1-\rho)}{2+\rho}$	0	0	0

The Greek expansion coefficients for Rayleigh scattering matrix are listed in Table 3.2.

Note that molecular scattering includes a strict elastic-scattering process (corresponding to the central Cabannes line) and the inelastic wavelength-distributed rotational Raman scattering (RRS). A physical model of inelastic scattering is not trivial. In fact, the term "molecular scattering" includes Rayleigh scattering and vibrational Raman scattering. In its turn, the Rayleigh scattering spectrum comprises rotational Raman lines and the central Cabannes line. The Cabannes line is composed of the Brillouin doublet and the Landau-Plachek [84]. The Cabannes line width is very small (e.g., around 0.002 at 500 nm) and can be neglected. The diatomic molecules, such as nitrogen and oxygen, in their electronic ground state can rotate and vibrate, and so, these motions lead to inelastic scattering and yield a characteristic scattering spectrum, which is important to take into account several applications, including trace gas retrieval. A detailed analysis of the inelastic scattering phenomenon in radiative transfer can be found in [85] and references therein. In the UV and visible spectral ranges, the Raman scattering cross-section is about 4 % of the Rayleigh cross-section.

The angular distributions for Cabannes scattering and RRS are described by phase functions [86]

$$p_{\text{Cab}}(\cos\Theta) = \frac{3}{8}\left[\frac{8-5\rho}{4-3\rho} + \frac{8-9\rho}{4-3\rho}\cos^2\Theta\right], \tag{3.206}$$

$$p_{\text{RSS}}(\cos\Theta) = \frac{3}{40}\left[13 + \cos^2\Theta\right], \tag{3.207}$$

respectively. The expansion coefficients for them are given as follows:

$$x_{2,\text{Cab}} = \frac{1}{20}\frac{8-9\rho}{4-3\rho} \tag{3.208}$$

and

$$x_{2,\text{RSS}} = \frac{1}{100}, \tag{3.209}$$

while $x_0 = 1$, $x_1 = 0$ and $x_k = 0$ for $x \geq 3$.

3.21 Geometrical Optics Approximation

Scattering and absorption of light by large spherical particles can be studied using both Maxwell equations (Mie theory) and also computationally easier geometrical optics approach. In practice, one can expect that the geometric optics approximation provides quite accurate results at the size parameter larger than 10^2–10^3. Interestingly, the formulas derived in the framework of geometrical optics can be obtained from the Mie solution (see [87]) for large particles. Obviously, in this case, the wave properties of the light are neglected. However, it is possible to design corresponding wave corrections for the geometrical optics approximation [88].

It is assumed in the framework of geometrical optics that the radius of particles r is much larger as compared to the wavelength of incident light, and the incident wave can be presented as rays striking the surface of the particle (see Fig. 3.17). These rays interact with the spherical particles according to the reflection and refraction laws for plane surfaces formulated by Fresnel [89]. It is assumed that the surface of the spherical particle is locally plane. The refracted ray is propagated along the direction specified by the refraction angle ψ, which is related to the incidence angle φ by the following equation:

$$\frac{\sin \varphi}{\sin \psi} = n, \tag{3.210}$$

where n is the real part of the relative complex refractive index $m = n - i\chi$ of a scatterer. It is assumed that $\chi \ll n$. The geometrical optics can be used to explain the main features of various optical phenomena, but with less computational efforts as compared to Mie's theory. For instance, it can be used to explain the existence and position of the rainbow [45]. It follows from Fig. 3.17 that the density of rays is highly close to the rainbow position (red line). It explains the enhanced scattered light intensity in the rainbow region. The sequence of colors can be explained by the dependence of the refraction angle on the refractive index n as given in Eq. (3.210). The refractive index of water changes with the wavelength. This leads to different positions of colors in the rainbow. A more detailed explanation of the scattered light intensity in the rainbow region still requires the wave optics approach [54, 90].

Geometrical optics (with account for diffraction effects in the forward scattering direction) leads to particularly simple expressions for integral light scattering characteristics of large spherical particles such as extinction efficiency factor Q_{ext}, scattering efficiency factor Q_{sca}, absorption efficiency factor Q_{abs}, and average cosine of scattering angle g. Namely for a large sphere with the size parameter x neglecting interference effects (i.e., $2x |m - 1| \gg 1$ and $x \gg 1$) the following relations are valid [91]:

$$\begin{cases} Q_{\text{ext}} = 2, \\ Q_{\text{abs}} = 1 - W, \\ Q_{\text{sca}} = 1 + W, \\ g = \frac{1+G}{1+W}, \end{cases} \tag{3.211}$$

Fig. 3.17 Interaction of geometrical optics rays with spherical particles

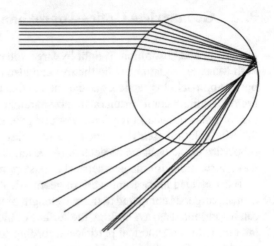

where

$$W = \frac{1}{2} \sum_{j=1}^{2} \int_0^{\pi/2} \left(R_j + \frac{(1 - R_j)^2 \, e^{-c\xi}}{1 - R_j e^{-c\xi}} \right) \sin 2\tau d\tau \qquad (3.212)$$

$$G = \frac{1}{2} \int_0^{\pi/2} (\varepsilon_1(c) + \varepsilon_2(c)) \sin 2\tau d\tau \qquad (3.213)$$

$$R_1 = \left| \frac{\sin \tau - m \sin \tau'}{\sin \tau + m \sin \tau'} \right|^2, \quad R_1 = \left| \frac{m \sin \tau - \sin \tau'}{m \sin \tau + \sin \tau'} \right|^2 \qquad (3.214)$$

$$\varepsilon_j(c) = \frac{(1 - R_j)^2 \cos 2(\tau - \tau') e^{-c\xi} + R_j \cos 2\tau (1 - e^{-2c\xi})}{1 - 2R_j e^{-c\xi} \cos 2\tau' + R_j^2 e^{-c\xi}}$$
$$+ \frac{2R_j^2 \cos 2\tau e^{-c\xi} (e^{-c\xi} - \cos 2\tau')}{1 - 2R_j e^{-c\xi} \cos 2\tau' + R_j^2 e^{-c\xi}}, \qquad (3.215)$$

while

$$\xi = \sqrt{1 - \left(\frac{\cos \tau}{n} \right)^2}, \quad \cos \tau' = \frac{\cos \tau}{n}, \quad c = 4\chi x. \qquad (3.216)$$

Therefore, the calculation of integral light scattering characteristics is reduced to the calculation of single integrals from elementary functions. The approximate evaluation of integrals given above is possible. In particular, it follows:

$$W = \rho + (1 - \rho) e^{-vc}, \qquad (3.217)$$

where

$$\rho = \frac{8n^4(1 + n^4) \ln n}{(1 + n^2)(1 - n^4)^2} + \frac{n^2 (1 - n^2) \ln [(n - 1)/(n + 1)]}{(1 + n^2)^3}$$

$$-\frac{\sum_{j=0}^{7} p_j n^j}{3(1+n)(1+n^2)(1-n^4)},$$

$$p = (-1, -1, -3, 7, -9, -13, -7, 3),$$

and

$$v = \frac{2n^2(1 - (1 - n^{-2})^{3/2})}{3(1-\rho)}.$$

Also it follows:

$$g = g_\infty - (g_\infty - g_0)e^{-\eta c}, \tag{3.218}$$

where

$$g_\infty \equiv g(c \to \infty) = \frac{1+K}{1+\rho},$$

$$g_0 = g(c \to 0) = \frac{1+M}{2},$$

$$K = \frac{8n^4(n^6 - 3n^4 + n^2 - 1)\ln n}{(1+n^2)^2(1-n^4)^2} + \frac{(1-n^2)^2(n^8 + 12n^6 + 54n^4 - 4n^2 + 1)\ln[(n-1)/(n+1)]}{16(1+n^2)^4}$$
$$+ \frac{\sum_{j=1}^{12} q_j n^j}{24(1+n)(1+n^2)(n^4-1)},$$

$$q = (-3, 13, -89, 151, 186, 138, -282, 22, 25, 25, 3, 3)$$

and

$$M = \frac{1}{2} \int_0^{\pi/2} (\varepsilon_1(0) + \varepsilon_2(0)) \sin 2\tau d\tau.$$

The values of g_0, g_∞ and η depend on the real part of the complex refractive index. In particular, we have found these values by parameterization of geometrical optics calculations of the asymmetry parameter:

$$g_0 = 1.006 - 0.3641(n - 1),$$

$$g_\infty = 1.008 - 0.11(n - 1),$$

$$\eta = 0.3639 - 1.676(n - 1) - 1.6284(n - 1)^2$$

for monodispersed spheres at $n = 1.2$–1.4. An important point is that Eqs. (3.217) and (3.218) can also be applied for the case of randomly oriented non-spherical particles. In this case, the radius a must be substituted by the ratio $3V/S$, where V is the volume

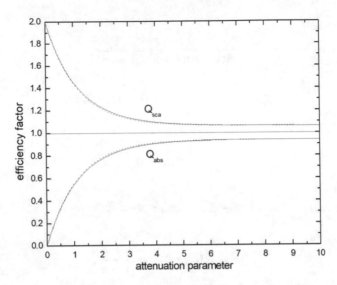

Fig. 3.18 The dependence of efficiency factors of monodispersed spherical particles on the attenuation parameter c calculated using Eq. (3.211) together with Eq. (3.212) (solid lines) and Eq. (3.211) together with Eq. (3.217) (dash lines). It is assumed that $n = 1.31$

and S is the surface area of particles. The parameters v, η and g_0 depend on the shape of particles and can be derived from geometrical optics calculations. Alternatively, they can be experimentally measured.

The dependence of Q_{abs} and Q_{sca} on the parameter c for monodispersed spheres at $n = 1.31$ as derived from geometrical optics calculations is shown in Fig. 3.18.

We also show the dependence of the asymmetry parameter of monodispersed spherical particles on the attenuation parameter c derived using Eqs. (3.211)–(3.213) and Eq. (3.218) in Fig. 3.19. The difference is below 0.5%. Therefore, simple Eq. (3.218) can be used to model the asymmetry parameter of monodispersed large spherical particles. Kokhanovsky and Macke [92] have applied Eq. (3.218) for modeling of asymmetry parameters of large non-spherical convex particles in random orientation. They have found that g_0 and η depend on the shape of particles. The value g_∞ corresponds to the case of strongly absorbing particles ($c \to \infty$). It depends just on the light reflection from the surface of particles. This process leads to the same result for g_∞ in the case of spherical particles and randomly oriented convex non-spherical particles. This important conclusion is confirmed by direct time-consuming geometrical optics calculations for hexagonal cylinders, plates and other types of non-spherical randomly oriented convex particles.

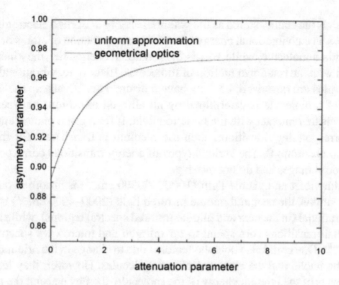

Fig. 3.19 The dependence of the asymmetry parameter of large monodispersed spherical particles on the attenuation parameter c calculated using geometrical optics results given by Eqs. (3.211)–(3.213) (solid line) and Eq. (3.218) (dash line). It is assumed that $n=1.31$

3.22 Absorption by Atmospheric Gases

Absorption of radiation is a inelastic wavelength-dependent process. The main parameter which describes absorption is the absorption cross-section (see Eq. (3.111)). We can define the total absorption thickness of the layer between altitudes z_1 and z_2 as follows:

$$\tau_{abs} = \sum_{i=1}^{N} \int_{z_1}^{z_2} C_{abs,i}(z)\, \varsigma_i(z)\, dz, \qquad (3.219)$$

where $C_{abs,i}$ is the ith gas absorption cross-section, N is the total number of gases present and $\varsigma_i(z)$ is the number density of the ith gas at a given height.

Accurate computations of the absorption cross-section dependence on the wavelength of light from first principles are quite complex. In our book we only outline the basic physical ideas behind the computational process.

A molecule has a discrete set of energy levels. A photon (or corresponding electromagnetic wave) can be absorbed, and its energy is transferred to the energy of a molecule. Therefore, the understanding of the structure of energy levels is crucial for computing absorption coefficients. The total energy W_{total} of a molecule can be represented as follows:

$$W_{total} = W_{elec} + W_{vib} + W_{rot}, \qquad (3.220)$$

where W_{elec} is the energy stored as the potential energy in excited electronic configurations, W_{vib} is the vibrational energy of the oscillatory motion of atoms or groups of atoms within a molecule, while W_{rot} is the rotational energy, i.e., the kinetic energy associated with the rotational motion of molecules. Electronic, vibrational and rotational energies are quantized, i.e., can have a discrete set of values. The absorption spectrum of a molecule is determined by all allowed transitions between pairs of energy levels that interact with the radiation field. If for a given photon energy there are no corresponding transitions, then the medium is transparent for this photon (there is no absorption). The various types of energy transitions correspond to different spectral ranges and do not overlap.

Near-ultraviolet and visible light ($10000–50000$ cm^{-1}) is absorbed due to electronic transitions, the near and middle infrared light ($2000–4000$ cm^{-1}) is absorbed due to vibrational (in the near and middle infrared spectral regions), while the energy of rotational transitions correspond to far-infrared and microwave spectral regions ($1–20$ cm^{-1}). One could also note the features due to change in the orientation of the spins of the nuclei and the electrons of the molecules. However, they lead to very small changes of the internal energy of the molecule, thereby causing the absorption of meter radiowave radiation [93]. The scheme of energy levels is shown in Fig. 3.20.

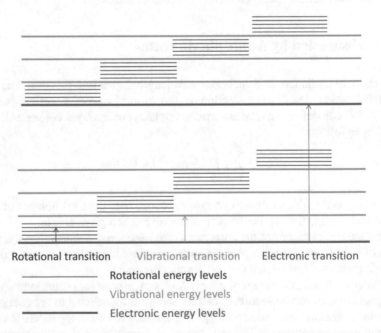

Fig. 3.20 The structure of energy levels due to electronic, vibrational and rotational transitions

A transition between two layers corresponds to a certain resonant frequency

$$\nu_0 = \Delta W / h, \tag{3.221}$$

where ΔW is the difference between energies of corresponding levels and h is Planck's constant. Consequently, a photon with energy $h\nu_0$ can be absorbed or emitted due to such a transition. However, the light is absorbed not only at ν_0, but also in a certain frequency range around. The absorption line is said to be broadened. Thus, to describe the line, we need to know the line position (i.e., at which frequency the absorption occurs), the line strength (i.e., how strong absorption is) and the line shape (how absorption is "distributed" across the spectrum).

Note that a set of absorption lines (and consequently, energy levels) is unique for a given chemical component. Therefore, the absorption spectrum can be used for retrieving quantitative information about absorbers (e.g., trace gases) and sometimes their distribution along a line of sight.

Several lines can overlap and contribute to the absorption process at a certain wavenumber ν. It is convenient to represent the differential absorption cross-section at a certain wavenumber ν as follows:

$$C_{\text{abs}}(\nu) = \sum_i S_i f_i(\nu), \tag{3.222}$$

where S_i is the line strength, while f_i is the line profile describing the line shape. Thus, the shape and the strength are decoupled. The profile f is normalized to the unit area:

$$\int_0^\infty f(\nu)\, d\nu = 1. \tag{3.223}$$

There are several mechanisms of line broadening, namely natural broadening, pressure broadening and Doppler broadening [94]. The natural broadening comes from one of the interpretations of Heisenberg's uncertainty principle:

$$\Delta W \Delta t \geq \frac{\hbar}{2}. \tag{3.224}$$

Here ΔW is the uncertainty in the energy and related to the line width, Δt is the uncertainty in time and is associated with the lifetime of an excited state, while $\hbar$ is the reduced Planck constant. Consequently, a short lifetime has a large energy uncertainty and a broad emission spectrum and vice versa. This broadening effect leads to the Lorentzian profile f_L given by

$$f_L(\nu, \nu_0, \alpha_L) = \frac{\alpha_L}{\pi\left[(\nu - \nu_0)^2 + \alpha_L^2\right]}, \tag{3.225}$$

here α_L specifies the half-width at half-maximum of the Lorentzian line profile. We should note that natural broadening is very small compared to other broadenings (about 10^{-8} cm^{-1}) and completely negligible in atmospheric spectroscopy.

The pressure broadening is caused by the collisions with other molecules [95]. If the gas pressure is high, the interval between atomic collisions can be much smaller than the emission time. These collisions cause the premature emission of a photon. The decreased lifetime results in the increased uncertainty in the emitted energy (the same mechanism as in natural broadening). Pressure broadening is described by a Lorentzian profile (3.225). It depends on the temperature and the pressure of the gas. On an average, the pressure broadening half-width is proportional to the pressure.

Doppler broadening is caused by the fact that the emitting atom has a certain velocity. Therefore, the emitted photon will be shifted in a lower or higher frequency direction depending on the atom velocity relative to the observer. Doppler broadening causes the Gaussian line shape f_G :

$$f_G\left(\nu, \nu_0, \alpha_D\right) = \frac{e^{-(\nu-\nu_0)^2/\alpha_D^2}}{\alpha_D\sqrt{\pi}}, \qquad (3.226)$$

where α_D is the half-width of the Gaussian profile. Note that α_D increases linearly with the atom velocity, which is proportional to the square root from the temperature. In the infrared frequency domain, α_D is about 10^{-3} cm^{-1}.

Given that the pressure decreases and the temperature increases with the altitude, we conclude that pressure broadening dominates at low altitudes, while Doppler broadening is dominant at high altitudes.

Assuming each broadening mechanism is independent, the resulting line shape is given by the convolution of the Lorentzian shape and the Gaussian shape. This is a Voigt profile. More sophisticated line shape functions (the so-called beyond-Voigt profiles) [96] can be used to account for details in the collisions that give rise to the line shape.

An example of the absorption cross-section for water vapor in 900–1000 nm spectral range at 300 K is shown in Fig. 3.21. The lines have the Voigt shape. As we can see, the water vapor has strong spectral signatures here; this spectral range can be used for water vapor retrieval. Figure 3.22 illustrates O_2 cross-sections in the near infrared spectral range. In particular, the absorption lines around 760 nm make the so-called O_2A absorption band, 690 nm–O_2B band, 630 nm–O_2C band. The zoomed O_2A band is shown in Fig. 3.23. This band is very important for retrieving information about clouds and aerosols. Indeed, since the total amount and the vertical distribution of O_2 in the atmosphere are well known, perturbations in this band can be associated with clouds and aerosols; this band is also sensible to the location of clouds making it possible to retrieve information about cloud-top height.

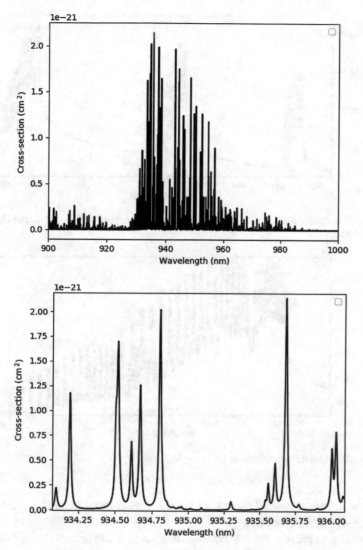

Fig. 3.21 Water vapor absorption cross-section: (top) in 900–1000 nm spectral range, (bottom) the same, but zoomed to 934–936 nm range

If the width of the lines is smaller than the spacing between the lines, the atmospheric spectrum is represented by individual lines. Complex molecules possess a lot of vibrational and rotational modes. A combination of these modes leads to a smearing of the discrete spectrum, and so, to appearing broad peaks rather than discrete lines. In Fig. 3.24 the ozone absorption cross-sections in visible and near-infrared spectral ranges is shown. We see a continuous function rather than a set of discrete lines. Moreover, the multispectral instruments cannot record a spectrum

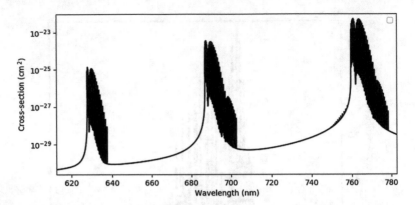

Fig. 3.22 Absorption cross-section of O_2. Three peaks correspond to O_2A, O_2B and O_2C bands (from right to left)

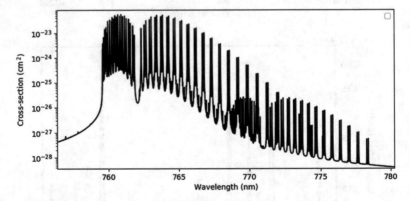

Fig. 3.23 Absorption cross-section of O_2 in the O_2A band

with so high resolution and resolve all the absorption lines. Instead, the instrument records a smoothed spectrum.

It is useful to introduce the direct transmission function (direct transmittance) for each gaseous component as follows:

$$T^{gas} = \exp(-\tau^{gas}_{abs}).$$ (3.227)

It shows which portion of radiation is absorbed when passing in the nadir direction through the whole atmospheric layer. Figure 3.25 illustrates transmission functions for vapor, ozone, carbon dioxide and oxygen. The transmission function equal to 1 shows that a gaseous component is transparent for the radiation at the corresponding spectral regions (atmospheric windows). If the transmission function is zero, then the radiation is completely absorbed.

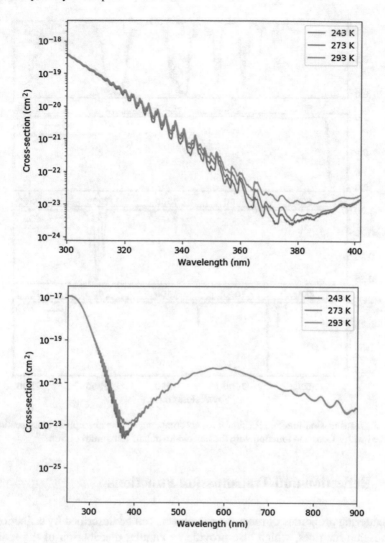

Fig. 3.24 Ozone absorption cross-section in the spectral range 300–400 nm (top) and 200–900 nm (bottom). The data is taken from [99]

The HITRAN database contains spectroscopic parameters of molecules. Accurate computations of the cross-sections from these data is not an easy task (for instance, due to long tales of the absorption lines and line overlapping). There are several open-source tools for reading the database and computing the absorption cross-sections [97, 98] (see references in section "Useful links" at the end of the book).

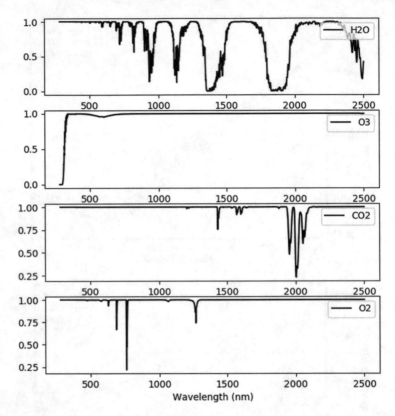

Fig. 3.25 Transmission function for four gaseous components. The absorption cross-sections are convolved with a Gaussian function with the half-width at half-maximum of $5\,cm^{-1}$

3.23 Reflection and Transmission Functions

The scattering properties of macroscopic objects can be described by reflection and transmission matrices, which also provide an angular distribution of the scattered light field given the known incoming radiation. Unlike the single-scattering phase function, which deals only with single-scattering, the reflection and transmission matrices take into account the multiple scattering process.

Let us consider a plane-parallel homogeneous layer of thickness z_0 illuminated from above as shown in Fig. 3.26. The reflectance and transmittance of light by layers can be defined in terms of reflection R and transmission T functions [19]. The R- and T-functions relate incident radiance with reflected and transmitted radiances. By definition, we have

$$L\left(0, \mu > 0, \varphi\right) = \frac{1}{\pi} \int_0^{2\pi} \int_{-1}^0 R\left(\mu, \varphi, \mu', \varphi'\right) L\left(0, \mu', \varphi'\right) \mu' d\mu' d\varphi', \quad (3.228)$$

Fig. 3.26 To the definition
of the reflection and
transmission functions

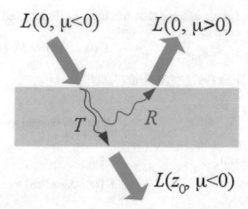

$$L(z_0, \mu < 0, \varphi) = \frac{1}{\pi} \int_0^{2\pi} \int_{-1}^{0} T\left(\mu, \varphi, \mu', \varphi'\right) L\left(0, \mu', \varphi'\right) \mu' d\mu' d\varphi'. \quad (3.229)$$

Note that R and T do not depend on the incoming radiance and characterize inherent
scattering properties of the layer. Next, we assume that the R and T functions are
axisymmetric, i.e., they depend on $\varphi - \varphi'$. For the azimuthally independent radiance
field and R and T functions we have

$$L\left(0, \mu > 0\right) = 2 \int_0^{2\pi} \int_{-1}^{0} R_0\left(\mu, \mu'\right) L\left(0, \mu'\right) \mu' d\mu' \quad (3.230)$$

and

$$L\left(z_0, \mu < 0\right) = 2 \int_{-1}^{0} T_0\left(\mu, \mu'\right) L\left(0, \mu'\right) \mu' d\mu', \quad (3.231)$$

respectively. Here R_0 and T_0 are the azimuthal averaged reflection and transmission
functions, respectively, defined as follows:

$$R_0\left(\mu, \mu'\right) = \frac{1}{2\pi} \int_0^{2\pi} R\left(\mu, \mu', \varphi - \varphi'\right) d\varphi' \quad (3.232)$$

and

$$T_0\left(\mu, \mu'\right) = \frac{1}{2\pi} \int_0^{2\pi} T\left(\mu, \mu', \varphi - \varphi'\right) d\varphi'. \quad (3.233)$$

Let us assume that the layer is illuminated by a plane-parallel beam. Then, the substitution of

$$L\left(0, \mu', \varphi'\right) = F\delta\left(\mu - \mu'\right)\delta\left(\varphi'\right) \quad (3.234)$$

into Eqs. (3.228) and (3.229) gives

$$R\left(\mu^{\uparrow}, \varphi, \mu_0, \varphi_0\right) = \frac{\pi L\left(0, \mu^{\uparrow}, \varphi\right)}{\mu_0 F} \quad (3.235)$$

and

$$T\left(\mu^{\downarrow}, \varphi, \mu_0, \varphi_0\right) = \frac{\pi L\left(z_0, \mu^{\downarrow}, \varphi\right)}{\mu_0 F}. \quad (3.236)$$

The plane albedo r, transmission t, the spherical albedo r_s and the spherical transmittance t_s of the layer are given by

$$r\left(\mu\right) = 2\int_0^1 R_0\left(\mu, \mu'\right)\mu' d\mu', \quad (3.237)$$

$$t\left(\mu\right) = 2\int_0^1 T_0\left(\mu, \mu'\right)\mu' d\mu', \quad (3.238)$$

$$r_s = 2\int_0^1 r\left(\mu\right)\mu d\mu \quad (3.239)$$

and

$$t_s = 2\int_0^1 t\left(\mu\right)\mu d\mu, \quad (3.240)$$

respectively.

3.24 Surface Bidirectional Reflectance Distribution Function

The surface bidirectional reflectance distribution function (BRDF) characterizes the reflection of the incident radiance by the surface. It describes how the surface reflectance depends on viewing and solar angles. The BRDF defines the lower boundary condition for the radiative transfer equation and has a strong influence on the surface energy balance. In addition, the BRDF has a strong impact on the reflected radiation measured by satellites, especially in atmospheric windows with weak atmospheric absorption.

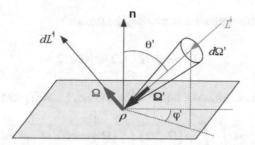

Fig. 3.27 To the definition of the bidirectional reflectance distribution function

Referring to Fig. 3.27, the BRDF ρ can be formally defined as the ratio of the reflected radiance to the incident irradiance:

$$\rho\left(\boldsymbol{\Omega}', \boldsymbol{\Omega}\right) = \frac{dL^\uparrow\left(\boldsymbol{\Omega}\right)}{dE^\downarrow\left(\boldsymbol{\Omega}'\right)}. \tag{3.241}$$

Here $dE^-\left(\boldsymbol{\Omega}'\right) = L^-\left(\boldsymbol{\Omega}'\right)\cos\theta' d\Omega'$ is the irradiance due to the incident radiance L' within a cone of solid angle $d\Omega'$ around the direction $\boldsymbol{\Omega}'$. From Eq. (3.241) the reflected radiance is computed as

$$L^\uparrow\left(\boldsymbol{\Omega}\right) = \int \rho\left(\boldsymbol{\Omega}', \boldsymbol{\Omega}\right) dE^\downarrow\left(\boldsymbol{\Omega}'\right) = \int_{\substack{\text{lower}\\\text{hemisphere}}} \rho\left(\boldsymbol{\Omega}', \boldsymbol{\Omega}\right) L^\downarrow\left(\boldsymbol{\Omega}'\right)\cos\theta' d\Omega'. \tag{3.242}$$

Comparing Eq. (3.242) and reflection matrix definition (3.228), we conclude that the BRDF can be regarded as the reflection matrix (divided by π) of the surface.

The BRDF is independent of the strength and geometry of the light source. The physically relevant BRDFs are positive for all angles and obey the Helmholtz reciprocity rule, i.e.,

$$\rho\left(-\boldsymbol{\Omega}', \boldsymbol{\Omega}\right) = \rho\left(\boldsymbol{\Omega}', -\boldsymbol{\Omega}\right). \tag{3.243}$$

In addition, due to the energy conservation law,

$$\int_{\substack{\text{lower}\\\text{hemisphere}}} \rho\left(\boldsymbol{\Omega}', \boldsymbol{\Omega}\right)\cos\theta' d\Omega' \le 1. \tag{3.244}$$

To prove it, we consider the irradiance due to incoming radiation, i.e.,

$$E^\downarrow = \int_{\Omega'} L^\downarrow\left(\boldsymbol{\Omega}'\right)\cos\theta' d\Omega' \tag{3.245}$$

and the radiosity (the radiant flux leaving the surface)

$$E^\uparrow = \int_\Omega L^\uparrow(\boldsymbol{\Omega}) \cos\theta d\Omega. \tag{3.246}$$

Substituting the substitution of Eq. (3.242) into Eq. (3.246) yields

$$E^\uparrow = \int_\Omega \int_{\Omega'} \rho(\boldsymbol{\Omega'}, \boldsymbol{\Omega}) L^\downarrow(\boldsymbol{\Omega'}) \cos\theta' d\Omega' \cos\theta d\Omega. \tag{3.247}$$

Using Eq. (3.245) in Eq. (3.247), one arrives at

$$E^\uparrow = E^\downarrow \int_\Omega \rho(\boldsymbol{\Omega'}, \boldsymbol{\Omega}) \cos\theta d\Omega. \tag{3.248}$$

Due to the energy conservation law, $E^\uparrow \leq E^\downarrow$. Therefore,

$$\int_\Omega \rho(\boldsymbol{\Omega'}, \boldsymbol{\Omega}) \cos\theta d\Omega \leq 1. \tag{3.249}$$

Finally, applying the reciprocity rule (i.e., Eq. (3.243)) in Eq. (3.249) proves Eq. (3.244).

Analogously to scalar BRDF, we can introduce the bi-directional reflection distribution matrix (BRDM) as follows:

$$\boldsymbol{\rho}(\boldsymbol{\Omega'}, \boldsymbol{\Omega}) = \frac{d\mathbf{L}^\uparrow(\boldsymbol{\Omega})}{d E^\downarrow(\boldsymbol{\Omega'})}. \tag{3.250}$$

In many textbooks, the reflection from the surface is described in terms of reflection functions (sometimes also referred to as the BRDF) analogously to Eq. (3.228). Thus,

$$L^\uparrow(\boldsymbol{\Omega}) = \frac{1}{\pi} \int_{\substack{\text{lower} \\ \text{hemisphere}}} R(\boldsymbol{\Omega'}, \boldsymbol{\Omega}) L^\downarrow(\boldsymbol{\Omega'}) \cos\theta' d\Omega'. \tag{3.251}$$

As we can see from Eqs. (3.242) and (3.251), two definitions differ by factor π :

$$\rho(\boldsymbol{\Omega'}, \boldsymbol{\Omega}) = \frac{1}{\pi} R(\boldsymbol{\Omega'}, \boldsymbol{\Omega}). \tag{3.252}$$

A surface is said to be Lambertian if its BRDF is independent of directions of incidence and observation. In this case,

$$\rho_L = \rho\left(-\boldsymbol{\Omega}', \boldsymbol{\Omega}\right) = \text{const}, \tag{3.253}$$

where ρ_L is the Lambert reflectance. In this case, Eq. (3.242) is transformed into

$$L^\uparrow\left(\boldsymbol{\Omega}\right) = \rho_L E^\downarrow. \tag{3.254}$$

From Eq. (3.244) it follows that $\rho_L \leq \pi^{-1}$. The diffuse reflectivity ρ_d (also referred to as the surface albedo) is defined as

$$\rho_d = \rho_L \pi. \tag{3.255}$$

Obviously, ρ_d varies from 0 to 1. Hence, ρ_d can be interpreted as the fraction of the incident sunlight that the surface reflects.

In the case of specular reflection (e.g., mirror), the reflected radiance is directed along the angle of reflection. The polar and azimuthal angles of reflection are

$$\theta = \theta', \tag{3.256}$$

$$\varphi = \varphi' + \pi \tag{3.257}$$

Hence, the specular BRDF reads as

$$\rho\left(\boldsymbol{\Omega}', \boldsymbol{\Omega}\right) = \frac{\rho_s\left(\theta\right) \delta\left(\cos\theta - \cos\theta'\right) \delta\left(\varphi - \left(\varphi' + \pi\right)\right)}{\cos\theta'}. \tag{3.258}$$

Here ρ_s is the specular reflectivity.

Real BRDFs may have a complex dependency on geometry and the wavelength λ. In the framework of the so-called kernel-based (or kernel-driven) approach, the BDRF is expressed as a linear sum of BRDF kernels:

$$R\left(\boldsymbol{\Omega}', \boldsymbol{\Omega}, \lambda\right) = \sum_k w_k\left(\lambda\right) K_k\left(\boldsymbol{\Omega}', \boldsymbol{\Omega}\right), \tag{3.259}$$

where $w_k\left(\lambda\right)$ are wavelength-dependent weights, while $K_k\left(\boldsymbol{\Omega}', \boldsymbol{\Omega}\right)$ are BRDF kernels. For kernels there are empirical [100] and semi-empirical [101, 102] models. Some BRDF models are summarized in Table 3.3.

Table 3.3 BRDF kernels

Kernel	Comments	Refs.
Lambertian	Isotropic reflectance	
Roujean	The reflectance of a random arrangement of rectangular blocks on a flat surface	[102]
Rahman	The model includes an explicit form for the phase function of the surface distribution of scatterers in the form of a Henyey–Greenstein function	[103]
Kokhanovsky-Breon	The reflectance for the snow surface	[104]
Ross-thick (-thin)	Volumetric scattering from horizontally homogeneous leaf canopies with a large (small) leaf area index	[101]
LiSparse-R	An analytical expression is derived assuming a sparse ensemble of tree crowns modeled as spheroids casting shadows on the background, which is assumed Lambertian	[105]
Hapke	Soil bidirectional reflectance. The surface is treated as a semi-infinite medium containing scatterers. The radiative transfer problem is solved for such medium by decomposing the reflected radiance into single-scattering term and multiple scattering radiance, which is taken into account assuming the isotropic phase function.	[106]
Cox-Munk	BRDF for the reflection from the oceanic surface with the wind-generated waves. A distribution of sea surface slopes is parametrized by a Gaussian function.	[107, 108]

3.25 Compilation of Atmospheric Parameters

In order to formulate the atmospheric radiative transfer problem, the optical parameters of the atmosphere should be defined. In particular, the physical parameters of the atmosphere (e.g., temperature profile, pressure profile, gas concentrations, aerosol and cloud parameters) should be transferred into optical parameters, i.e., optical thicknesses, single-scattering albedos, and phase functions. The properties of the surface should be provided in terms of BDRF or BRDM.

The scheme of computing optical parameters of the atmosphere is outlined in Fig. 3.28. The computations are started by defining the atmospheric model. It includes information on gas profiles, as well as temperature and pressure profiles. The representation of the non-homogeneous atmosphere along the vertical direction as a system of homogeneous layers is called "layering" and is an important step in preparation of the input data for radiative transfer solvers. For an inhomogeneous atmosphere, we consider a spatial discretization with $N + 1$ levels: $z_1 < z_2 < \ldots < z_N < z_{N+1}$. A layer j is bounded above by the level z_{j+1} and below by the level z_j; the number of layers is N.

The information about gaseous absorption can be retrieved from databases, such as HITRAN, or is available in the form of cross-sections (for some gases in the ultraviolet and visible spectral ranges). Then, for given temperature/pressure profiles, the absorption spectrum can be computed by means of line-by-line models.

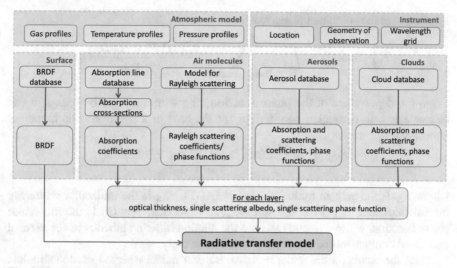

Fig. 3.28 Computations of optical parameters of the atmosphere, which serve as inputs to the radiative transfer model

Depending on location/time/season of measurement, corresponding aerosol, cloud and BRDF models can be chosen. Usually, aerosol/cloud models are defined as mixtures of pre-computed basic aerosol/cloud particle types with certain size distributions.

The absorption, scattering and extinction coefficients and the corresponding optical thicknesses have the additive property, meaning that the resulting coefficient in a certain layer is a sum of corresponding coefficients of all components:

$$\sigma_{abs}^{tot}(z) = \sigma_{abs}^{mol}(z) + \sigma_{abs}^{aer}(z) + \sigma_{abs}^{cloud}(z), \tag{3.260}$$

$$\sigma_{sca}^{tot}(z) = \sigma_{sca}^{mol}(z) + \sigma_{sca}^{aer}(z) + \sigma_{sca}^{cloud}(z) \tag{3.261}$$

and

$$\sigma_{ext}^{tot}(z) = \sigma_{ext}^{mol}(z) + \sigma_{ext}^{aer}(z) + \sigma_{ext}^{cloud}(z). \tag{3.262}$$

Note that σ_{abs}^{mol} is due to gaseous absorption, while σ_{sca}^{mol} is equal to due to Rayleigh (molecular) scattering coefficient. The optical thickness of the jth layer is given by

$$\tau_{ext,j}^{tot} = \int_{z_j}^{z_{j+1}} \sigma_{ext}(z)\, dz. \tag{3.263}$$

The phase function is expressed as the weighted sum of phase functions of all components:

$$P_{tot,j}(\cos\Theta) = \int_{z_j}^{z_{j+1}} \sum_{k=mol,aer,cloud} \frac{\sigma_{sca}^{k}(z)\, p^{k}(z,\cos\Theta)}{\sigma_{sca}^{tot}(z)} dz. \tag{3.264}$$

It follows for the combined phase matrix:

$$F_{\text{tot},j}\left(\cos\Theta\right) = \int_{z_j}^{z_{j+1}} \sum_{k=\text{mol,aer,cloud}} \frac{\sigma_{\text{sca}}^k\left(z\right) F^k\left(z, \cos\Theta\right)}{\sigma_{\text{sca}}^{\text{tot}}\left(z\right)} dz.$$

Often the dependence of the phase function on z within a layer is weak and can be neglected. In particular, one derives for the case of a homogeneous cloudless atmosphere:

$$p\left(\cos\Theta\right) = \frac{\tau_{\text{aer}}\omega_{\text{aer}} p_{\text{aer}}\left(\theta\right) + \tau_{\text{mol}}\omega_{\text{mol}} p_{\text{mol}}\left(\theta\right)}{\tau_{\text{aer}} + \tau_{\text{mol}}},$$

where τ_{aer} is the aerosol optical thickness (AOT), τ_{mol} is the molecular scattering optical thickness, $p_{\text{aer}}\left(\theta\right)$ is the aerosol phase function, $p_{\text{mol}}\left(\theta\right)$ is the molecular phase function, while ω_{aer} and ω_{mol} are the single-scattering albedos of the aerosol and cloud components of the atmosphere, respectively.

Thus, the inputs for the scalar radiative transfer model which is used to simulate the scattered radiation are the following ones:

1. The optical thickness for each layer,
2. The single-scattering albedo for each layer,
3. The single-scattering phase function for each layer,
4. The surface BRDF,
5. The definition of sources of radiation (thermal emission, solar radiation and artificial source of radiation) and
6. The temperature profile (if thermal emission is taken into account).

Appendix 1: On the Use of Complex Electromagnetic Fields for Computing Time-Averaged Poynting Vector

In electrodynamics, the complex representation of electric and magnetic fields is used quite extensively for mathematical convenience. Consider time-harmonic complex fields

$$\hat{\mathbf{E}}\left(\mathbf{r}, t\right) = \hat{\mathbf{E}}\left(\mathbf{r}\right) e^{-j\omega t}, \tag{3.265}$$

$$\hat{\mathbf{H}}\left(\mathbf{r}, t\right) = \hat{\mathbf{H}}\left(\mathbf{r}\right) e^{-j\omega t}. \tag{3.266}$$

Here the "hat" sign is used for complex quantities, and $j^2 = -1$. Note that time-independent amplitudes $\hat{\mathbf{E}}\left(\mathbf{r}\right)$ and $\hat{\mathbf{H}}\left(\mathbf{r}\right)$ correspond to perfectly monochromatic radiation, while slowly varying $\hat{\mathbf{E}}\left(\mathbf{r}\right)$ and $\hat{\mathbf{H}}\left(\mathbf{r}\right)$ represent the so-called quasi-monochromatic radiation.

In practice, only the real parts, i.e., $\mathbf{E} = \text{Re}\{\hat{\mathbf{E}}\}$ and $\mathbf{H} = \text{Re}\{\hat{\mathbf{H}}\}$, have the physical meaning. Furthermore, letting * denote complex conjugate, we can express $\mathbf{E}$ as follows:

$$\mathbf{E}\left(\mathbf{r}, t\right) = \text{Re}\left\{\hat{\mathbf{E}}\left(\mathbf{r}, t\right)\right\} = \frac{1}{2}\left[\hat{\mathbf{E}}\left(\mathbf{r}\right) e^{-j\omega t} + \hat{\mathbf{E}}^*\left(\mathbf{r}\right) e^{j\omega t}\right] \tag{3.267}$$

and similarly for $\mathbf{H}$:

$$\mathbf{H}\left(\mathbf{r}, t\right) = \text{Re}\left\{\hat{\mathbf{H}}\left(\mathbf{r}, t\right)\right\} = \frac{1}{2}\left[\hat{\mathbf{H}}\left(\mathbf{r}\right) e^{-j\omega t} + \hat{\mathbf{H}}^*\left(\mathbf{r}\right) e^{j\omega t}\right]. \tag{3.268}$$

Substituting Eqs. (3.267) and (3.268) into Eq. (3.39) yields the expression for the complex Poynting vector:

$$\begin{aligned}\hat{\mathbf{S}} = &\frac{1}{4}\left[\hat{\mathbf{E}}\left(\mathbf{r}\right) \times \hat{\mathbf{H}}\left(\mathbf{r}\right) e^{-2j\omega t} + \hat{\mathbf{E}}^*\left(\mathbf{r}\right) \times \hat{\mathbf{H}}^*\left(\mathbf{r}\right) e^{2j\omega t}\right] \\ &+ \frac{1}{4}\left[\hat{\mathbf{E}}^*\left(\mathbf{r}\right) \times \hat{\mathbf{H}}\left(\mathbf{r}\right) + \hat{\mathbf{E}}\left(\mathbf{r}\right) \times \hat{\mathbf{H}}^*\left(\mathbf{r}\right)\right].\end{aligned} \tag{3.269}$$

Then, taking into account that

$$\mathbf{S} = \text{Re}\{\hat{\mathbf{S}}\}, \tag{3.270}$$

$$\text{Re}\{\hat{\mathbf{E}}^*\left(\mathbf{r}\right) \times \hat{\mathbf{H}}\left(\mathbf{r}\right)\} = \text{Re}\{\hat{\mathbf{E}}\left(\mathbf{r}\right) \times \hat{\mathbf{H}}^*\left(\mathbf{r}\right)\} \tag{3.271}$$

and

$$\left\langle e^{\pm 2j\omega t}\right\rangle = 0, \tag{3.272}$$

we obtain the expression for the time-averaged Poynting vector in the form of Eq. (3.44).

Appendix 2: Computations of Legendre Polynomials

The associated Legendre polynomials $P_l^m\left(\mu\right)$ (sometimes referred to as Ferrers' functions) can be computed recurrently. The base of the recurrence is given by two formulas for $m = l$ and $l = m + 1$, respectively,

$$P_m^m\left(\mu\right) = (-1)^m (2m - 1)!!(1 - \mu^2)^{m/2} \tag{3.273}$$

and

$$P_{m+1}^m(\mu) = \mu(2m + 1) P_m^m\left(\mu\right) \tag{3.274}$$

with !! the double factorial ($n!!$ equals to the product of all the integers from 1 up to n that have the same parity (odd or even) as n — it should not be confused with the factorial iterated twice, which would be written as $(n!)!$). Polynomials $P_l^m\left(\mu\right)$ for

Table 3.4 The first few associated Legendre polynomials

	$l = 0$	$l = 1$	$l = 2$	$l = 3$
$m = 0$	1	μ	$\dfrac{1}{2}\left(3\mu^2 - 1\right)$	$\dfrac{1}{2}\mu\left(5\mu^2 - 3\right)$
$m = 1$		$-\left(1 - \mu^2\right)^{1/2}$	$-3\mu\left(1 - \mu^2\right)^{1/2}$	$\dfrac{3}{2}\left(1 - 5\mu^2\right)\left(1 - \mu^2\right)^{1/2}$
$m = 2$			$3\left(1 - \mu^2\right)$	$15\mu\left(1 - \mu^2\right)$
$m = 3$				$-15\left(1 - \mu^2\right)^{3/2}$

$l \geq m + 2$ are computed by using three-term recurrence relation with respect to the degree (for $m = 0$ known is Bonnet's recursion formula):

$$(2l + 1)\mu P_l^m(\mu) = (l + m)P_{l-1}^m(\mu) + (l - m + 1)P_{l+1}^m(\mu). \qquad (3.275)$$

For validation of the recurrent formula implementation, it is convenient to use analytical forms of the first few associated Legendre polynomials are listed in Table 3.4.

Note that polynomials $P_l^m(\mu)$ satisfy the following normalization condition

$$\int_{-1}^{1} P_l^m(\mu) P_{l'}^m(\mu) d\mu = \frac{2}{2l + 1} \frac{(l + m)!}{(l - m)!} \delta_{ll'} \qquad (3.276)$$

and the maximum value of $P_l^m(\mu)$ rapidly increases with m. To avoid overflow during the recurrent process, it is recommended to compute normalized Legendre polynomials $\bar{P}_l^m(\mu)$ defined as

$$\bar{P}_l^m(\mu) = \sqrt{(2l + 1)\frac{(l - m)!}{(l + m)!}} P_l^m(\mu) \qquad (3.277)$$

instead of $P_l^m(\mu)$. For $\bar{P}_l^m(\mu)$ the following normalization condition holds:

$$\int_{-1}^{1} \bar{P}_l^m(\mu) \bar{P}_{l'}^m(\mu) d\mu = 2\delta_{ll'}. \qquad (3.278)$$

Appendix 3: Computations of Generalized Spherical Functions

The generalized spherical functions $R_l^m(\mu)$ and $T_l^m(\mu)$ can be expressed as follows [67, 109–111]:

$$R_l^m (\mu) = -\frac{1}{2} (i)^m \sqrt{\frac{(l+m)!}{(l-m)!}} \left\{ P_{m,2}^l (\mu) + P_{m,-2}^l (\mu) \right\}, \tag{3.279}$$

$$T_l^m (\mu) = -\frac{1}{2} (i)^m \sqrt{\frac{(l+m)!}{(l-m)!}} \left\{ P_{m,2}^l (\mu) - R_{m,-2}^l (\mu) \right\}, \tag{3.280}$$

where for $l \geq \max(m, 2)$,

$$\begin{aligned} P_{m,n}^l (\mu) &= A_{m,n}^l (1-\mu)^{-(n-m)/2} (1+\mu)^{-(n+m)/2} \\ &\times \frac{d^{l-n}}{d\mu^{l-n}} \left[(1-\mu)^{l-m} (1+\mu)^{l+m} \right], \end{aligned} \tag{3.281}$$

with

$$A_{m,n}^l = \frac{(-1)^{l-m} (i)^{n-m}}{2^l (l-m)!} \sqrt{\frac{(l-m)!(l+n)!}{(l+m)!(l-n)!}}. \tag{3.282}$$

For $P_{m,n}^l$ the recursion formula was derived in [109]:

$$e_{m,n}^l P_{m,n}^{l+1} (\mu) = (2l+1)\mu P_{m,n}^l (\mu) - f_{m,n}^l P_{m,n}^{l-1} (\mu) - \frac{mn(2l+1)}{l(l+1)} P_{m,n}^l (\mu) \tag{3.283}$$

with

$$e_{m,n}^l = \frac{1}{l+1} \sqrt{(l+m+1)(l-m+1)(l+n+1)(l-n+1)} \tag{3.284}$$

and

$$f_{m,n}^l = \frac{1}{l} \sqrt{(l+m)(l-m)(l+n)(l-n)}. \tag{3.285}$$

Given Eqs. (3.279)-(3.285), the recurrence formulas for R_l^m and T_l^m can be derived for $l > \max(m, 2)$ and $m \geq 0$:

$$\begin{aligned} &\tfrac{l+1-m}{l+1} \sqrt{(l+3)(l-1)} R_{l+1}^m (\mu) = \\ &(2l+1)\mu R_l^m (\mu) - \tfrac{l+m}{l} \sqrt{(l+2)(l-2)} R_{l-1}^m (\mu) - \tfrac{2m(2l+1)}{l(l+1)} T_l^m (\mu), \end{aligned} \tag{3.286}$$

$$\begin{aligned} &\tfrac{l+1-m}{l+1} \sqrt{(l+3)(l-1)} T_{l+1}^m (\mu) = \\ &(2l+1)\mu T_l^m (\mu) - \tfrac{l+m}{l} \sqrt{(l+2)(l-2)} T_{l-1}^m (\mu) - \tfrac{2m(2l+1)}{l(l+1)} R_l^m (\mu). \end{aligned} \tag{3.287}$$

The base of recurrence computation is

$$R_j^j (\mu) = \frac{(2j)!}{2^j j!} \sqrt{\frac{j(j-1)}{(j+1)(j+2)}} \sqrt{1-\mu^2} \frac{1+\mu^2}{1-\mu^2}, \tag{3.288}$$

$$T_j^j(\mu) = \frac{(2j)!}{2^j j!} \sqrt{\frac{j(j-1)}{(j+1)(j+2)}} \sqrt{1-\mu^2} \frac{2\mu}{1-\mu^2}, \tag{3.289}$$

while

$$R_l^m(\mu) = T_l^m(\mu) = 0, \quad l < m. \tag{3.290}$$

In particular, it follows:

$$R_2^2(\mu) = \frac{\sqrt{6}}{2}\left(1+\mu^2\right), \tag{3.291}$$

$$T_2^2(\mu) = \sqrt{6}\mu. \tag{3.292}$$

For $m = 1$, the initial values are

$$R_2^1(\mu) = -\frac{1}{2}\mu\sqrt{6}\sqrt{1-\mu^2}, \tag{3.293}$$

$$T_2^1(\mu) = -\frac{1}{2}\sqrt{6}\sqrt{1-\mu^2}. \tag{3.294}$$

Finally, for $m = 0$, we have

$$R_2^0(\mu) = \frac{\sqrt{6}}{4}(1-\mu^2), \tag{3.295}$$

$$T_2^0(\mu) = 0. \tag{3.296}$$

Note that R_l^m and T_l^m satisfy the following normalization condition:

$$\int_{-1}^{+1}\left[R_l^m(\mu)\,R_r^m(\mu) + T_l^m(\mu)\,T_r^m(\mu)\right]d\mu = 1, \tag{3.297}$$

where $l, r \geq \max(m, 2)$.

References

1. O. Chwolson, Grundzüge einer matimatischen Theorie der inneren Diffusion des Licht. Bull. Acad. Imp. Sci. St. Petersbourg **33**, 221–256 (1889). [in German]
2. E. Lommel, Die Photometric der diffusen Zuruckwerfung. Sitzber. Acad. Wissensch. Munchen **17**, 95–124 (1887). [in German]
3. A. Schuster, The influence of radiation of the transmission of heat. Phil.Mag. **5**, 243–257 (1903)
4. A. Schuster, Radiation through a foggy atmosphere. The Astrophysical Journal **21**, 1 (1905). https://doi.org/10.1086/141186

5. K. Schwarzschild, Über das Gleichgewicht der Sonnenatmosphäre. Nachr. Konig. Gesel. der Wiss., Gottingen **1**, 41–53 (1906). [in German]
6. E.A. Milne, The reflection effect in eclipsing binaries. Monthly Notices of the Royal Astronomical Society **87**(1), 43–55 (1926). https://doi.org/10.1093/mnras/87.1.43
7. V. Ambartsumian, On the problem of diffuse reflection of light. J. Phys. USSR **8**, 65–75 (1944). [in Russian]
8. V. Sobolev, *The transfer of radiant energy in the atmospheres of stars and planets*. Moscow: GITTL (1956). [in Russian]
9. S. Chandrasekhar, *Radiative trasnfer*. Dover publications, inc. New York (1950)
10. G. Rozenberg, Vector-parameter of stokes (matrix methods for accounting for radiation polarization in the ray optics approximation). Uspekhi Fizicheskih Nauk **56**(5), 77–110 (1955). https://doi.org/10.3367/ufnr.0056.195505c.0077. [in Russian]
11. Y. Kravtsov, L. Apresyan, IV Radiative transfer: New aspects of the old theory. in *Progress in Optics*, pp. 179–244. Elsevier (1996). https://doi.org/10.1016/s0079-6638(08)70315-9
12. V. Afanas'ev, V. Budak, D. Efremenko, P. Kaplya, Application of the photometric theory of the radiance field in the problems of electron scattering. Light & Engineering pp. 88–96 (2019). https://doi.org/10.33383/2018-034
13. Y.N. Barabanenkov, Y.A. Kravtsov, V. Ozrin, A. Saichev II enhanced backscattering in optics.in *Progress in Optics*, pp. 65–197. Elsevier (1991). https://doi.org/10.1016/s0079-6638(08)70006-4
14. E. Akkermans, P.E. Wolf, R. Maynard, Coherent backscattering of light by disordered media: Analysis of the peak line shape. Physical Review Letters **56**(14), 1471–1474 (1986). https://doi.org/10.1103/physrevlett.56.1471
15. G. Thomas, K. Stamnes, *Radiative Transfer in the Atmosphere and Ocean*. Cambridge University Press (1999). https://doi.org/10.1017/cbo9780511613470
16. A. Peraiah, *An Introduction to Radiative Transfer: Methods and Applications in Astrophysics*. Cambridge University Press (2001)
17. E. Zege, A. Ivanov, I. Katsev, Image Transfer Through a Scattering Medium. Springer Berlin Heidelberg (1991). https://doi.org/10.1007/978-3-642-75286-5
18. G.C. Pomraning, *The Equations of Radiation Hydrodynamics*. Oxford: Pergamon Press (1973)
19. E. Yanovitskij, Light Scattering in Inhomogeneous Atmospheres. Springer Berlin Heidelberg (1997). https://doi.org/10.1007/978-3-642-60465-2
20. J. Poynting, XV. On the transfer of energy in the electromagnetic field. Philosophical Transactions of the Royal Society of London **175**, 343–361 (1884). https://doi.org/10.1098/rstl.1884.0016
21. E. Wolf, Coherence and radiometry. Journal of the Optical Society of America **68**(1), 6 (1978). https://doi.org/10.1364/josa.68.000006
22. A. Ishimaru, *Wave Propagation and Scattering in Random Media*. Elsevier (1978). https://doi.org/10.1016/c2013-0-10906-3
23. M. Mishchenko, 125 years of radiative transfer: Enduring triumphs and persisting misconceptions. in R. Cahalan, J. Fischer (eds.) *Radiation Processes in the Atmosphere and Ocean (IRS2012): Proceedings of the International Radiation Symposium (IRC/IAMAS), 6-10 Aug. 2012, Berlin, AIP Conference Proceedings*, vol. 1531, pp. 11–18. AIP, Melville, N.Y. (2013)
24. M. Mishchenko, Multiple scattering by particles embedded in an absorbing medium. 1. Foldy–Lax equations, order-of-scattering expansion, and coherent field. Optics Express **16**(3), 2288 (2008). https://doi.org/10.1364/oe.16.002288
25. A. Doicu, M. Mishchenko, Overview of methods for deriving the radiative transfer theory from the Maxwell equations. I: Approach based on the far-field Foldy equations. Journal of Quantitative Spectroscopy and Radiative Transfer **220**, 123–139 (2018). https://doi.org/10.1016/j.jqsrt.2018.09.004
26. A. Doicu, M. Mishchenko, An overview of methods for deriving the radiative transfer theory from the Maxwell equations. II: Approach based on the Dyson and Bethe-Salpeter equations. Journal of Quantitative Spectroscopy and Radiative Transfer **224**, 25–36 (2019). https://doi.org/10.1016/j.jqsrt.2018.10.032

27. L. Foldy, The multiple scattering of waves. I. General theory of isotropic scattering by randomly distributed scatterers. Physical Review **67**(3-4), 107–119 (1945). https://doi.org/10.1103/physrev.67.107
28. M. Lax, Multiple scattering of waves. Reviews of Modern Physics **23**(4), 287–310 (1951). https://doi.org/10.1103/revmodphys.23.287
29. M. Mishchenko, *Electromagnetic Scattering by Particles and Particle Groups: An Introduction*. Cambridge (2014)
30. V. Twersky, On propagation in random media of discrete scatterers. Proc Symp Appl Math **16**, 84–116 (1964)
31. L. Dolin, Scattering of light in a layer of turbid medium. Izv. VUZ Radiofiz **7**(2), 380–382 (1964)
32. L. Apresyan, Y. Kravtsov, *Radiation Transfer: Statistical and Wave Aspects*. Gordon and Breach, London, NY (1996)
33. M. Kuzmina, L. Bass, O. Nikolaeva, Polarized radiative transfer in optically active light scattering media. in *Springer Series in Light Scattering*, pp. 1–53. Springer International Publishing (2017). https://doi.org/10.1007/978-3-319-70808-9_1
34. J. Aizenberg, V. Budak, The science of light engineering, fields of application and theoretical foundations. Light & Engineering pp. 4–6 (2018). https://doi.org/10.33383/2018-045
35. J. Aizenberg, V. Budak, Light science is not only science of lighting: theoretical bases and application area. *in Proceedings of the 29th Quadrennial Session of the CIE*. International Commission on Illumination, CIE (2019). https://doi.org/10.25039/x46.2019.po120
36. G. Konnen, *Polarized Light in Nature*. Cambridge University Press (1985)
37. T. Ramskou, Solstenen. Skalk **2**, 16–17 (1967)
38. C. Roslund, C. Beckman, Disputing Viking navigation by polarized skylight. Applied Optics **33**(21), 4754 (1994). https://doi.org/10.1364/ao.33.004754
39. E. Bartholin, *Experiments with the double refracting Iceland crystal which led to the discovery of a marvelous and strange refraction*. Copenhagen, Denmark (1669)
40. É.L. Malus, Mémoire sur la mesure du pouvoir réfringent des corps opaques. Nouveau bulletin des sciences de la Société philomathique de Paris **1**, 77–81 (1807). [in French]
41. J. Tyndall, IV. On the blue colour of the sky, the polarization of skylight, and on the polarization of light by cloudy matter generally. Proceedings of the Royal Society of London **17**, 223–233 (1869). https://doi.org/10.1098/rspl.1868.0033
42. D. Brewster, On the laws which regulate the polarisation of light by reflexion from transparent bodies. Philosophical Transactions of the Royal Society of London **105**, 125–159 (1815).https://doi.org/10.1098/rstl.1815.0010
43. A. Fresnel, Note sur le calcul des teintes que la polarisation développe dans les lames cristallisées. Annales de Chimie et de Physique **17**, 102–12 (1821). [in French]
44. G. Stokes, On the composition and resolution of streams of polarized light from different sources. Transactions of the Cambridge Philosophical Society **9**, 399–416 (1852)
45. C. Bohren, D. Huffman, *Absorption and Scattering of Light by Small Particles*. Wiley (1998). https://doi.org/10.1002/9783527618156
46. M. Born, E. Wolf, *Principles of Optics*. Electromagnetic Theory of Propagation, Interference and Diffraction of Light, Sixth Edition. Pergamon Press, Oxford (1993)
47. A.Z. Dolginov, Y. Gnedin, N. Silant'Ev, *Propagation and Polarization of Radiation in Cosmic Media*. Gordon & Breach Science Pub (1995)
48. F. Fedorov, *Optics of Anisotropic media*. Academy of Sciences of the Belarus Soviet Socialist Republic, Minsk (1958)
49. A.F. Konstantinova, A.Y. Tronin, B.V. Nabatov, Developments of Fedorov covariant methods and applications to optically active crystals. in *Advances in Complex Electromagnetic Materials*, pp. 19–32. Springer Netherlands (1997). https://doi.org/10.1007/978-94-011-5734-6_2
50. A. Kokhanovsky, The tensor radiative transfer equation. Journal of Physics A: Mathematical and General **33**(22), 4121–4128 (2000). https://doi.org/10.1088/0305-4470/33/22/314
51. G. Goldstein, *Polarized Light*. CRC Press (2003)

52. I. Kuščer, M. Ribarič, Matrix formalism in the theory of diffusion of light. Optica Acta: International Journal of Optics **6**(1), 42–51 (1959). https://doi.org/10.1080/713826264
53. Z. Sekera, Scattering matrices and reciprocity relationships for various representations of the state of polarization. Journal of the Optical Society of America **56**(12), 1732 (1966). https://doi.org/10.1364/josa.56.001732
54. H. van de Hulst, Light scattering by small particles. Dover Publications, Inc. New York (1957)
55. A.M. Legendre, Recherches sur l'attraction des sphéroïdes homogènes. Mémoires de Mathématiques et de Physique, présentés à l'Académie Royale des Sciences, par divers savans, et lus dans ses Assemblées **X**, 411–435 (1785). [in French]
56. J. Dave, B. Armstrong, Computations of high-order associated legendre polynomials. Journal of Quantitative Spectroscopy and Radiative Transfer **10**(6), 557–562 (1970). https://doi.org/10.1016/0022-4073(70)90073-7
57. B. Hapke, *Theory of Reflectance and Emittance Spectroscopy*. Cambridge University Press (2009). https://doi.org/10.1017/cbo9781139025683
58. V. Sobolev, *Light Scattering in Planetary Atmospheres* (Pergamon Press, 1975)
59. L. Henyey, J. Greenstein, Diffuse radiation in the galaxy. Astrophys J **93**, 70–83 (1941)
60. W. Cornette, J. Shanks, Physically reasonable analytic expression for the single-scattering phase function. Applied Optics **31**(16), 3152 (1992). https://doi.org/10.1364/ao.31.003152
61. H. Kagiwada, R. Kalaba, RAND Report RM-5537-PR RAND Corporation. Santa Monica, California (1967)
62. G. Hunt, A review of computational techniques for analysing the transfer of radiation through a model cloudy atmosphere. Journal of Quantitative Spectroscopy and Radiative Transfer **11**(6), 655–690 (1971). https://doi.org/10.1016/0022-4073(71)90046-x
63. M.I. Mishchenko, L.D. Travis, A.A. Lacis, *Scattering, Absorption, and Emission of Light by Small Particles*. New York (2004)
64. J. Hovenier, The polarization of light scattered by small particles: A personal review. Journal of Quantitative Spectroscopy and Radiative Transfer **113**(18), 2280–2291 (2012). https://doi.org/10.1016/j.jqsrt.2012.03.029
65. V.P. Budak, S.V. Korkin, The aerosol influence upon the polarization state of the atmosphere solar radiation. International Journal of Remote Sensing **29**(9), 2469–2506 (2008). https://doi.org/10.1080/01431160701767542
66. R. Garcia, C. Siewert, A generalized spherical harmonics solution for radiative transfer models that include polarization effects. Journal of Quantitative Spectroscopy and Radiative Transfer **36**(5), 401–423 (1986). https://doi.org/10.1016/0022-4073(86)90097-x
67. J. Lenoble, M. Herman, J. Deuzé, B. Lafrance, R. Santer, D. Tanré, A successive order of scattering code for solving the vector equation of transfer in the earth's atmosphere with aerosols. Journal of Quantitative Spectroscopy and Radiative Transfer **107**(3), 479–507 (2007). https://doi.org/10.1016/j.jqsrt.2007.03.010
68. G. Mie, Beiträge zur optik trüber medien, speziell kolloidaler metallösungen. Annalen der Physik **330**(3), 377–445 (1908). https://doi.org/10.1002/andp.19083300302. [in German]
69. J. Hansen, L. Travis, Light scattering in planetary atmospheres. Space Science Reviews **16**(4), 527–610 (1974). https://doi.org/10.1007/bf00168069
70. Waterman, P.: Matrix formulation of electromagnetic scattering. Proceedings of the IEEE **53**(8), 805–812 (1965). https://doi.org/10.1109/proc.1965.4058
71. M. Mishchenko, L. Travis, D. Mackowski, T-matrix computations of light scattering by nonspherical particles: A review. Journal of Quantitative Spectroscopy and Radiative Transfer **55**(5), 535–575 (1996). https://doi.org/10.1016/0022-4073(96)00002-7
72. T. Rother, *Sound Scattering on Spherical Objects*. Springer International Publishing (2020). https://doi.org/10.1007/978-3-030-36448-9
73. A. Doicu, A. Vasilyeva, D. Efremenko, C. Wirth, T. Wriedt, A light scattering model for total internal reflection microscopy of geometrically anisotropic particles. Journal of Modern Optics **66**(10), 1139–1151 (2019). https://doi.org/10.1080/09500340.2019.1605005
74. W.M. Grundy, S. Douté, B. Schmitt, A Monte Carlo ray-tracing model for scattering and polarization by large particles with complex shapes. Journal of Geophysical Research: Planets **105**(E12), 29291–29314 (2000). https://doi.org/10.1029/2000je001276

75. P. Yang, K. Liou, K. Wyser, D. Mitchell, Parameterization of the scattering and absorption properties of individual ice crystals. Journal of Geophysical Research: Atmospheres **105**(D4), 4699–4718 (2000). https://doi.org/10.1029/1999jd900755

76. P. Yang, K. Liou, Finite-difference time domain method for light scattering by small ice crystals in three-dimensional space. Journal of the Optical Society of America A **13**(10), 2072 (1996). https://doi.org/10.1364/josaa.13.002072

77. H. DeVoe, Optical properties of molecular aggregates. I. Classical model of electronic absorption and refraction. The Journal of Chemical Physics **41**(2), 393–400 (1964). https://doi.org/10.1063/1.1725879

78. M. Yurkin, A. Hoekstra, The discrete dipole approximation: An overview and recent developments. Journal of Quantitative Spectroscopy and Radiative Transfer **106**(1-3), 558–589 (2007). https://doi.org/10.1016/j.jqsrt.2007.01.034

79. B. Bodhaine, N. Wood, E. Dutton, J. Slusser, On Rayleigh optical depth calculations. Journal of Atmospheric and Oceanic Technology **16**(11), 1854–1861 (1999). https://doi.org/10.1175/1520-0426(1999)016<1854:orodc>2.0.co;2

80. D. Bates, Rayleigh scattering by air. Planetary and Space Science **32**(6), 785 – 790 (1984). https://doi.org/10.1016/0032-0633(84)90102-8

81. A. Bucholtz, Rayleigh-scattering calculations for the terrestrial atmosphere. Applied Optics **34**(15), 2765 (1995). https://doi.org/10.1364/ao.34.002765

82. E. Dutton, P. Reddy, S. Ryan, J. DeLuisi, Features and effects of aerosol optical depth observed at Mauna Loa, Hawaii: 1982–1992. Journal of Geophysical Research **99**(D4), 8295 (1994). https://doi.org/10.1029/93jd03520

83. V. Rozanov, A. Rozanov, A. Kokhanovsky, J. Burrows, Radiative transfer through terrestrial atmosphere and ocean: Software package SCIATRAN. Journal of Quantitative Spectroscopy and Radiative Transfer **133**, 13–71 (2014). https://doi.org/10.1016/j.jqsrt.2013.07.004

84. A. Young, Rayleigh scattering. Applied Optics **20**(4), 533 (1981). https://doi.org/10.1364/ao.20.000533

85. V. Rozanov, M. Vountas, Radiative transfer equation accounting for rotational Raman scattering and its solution by the discrete-ordinates method. Journal of Quantitative Spectroscopy and Radiative Transfer **133**, 603–618 (2014). https://doi.org/10.1016/j.jqsrt.2013.09.024

86. R. Spurr, J. de Haan, R. van Oss, A. Vasilkov, Discrete-ordinate radiative transfer in a stratified medium with first-order rotational Raman scattering. Journal of Quantitative Spectroscopy and Radiative Transfer **109**(3), 404–425 (2008). https://doi.org/10.1016/j.jqsrt.2007.08.011

87. P. Debye, Der Lichtdruck auf Kugeln von beliebigem Material. Annalen der Physik **335**(11), 57–136 (1909). https://doi.org/10.1002/andp.19093351103. [in German]

88. H.M. Nussenzveig, W.J. Wiscombe, Efficiency factors in Mie scattering. Physical Review Letters **45**(18), 1490–1494 (1980). https://doi.org/10.1103/physrevlett.45.1490

89. A. Fresnel, Mémoire sur la loi des modifications que la réflexion imprime á la lumiére polarisée ("memoir on the law of the modifications that reflection impresses on polarized light") (1823). Read 7 January 1823; reprinted in Fresnel, 1866-70, vol. 1, pp. 767–799 (full text, published 1831), pp. 753–762 (extract, published 1823)

90. K.S. Shifrin, *Scattering of Light in a Turbid Medium*. NASA, Washington, DC (1968)

91. A. Kokhanovsky, E. Zege, Local optical parameters of spherical polydispersions: simple approximations. Applied Optics **34**(24), 5513 (1995). https://doi.org/10.1364/ao.34.005513

92. A. Kokhanovsky, A. Macke, Integral light-scattering and absorption characteristics of large, nonspherical particles. Applied Optics **36**(33), 8785 (1997). https://doi.org/10.1364/ao.36.008785

93. Y. Timofeyev, A. Vasil'ev, *Theoretical fundamentals of Atmospheric Optics*. Cambridge (2008)

94. W.G. Rees, *Physical Principles of Remote Sensing*. Cambridge University Press (2012). https://doi.org/10.1017/cbo9781139017411

95. G. Peach Theory of the pressure broadening and shift of spectral lines. Advances in Physics **30**(3), 367–474 (1981). https://doi.org/10.1080/00018738100101467

96. N. Ngo, D. Lisak, H. Tran, J.M. Hartmann, An isolated line-shape model to go beyond the Voigt profile in spectroscopic databases and radiative transfer codes. Journal of Quantitative Spectroscopy and Radiative Transfer **129**, 89–100 (2013). https://doi.org/10.1016/j.jqsrt.2013.05.034

97. F. Schreier, S. Gimeno García, P. Hochstaffl, S. Städt, Py4cats –PYthon for computational Atmospheric spectroscopy. Atmosphere **10**(5), 262 (2019). https://doi.org/10.3390/atmos10050262

98. R. Kochanov, I. Gordon, L. Rothman, P. Wcisło, C. Hill, J. Wilzewski, HITRAN application programming interface (HAPI): A comprehensive approach to working with spectroscopic data. Journal of Quantitative Spectroscopy and Radiative Transfer **177**, 15–30 (2016). https://doi.org/10.1016/j.jqsrt.2016.03.005

99. K. Bogumil, J. Orphal, T. Homann, S. Voigt, P. Spietz, O. Fleischmann, A. Vogel, M. Hartmann, H. Kromminga, H. Bovensmann, J. Frerick, J. Burrows, Measurements of molecular absorption spectra with the SCIAMACHY pre-flight model: instrument characterization and reference data for atmospheric remote-sensing in the 230–2380 nm region. Journal of Photochemistry and Photobiology A: Chemistry **157**(2-3), 167–184 (2003). https://doi.org/10.1016/s1010-6030(03)00062-5

100. C.L. Walthall, J.M. Norman, J.M. Welles, G. Campbell, B.L. Blad, Simple equation to approximate the bidirectional reflectance from vegetative canopies and bare soil surfaces. Applied Optics **24**(3), 383 (1985). https://doi.org/10.1364/ao.24.000383

101. W. Wanner, X. Li, A.H. Strahler, On the derivation of kernels for kernel-driven models of bidirectional reflectance. Journal of Geophysical Research **100**(D10), 21077 (1995). https://doi.org/10.1029/95jd02371

102. J.L. Roujean, M. Leroy, P.Y. Deschamps, A bidirectional reflectance model of the Earth's surface for the correction of remote sensing data. Journal of Geophysical Research **97**(D18), 20455 (1992). https://doi.org/10.1029/92jd01411

103. H. Rahman, B. Pinty, M. Verstraete, Coupled surface-atmosphere reflectance (CSAR) model: 2. semiempirical surface model usable with NOAA advanced very high resolution radiometer data. Journal of Geophysical Research **98**(D11), 20791 (1993). https://doi.org/10.1029/93jd02072

104. A.A. Kokhanovsky, F.M. Breon, Validation of an analytical snow BRDF model using PARASOL multi-angular and multispectral observations. IEEE Geoscience and Remote Sensing Letters **9**(5), 928–932 (2012). https://doi.org/10.1109/lgrs.2012.2185775

105. X. Li, A.H. Strahler, Geometric-optical bidirectional reflectance modeling of the discrete crown vegetation canopy: effect of crown shape and mutual shadowing. IEEE Transactions on Geoscience and Remote Sensing **30**(2), 276–292 (1992)

106. B. Hapke, Scattering and diffraction of light by particles in planetary regoliths. Journal of Quantitative Spectroscopy and Radiative Transfer **61**(5), 565–581 (1999). https://doi.org/10.1016/s0022-4073(98)00042-9

107. J. Chowdhary, P.W. Zhai, E. Boss, H. Dierssen, R. Frouin, A. Ibrahim, Z. Lee, L. Remer, M. Twardowski, F. Xu, X. Zhang, M. Ottaviani, W. Espinosa, D. Ramon, Modeling atmosphere-ocean radiative transfer: A PACE mission perspective. Frontiers in Earth Science **7** (2019). https://doi.org/10.3389/feart.2019.00100

108. C. Cox, W. Munk, Measurement of the roughness of the sea surface from photographs of the sun's glitter. Journal of the Optical Society of America **44**(11), 838 (1954). https://doi.org/10.1364/josa.44.000838

109. I. Gelfand, Z. Sapiro, Representations of the group of rotation in three-dimensional space and their applications. Amer. Math. Soc. Translations **2**, 207–316 (1956)

110. J. Hovenier, C. Van Der Mee, H. Domke, Transfer of Polarized Light in Planetary Atmospheres. Springer Netherlands (2004). https://doi.org/10.1007/978-1-4020-2856-4

111. C. Siewert, On the phase matrix basic to the scattering of polarized light. Astron. Astrophys. **109**, 195–200 (1982)

Chapter 4
Radiative Transfer Models

4.1 Radiative Transfer Equation

As the thickness of the terrestrial atmosphere is much less than the radius of the Earth, the sphericity of the atmosphere is often neglected and the atmospheric radiative transfer is described in the framework of one-dimensional models, assuming that the properties of the atmosphere (such as density, temperature, pressure and concentration of atmospheric gases/particulate matter) change along the vertical direction only, while the solar radiation at the top of the atmosphere is described as a plane-parallel source of radiation. The horizontal inhomogeneity of the atmosphere and underlying surface cannot be ignored in a number of atmospheric optics and remote sensing problems, including radiative transfer/atmospheric remote sensing in the presence of broken cloud fields.

Let us consider a one-dimensional problem of scalar radiative transfer for a single atmospheric layer. It is assumed that the albedo of the underlying surface is equal to zero (an absolutely black underlying surface). Our goal is to formulate an equation for the radiance field $L(z, \mu, \varphi)$, which depends on the vertical coordinate z, the cosine μ of the polar angle θ and the azimuthal angle φ, as shown in Fig. 4.1. The principal plane is perpendicular to the layer boundaries and comprises the incident radiation. The polar angle is defined with respect to the vertical axis Oz, while the azimuthal angle is defined with respect to the principal plane. We use $\mu > 0$ for the downwelling radiation and $\mu < 0$ for the upwelling radiation. It is assumed that the frequency of the light beam does not change while it propagates in the medium. The relative azimuthal angle is equal to zero in the specular reflection direction and π in the backward observation direction (at equal incidence and viewing zenith angles). It should be pointed out that the relative azimuthal angle is often defined in the opposite way with φ equal to π in the direction of specular reflection and zero in the backward observation direction (e.g., as in some satellite datasets). The vertical coordinate z is equal to 0 at the top of the atmosphere, while at the bottom of the medium it equals to z_L, where z_L is the geometrical layer thickness. The extinction coefficient $\sigma_{ext}(z)$ as a function of z is assumed to be known.

© Springer Nature Switzerland AG 2021
D. Efremenko and A. Kokhanovsky, *Foundations of Atmospheric Remote Sensing*,
https://doi.org/10.1007/978-3-030-66745-0_4

Fig. 4.1 Formulation of the radiative transfer equation. The radiance L depends on the vertical coordinate z, cosine μ of polar angle θ and the azimuthal angle φ. The principal plane is perpendicular to the layer boundaries and comprises the incident radiation direction. The polar angle θ is defined with respect to the axis Oz, while the azimuthal angle φ is defined with respect to the principal plane

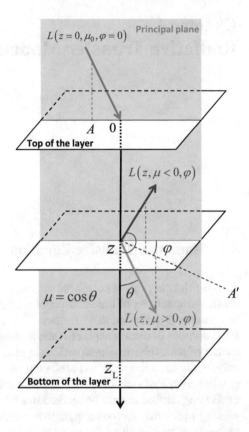

The radiation inside the layer and escaping from the layer is due to scattering, absorption and emission processes occurring inside an atmospheric layer under consideration. A light beam propagating in a given direction in the absence of internal radiation sources is attenuated due to absorption and scattering processes. The light beam is also enhanced due to scattering from all other directions to a propagation direction. Let us first neglect this enhancement process and consider the extinction process only. The light extinction process is characterized by the extinction coefficient equal to the sum of absorption and scattering coefficients of a given atmospheric layer.

Let us consider a plane-parallel beam of radiation which strikes the plane-parallel turbid layer with the extinction coefficient σ_{ext} and the geometrical thickness z_L, as shown in Fig. 4.2. The incidence angle cosine is μ. The radiance traversing the layer is attenuated due to the extinction process (i.e., absorption and scattering). For a sufficiently thin layer, the attenuation scales linearly with the layer thickness, i.e.,

$$dL\,(z, \mu, \varphi) = L\,(z + dz, \mu, \varphi) - L\,(z, \mu, \varphi) = -L\,(z, \mu, \varphi)\,\sigma_{ext}\,(z)\,\frac{dz}{\mu}. \quad (4.1)$$

Thus, we obtain the following equation:

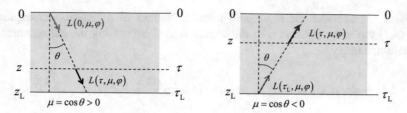

Fig. 4.2 Attenuation of the pencil of radiation passing through a slab

$$\mu \frac{dL(z, \mu, \varphi)}{dz} = -\sigma_{\text{ext}}(z) L(z, \mu, \varphi). \tag{4.2}$$

The minus sign on the right-hand side of Eq. (4.1) shows that the radiance will be weakened due to interactions with the slab. Equation (4.2) can be readily solved and we get the well-known exponential decay law:

$$L(z, \mu, \varphi) = L(0, \mu, \varphi) e^{-\int_0^z \frac{\sigma_{\text{ext}}(z')dz'}{\mu}}. \tag{4.3}$$

We introduce the optical depth $\tau(z_1, z_2)$ of the layer with top and bottom boundaries at z_1 and z_2, respectively, as follows:

$$\tau(z_1, z_2) = \int_{z_1}^{z_2} \sigma_{\text{ext}}(z') dz'. \tag{4.4}$$

If the integration is performed from the top of the atmosphere, then z_1 is set to zero, i.e.,

$$\tau(z) = \int_0^z \sigma_{\text{ext}}(z') dz'. \tag{4.5}$$

Equation (4.5) allows us to work with the unitless coordinate τ. Then Eq. (4.2) becomes

$$\mu \frac{dL(\tau, \mu, \varphi)}{d\tau} = -L(\tau, \mu, \varphi) \tag{4.6}$$

and Eq. (4.3) can be written as

$$L(\tau, \mu > 0, \varphi) = L(0, \mu > 0, \varphi) e^{-\frac{\tau}{\mu}}. \tag{4.7}$$

Analogously, for the radiance striking a slab from the bottom, we have

$$L(\tau, \mu < 0, \varphi) = L(\tau_{\text{L}}, \mu < 0, \varphi) e^{-\frac{\tau_{\text{L}} - \tau}{|\mu|}}, \tag{4.8}$$

where

$$\tau_{\text{L}} = \int_0^{z_{\text{L}}} \sigma_{\text{ext}}(z') dz' \tag{4.9}$$

is the optical thickness of the slab.

Now we consider the layer with internal sources of radiation. Consequently, Eq. (4.1) has to be updated with the source term J expressing the enhancement of radiation, namely

$$L\left(z+dz,\mu,\varphi\right)=L\left(z,\mu,\varphi\right)\left(1-\sigma_{\text{ext}}\left(z\right)\frac{dz}{\mu}\right)+J\left(z,\mu,\varphi\right)\frac{dz}{\mu}. \qquad (4.10)$$

This leads us to the following radiative transfer equation:

$$\mu\frac{dL\left(z,\mu,\varphi\right)}{dz}=-\sigma_{\text{ext}}\left(z\right)L\left(z,\mu,\varphi\right)+J\left(z,\mu,\varphi\right), \qquad (4.11)$$

or dividing by σ_{ext}:

$$\mu\frac{dL\left(\tau,\mu,\varphi\right)}{d\tau}=-L\left(\tau,\mu,\varphi\right)+J\left(\tau,\mu,\varphi\right). \qquad (4.12)$$

The formal solution of Eq. (4.12) can be derived by integration. The result reads as follows [1]:

$$L\left(\tau,\mu>0,\varphi\right)=L\left(0,\mu>0,\varphi\right)e^{-\frac{\tau}{\mu}}+\int_{0}^{\tau}J\left(\tau',\mu,\varphi\right)e^{-\frac{\tau-\tau'}{\mu}}\frac{d\tau'}{\mu} \qquad (4.13)$$

for the downwelling radiation, and

$$L\left(\tau,\mu<0,\varphi\right)=L\left(\tau_{\text{L}},\mu<0,\varphi\right)e^{-\frac{\tau_{\text{L}}-\tau}{|\mu|}}+\int_{\tau}^{\tau_{\text{L}}}J\left(\tau',\mu,\varphi\right)e^{-\frac{\tau'-\tau}{|\mu|}}\frac{d\tau'}{|\mu|} \qquad (4.14)$$

for the upwelling radiation. These solutions correspond to a physical picture, illustrated in Fig. 4.3. The radiance striking the boundaries of the slab is attenuated exponentially on its way from the top or bottom of the layer to the coordinate τ. For the downward radiation, the path is τ/μ, while for the upward radiation the path is $(\tau_{\text{L}}-\tau)/|\mu|$ (we use the absolute value of μ since for the upward radiation μ is negative). The emitted radiation by the source J at the coordinate τ' is attenuated exponentially when it passes through the medium from τ' to τ. Note that we should take into account the radiation sources at all interior points reduced by corresponding exponential decay factors.

Equations (4.13) and (4.14) are the formal solutions of Eq. (4.12). In particular, for the radiation transmitted through the layer we have

$$L\left(\tau_{\text{L}},\mu>0,\varphi\right)=L\left(0,\mu>0,\varphi\right)e^{-\frac{\tau_{\text{L}}}{\mu}}+\int_{0}^{\tau_{\text{L}}}J\left(\tau',\mu,\varphi\right)e^{-\frac{\tau_{\text{L}}-\tau'}{\mu}}\frac{d\tau'}{\mu}, \qquad (4.15)$$

while the reflected light intensity is given by

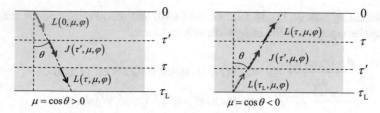

Fig. 4.3 Attenuation of the radiation and internal sources

$$L\left(0, \mu < 0, \varphi\right) = L\left(\tau_{\mathrm{L}}, \mu < 0, \varphi\right) e^{-\frac{\tau_L}{|\mu|}} + \int_0^{\tau_L} J\left(\tau', \mu, \varphi\right) e^{-\frac{\tau'}{|\mu|}} \frac{d\tau'}{|\mu|} \qquad (4.16)$$

under assumption of the absolutely black underlying surface.

4.2 Multiple Scattering Pseudo-source and Thermal Emission Source

For a scattering atmosphere, J is the sum of the multiple scattering term $J_{\mathrm{ms}}(z, \mu, \varphi)$ describing the multiple scattering of the solar radiation and the thermal emission term $J_{\mathrm{th}}(z, \mu, \varphi)$ governed by the Planck function B:

$$J(z, \mu, \varphi) = J_{\mathrm{ms}}(z, \mu, \varphi) + J_{\mathrm{th}}(z, \mu, \varphi) \qquad (4.17)$$

with

$$J_{\mathrm{ms}}(z, \mu, \varphi) = \frac{\sigma_{\mathrm{sca}}}{4\pi} \int_0^{2\pi} \int_{-1}^{1} p\left(z, \mu', \mu, \varphi - \varphi'\right) L\left(z, \mu', \varphi'\right) d\mu' d\varphi' \qquad (4.18)$$

and assuming local thermodynamic equilibrium

$$J_{\mathrm{th}}(z, \mu, \varphi) = \sigma_{\mathrm{abs}}(z) B(T(z)), \qquad (4.19)$$

where $T(z)$ is the temperature profile. According to Eq. (4.18), the radiation which goes along the direction (μ', φ') is scattered into the direction (μ, φ), thereby increasing $L(z, \mu, \varphi)$. Thus, J_{ms} refers to a pseudo-source, i.e., not a physical source, yet enhancing the radiation in a given direction.

Note that J_{ms} (and consequently J) depends on the radiance L. Therefore, the formal solution (4.13)–(4.14) cannot be readily applied to the radiative transfer equation with the multiple scattering term (we are confronted with an "egg-and-chicken problem"). However, as we will see further, these equations are used in several numerical techniques.

Substituting Eqs. (4.17)–(4.19) into Eq. (4.10) and rearranging, we get the radiative transfer equation in the integro-differential form:

$$\mu \frac{dL\,(z, \mu, \varphi)}{dz} = -\sigma_{\text{ext}}\,(z)\,L\,(z, \mu, \varphi) + \sigma_{\text{abs}}\,(z)\,B\,(T\,(z))$$
$$+ \frac{\sigma_{\text{sca}}\,(z)}{4\pi} \int_0^{2\pi} \int_{-1}^{1} P\,(z, \mu', \mu, \varphi - \varphi')\,L\,(z, \mu', \varphi')\,d\mu' d\varphi'. \tag{4.20}$$

Dividing Eq. (4.20) by σ_{ext} we obtain the radiative transfer equation, which depends on the unitless parameters, i.e., the optical depth τ and the single-scattering albedo ω:

$$\mu \frac{dL\,(\tau, \mu, \varphi)}{d\tau} = -L\,(\tau, \mu, \varphi) + (1 - \omega\,(\tau))\,B\,(T\,(\tau))$$
$$+ \frac{\omega\,(\tau)}{4\pi} \int_0^{2\pi} \int_{-1}^{1} P\,(\tau, \mu', \mu, \varphi - \varphi')\,L\,(\tau, \mu', \varphi')\,d\mu' d\varphi'. \tag{4.21}$$

Note that the multiple scattering term changes the type of the radiative transfer equation from differential to integro-differential, thereby causing mathematical complications. The thermal source is negligible as compared to the multiple scattering term at wavelengths smaller than approximately $2.5\,\mu$m.

4.3   Rotational Raman Scattering

So far we considered only elastic processes, i.e., those in which the energy of the photon (or the wavelength) does not change (unless it is absorbed). However, the processes of inelastic scattering, in which the photon either gets or gives the energy, may be important. There are several inelastic processes which may lead to the energy change, such as, for instance, the fluorescence [2], which can be used for distinguishing organic and inorganic compounds of aerosol particles. To take into account inelastic scattering processes, the radiative transfer equation has to be equipped with additional source terms and absorption terms. The former ones correspond to the processes which lead to the energy gain for a given wavelength, while the latter one describes the additional energy leakage from a given wavelength to other wavelengths. In principle, the radiative transfer model with inelastic scattering is computationally expensive as the problem is not monochromatic anymore and so, the radiative transfer equation should be considered across several wavelengths.

In this section we consider the so-called Ring effect [3] (sometimes referred to as "filling-in effect"), which is important to take into account in trace gas retrievals (interested readers are encouraged to read an excellent didactic tutorial of McLinden [4]). It was discovered that the depth of solar Fraunhofer lines in scattered light is a few percent less than that observed in the direct sunlight [5]. Several mechanisms were proposed to explain this phenomenon, including daytime air glow and aerosol fluorescence [6] (early suggestions are outlined in [7]). Brinkmann [8] pointed out that the observed filling-in of the solar Fraunhofer lines in the Earth's atmosphere can

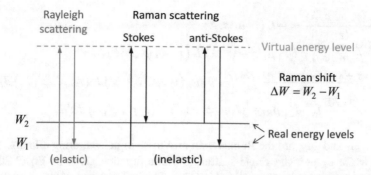

Fig. 4.4 Schematic representation of Rayleigh (elastic) and Raman (inelastic) scattering processes

be explained by Raman scattering, which was predicted theoretically by Smekal [9] and experimentally observed by Raman [10] as well as Landsberg and Mandelstam. Figure 4.4 gives a schematic representation of Rayleigh and Raman scattering. Two energy levels of a molecule with energies W_1 and W_2 are shown. The photon interacts with a molecule in a very short time Δt. Following the time–energy uncertainty relation we have

$$\Delta t \Delta \tilde{W} \geq \frac{\hbar}{2}, \tag{4.22}$$

where $\Delta \tilde{W}$ is the energy difference between the virtual and nearest real energy levels (the so-called detuning energy) [11], while $\hbar$ is the reduced Planck constant. After this virtual absorption, the molecule returns back to its ground state by emitting a photon. The energy of the final state of the photon can be larger or smaller (corresponding to Stokes Raman scattering and anti-Stokes scattering, respectively). The change in the photon energy (the so-called Raman shift) is determined by the structure of molecular energy levels, and is unique for every molecule.

In recent studies (e.g., [12–14]), the Ring effect is explained by Raman scattering. Note that in ultraviolet and visible spectral ranges, only rotational Raman scattering (RRS) is considered (the vibrational transitions correspond to larger shifts in wavelength space and usually are not considered) [15]. As shown in numerous studies (see [16] and references therein), the impact of RRS can reach 2–3 % in the radiance.

Let us consider the radiance $L(\ldots, \lambda)$ explicitly dependent on the wavelength λ. The radiative transfer equation should be equipped with two processes governing the energy exchange. The first one is the energy loss. Here the energy is distributed from the wavelength λ across all Raman-shifted lines λ_s. The second process is the energy gain from $L(\ldots, \lambda_s)$ for which λ is itself a Raman-shifted wavelength.

RRS can be included by adding additional source terms in the radiative transfer equation:

$$
\mu \frac{dL\left(\tau, \mu, \varphi, \lambda\right)}{d\tau} = -L\left(\tau, \mu, \varphi, \lambda\right)
$$
$$
+ \frac{\omega\left(\lambda\right)}{4\pi} \int_0^{2\pi} \int_{-1}^1 p\left(\mu', \mu, \varphi - \varphi', \lambda\right) L\left(\tau, \mu', \varphi', \lambda\right) d\mu' d\varphi'
$$
$$
+ \sum_{s=1}^{NS} \frac{\omega_{RG}\left(\tau, \lambda_s \to \lambda\right)}{4\pi} \int_0^{2\pi} \int_{-1}^1 p_{RRS}\left(\mu', \mu, \varphi - \varphi'\right) L\left(\tau_s, \mu', \varphi', \lambda_s\right) d\mu' d\varphi'
$$
$$
- \frac{\omega_{RL}\left(\tau, \lambda\right)}{4\pi} \int_0^{2\pi} \int_{-1}^1 p_{RRS}\left(\mu', \mu, \varphi - \varphi'\right) L\left(\tau, \mu', \varphi', \lambda\right) d\mu' d\varphi',
$$

$$(4.23)$$

where ω_{RG} and ω_{RL} are the Raman gain and loss single-scattering albedos, respectively, while p_{RRS} is the single-scattering phase function given by Eq. (3.207) (the computational formulas are outlined in [17]). The first and second terms on the right-hand side are the same as in the radiative transfer equation without Raman scattering. The third term is the Raman gain term. The summation is performed across all NS Raman-shifted lines. Note that the molecular scattering coefficient computed as in Sect. 3.20 comprises both Cabannes and RRS scattering. Therefore, the RRS contribution has to be subtracted—that is the fourth term.

In Eq. (4.23), $L\left(\ldots, \lambda\right)$ and $L\left(\ldots, \lambda_s\right)$ are unknown. Therefore, the radiative transfer equations for all Raman-shifted wavelengths have to be solved in one system. This approach is computationally expensive. In the first-order RRS scattering model [17, 18], this equation is solved in two steps. At the first step, the radiative transfer equation without the Raman loss and gain terms provides the elastic scattering field L_E. At the second step, L_E is used for computing the Raman loss and gain terms:

$$
\mu \frac{dL\left(\tau, \mu, \varphi, \lambda\right)}{d\tau} = -L\left(\tau, \mu, \varphi, \lambda\right)
$$
$$
+ \frac{\omega\left(\lambda\right)}{4\pi} \int_0^{2\pi} \int_{-1}^1 p_E\left(\mu', \mu, \varphi - \varphi', \lambda\right) L\left(\tau, \mu', \varphi', \lambda\right) d\mu' d\varphi'
$$
$$
+ \sum_{s=1}^{NS} \frac{\omega_{RG}\left(\tau, \lambda_s\right)}{4\pi} \int_0^{2\pi} \int_{-1}^1 p_{RRS}\left(\mu', \mu, \varphi - \varphi'\right) L_E\left(\tau_s, \mu', \varphi', \lambda_s\right) d\mu' d\varphi'
$$
$$
- \frac{\omega_{RL}\left(\tau, \lambda\right)}{4\pi} \int_0^{2\pi} \int_{-1}^1 p_{RRS}\left(\mu', \mu, \varphi - \varphi'\right) L_E\left(\tau, \mu', \varphi', \lambda\right) d\mu' d\varphi'.
$$

$$(4.24)$$

Equation (4.24) is similar to Eq. (4.21), but with additional source terms, which are already known. Note that the boundary conditions should be imposed for radiances at each wavelength considered. The accuracy of the first-order RRS model was analyzed in several works (e.g., [19, 20]) revealing the insignificance of higher-order RRS terms. The first-order RRS for the Stokes vector is considered in [16].

In principle, the Raman effect is quite weak. The RRS spectrum is temperature-dependent. Thus, the RRS effects can be considered not only as a distortion (as in trace gas retrievals), but also as a valuable source of information on the atmospheric composition and structure. Raman scattering is studied in the context of probing the atmospheres of solar system planets [21] and exoplanets [22]. Further, we will consider only the equation without Raman scattering. In practice, we can use a scheme corresponding to Eq. (4.24).

4.4 Boundary Value Problems in Radiative Transfer

A radiative transfer equation has to be equipped with boundary conditions to derive a unique solution. We consider a plane-parallel atmosphere illuminated by the plane-parallel beam of radiation given as $F_\odot \delta (\mu_0 - \mu) \delta (\varphi)$, where μ_0 are the cosine of the zenith angle, $F_\odot = E_\odot / \pi$ with $E_\odot$ being the downwelling solar irradiance at the top of the atmosphere. It is assumed that the azimuth angle of the incident light beam $\varphi_0 = 0$. The viewing direction is characterized by the cosine of the zenith angle $\mu = \cos \theta$ and the azimuthal angle φ (which is measured with respect to $\varphi_0 = 0$).

In this formulation, the monodirectional non-scattered solar radiation at the top of the atmosphere decays exponentially as it goes through the atmosphere. Following [1, 23], the radiance field (referred to as the total radiance L_t) is represented as a sum of the direct radiance $L_\odot$ and the diffuse radiance (hereinafter denoted as L):

$$L_t (z, \mu, \varphi) = L (z, \mu, \varphi) + L_\odot (z, \mu, \varphi) . \tag{4.25}$$

Note that the diffuse part refers to the radiation field which has been scattered at least once, while the direct radiation corresponds to the unscattered part. In a plane-parallel atmosphere with a given bidirectional reflectance distribution function (BRDF) ρ at the underlying surface, the integro-differential equation for the total radiance $L_t (z, \mu, \varphi)$ should be equipped with boundary conditions:

$$\begin{cases} \mu \dfrac{dL_t}{dz} (z, \mu, \varphi) = -\sigma_{ext} (z) L_t (z, \mu, \varphi) + J_{ms} (z, \mu, \varphi), \\ L_t (0, \mu > 0, \varphi) = F_\odot \delta (\mu_0 - \mu) \delta (\varphi), \\ L_t (z_s, \mu < 0, \varphi) = \dfrac{1}{\pi} \int_0^{2\pi} \int_{-1}^0 \rho (\mu, \mu', \varphi - \varphi') L_t (z_s, -\mu', \varphi') \mu' d\mu' d\varphi' \\ \qquad + \varepsilon (\mu) B (T_s), \end{cases} \tag{4.26}$$

where

$$J_{ms} (z, \mu, \varphi) = \frac{\sigma_{sca} (z)}{4\pi} \int_0^{2\pi} \int_{-1}^1 p (z, \mu, \mu', \varphi - \varphi') L_t (z, \mu', \varphi') d\mu' d\varphi', \tag{4.27}$$

ε is the surface emissivity, while T_s is the surface temperature. The points $z_{TOA} = 0$ and z_s correspond to the top of the atmosphere and the reflecting surface, respectively. Note that a half of boundary conditions are given at the top, and a half at the bottom. System (4.26) is called a two-point boundary value problem.

The direct light satisfies the following boundary value problem

$$\begin{cases} \mu \dfrac{dL_\odot (z, \mu, \varphi)}{dz} = -\sigma_{ext} (z) L_\odot (z, \mu, \varphi), \\ L_\odot (z_{TOA}, \mu, \varphi) = F_\odot \delta (\mu_0 - \mu) \delta (\varphi) \end{cases} \tag{4.28}$$

and is expressed as follows:

$$L_\odot(z, \mu, \varphi) = F_\odot \delta(\mu_0 - \mu)\, \delta(\varphi) \exp\left(-\frac{\tau(z)}{\mu_0}\right). \tag{4.29}$$

From Eqs. (4.26) and (4.25) the boundary value problem for the diffuse radiance L can be derived in the following form:

$$
\begin{cases}
\mu\dfrac{dL}{dz}(z, \mu, \varphi) = -\sigma_{\text{ext}}(z) L(z, \mu, \varphi) + J^*(z, \mu, \varphi), \\[2mm]
L(0, \mu > 0, \varphi) = 0, \\[2mm]
L(z_s, \mu < 0, \varphi) = \dfrac{1}{\pi} \int_0^{2\pi} \int_{-1}^0 L(z_s, -\mu, \varphi)\, \rho(\mu, \mu', \varphi - \varphi')\, \mu' d\mu' d\varphi' \\[2mm]
+ \dfrac{\mu_0 F_\odot}{\pi} \exp\left(-\dfrac{\tau(z_s, z_{\text{TOA}})}{\mu_0}\right) \rho(\mu, \mu_0, \varphi) + \varepsilon(\mu) B(T_s)
\end{cases} \tag{4.30}
$$

with

$$J^*(z, \mu, \varphi) = J_{\text{ss}}(z, \mu, \varphi) + J_{\text{ms}}(z, \mu, \varphi) + J_{\text{th}}(z, \mu, \varphi). \tag{4.31}$$

Here $J_{\text{ss}}(z, \Omega)$ is the single-scattering term given as

$$J_{\text{ss}}(z, \mu, \varphi) = \frac{\sigma_{\text{sca}}(z)}{4\pi} F_\odot p(z, \mu_0, \mu, \varphi) \exp\left(-\frac{\tau(z)}{\mu_0}\right). \tag{4.32}$$

Unlike BVP (4.26), BVP (4.30) does not contain delta-functions at the boundaries and hence can be solved numerically. Note that in ultraviolet and visible spectral ranges, the thermal emission term in Eq. (4.31) can be neglected, while the multiple scattering term has a strong impact. Finally, dividing Eq. (4.30) by σ_{ext} and taking into account Eqs. (4.19), (4.27), (4.31) and (4.32) we obtain the boundary value problem for the diffuse radiance:

$$
\begin{cases}
\mu\dfrac{dL(\tau, \mu, \varphi)}{d\tau} = -L(\tau, \mu, \varphi) + (1 - \omega(\tau)) B(T(\tau)) \\[2mm]
+ \dfrac{\omega(\tau)}{4\pi} \int_0^{2\pi} \int_{-1}^1 p(\tau, \mu', \mu, \varphi - \varphi')\, L(\tau, \mu', \varphi')\, d\mu' d\varphi' \\[2mm]
+ \dfrac{\omega(\tau)}{4\pi} p(\tau, \mu_0, \mu, \varphi) \exp\left(-\dfrac{\tau}{\mu_0}\right), \\[2mm]
L(0, \mu > 0, \varphi) = 0, \\[2mm]
L(z_s, \mu < 0, \varphi) = \dfrac{1}{\pi} \int_0^{2\pi} \int_{-1}^0 L(z_s, -\mu, \varphi)\, \rho(\mu, \mu', \varphi - \varphi')\, \mu' d\mu' d\varphi' \\[2mm]
+ \dfrac{\mu_0 F_\odot}{\pi} \exp\left(-\dfrac{\tau(z_s, z_{\text{TOA}})}{\mu_0}\right) \rho(\mu, \mu_0, \varphi) + \varepsilon(\mu) B(T_s).
\end{cases} \tag{4.33}
$$

4.5 Exact and Approximate Solution Techniques

The analytical solutions of the boundary value problem for the diffuse radiance (Eq. (4.30)) are known only for special cases. One of them is the so-called Milne problem corresponding to a non-absorbing semi-infinite medium with a source at infinity and isotropic scattering [24, 25]. For a semi-infinite homogeneous plane-parallel atmosphere, a solution is given in terms of H-functions [1, 26]. This approach is reviewed in [27, 28]. In atmospheric remote sensing applications, the numerical solution techniques can be applied. Essentially, they are based on certain approximations. In this regard, solution techniques can be broadly categorized into exact numerical methods and approximate methods. The examples of exact methods are the discrete ordinate method, the adding-doubling technique and successive orders of scattering. These exact methods converge to a "true" solution for any values of optical characteristics (τ, ω, p) when a certain parameter, which governs the accuracy, is sufficiently large (e.g., number of discrete ordinates or number of orders of scattering etc.). The approximate models are close to the true solution only under certain conditions imposed on τ, ω and p. The examples are the single-scattering approximation, the two-stream approximation and the asymptotic theory. Usually, the approximate models provide analytical solutions to the radiative transfer equation in a closed form (which are beautifully elegant and lead to semi-analytical retrieval algorithms), while the exact methods provide a computational algorithm rather than a final ready-to-calculate expression [29]. Some references to open-source radiative transfer models can be found in the section "Useful links" at the end of the book.

4.6 Azimuthal Expansion of the Radiative Transfer Equation

The numerical solution of the monochromatic radiative transfer equation is complicated by the fact that the desired function L depends on three variables (τ, μ, φ) for a given single-scattering albedo and a single-scattering phase function. To make the problem easier, the φ-dependence can be factored out. To do that, we consider Fourier cosine series for the diffuse radiance, i.e.,

$$L(\tau, \mu, \varphi) = \sum_{m=0}^{M_{max}} L_m(\tau, \mu) \cos m\varphi, \qquad (4.34)$$

and the expansion of the single-scattering phase function into Legendre series:

$$p(\tau, \mu, \mu', \varphi - \varphi') = \sum_{m=0}^{M_{max}} (2 - \delta_{m0}) p_m(\tau, \mu, \mu') \cos[m(\varphi - \varphi')], \qquad (4.35)$$

where δ is the Kronecker delta, i.e.,

$$\delta_{ij} = \begin{cases} 0 & \text{if } i \neq j, \\ 1 & \text{if } i = j, \end{cases} \tag{4.36}$$

and

$$p_m\left(\tau, \mu, \mu'\right) = \sum_{n=m}^{N_{\max}} \chi_n\left(\tau\right) \bar{P}_n^m(\mu) \bar{P}_n^m(\mu'). \tag{4.37}$$

Two remarks are in order:

1. Actually, Eqs. (4.34) and (4.35) are valid only if the infinite number of terms is taken, i.e., $M_{\max} \to \infty$ and $N_{\max} \to \infty$. In practice, however, we take a finite number of terms.
2. In Eq. (4.34) we neglect sine terms of the Fourier expansion because it is assumed that $L\left(z, \mu, \varphi\right)$ exhibits the azimuthal symmetry with respect to the incidence plane and, thus, L is an even function of φ.

Substituting Eqs. (4.34) and (4.35) into Eq. (4.30) and applying the orthogonality property of cosine functions, the integro-differential equation for the mth azimuthal harmonic can be derived (see Appendix 1 for details):

$$\mu \frac{d L_m(\tau, \mu)}{d\tau} = -L_m(\tau, \mu) + J_{ss,m}\left(\tau, \mu\right) + J_{ms,m}\left(\tau, \mu\right) + J_{th,m}\left(\tau\right) \tag{4.38}$$

with

$$J_{ss,m}\left(\tau, \mu\right) = (2 - \delta_{m0}) \frac{\omega\left(\tau\right)}{4\pi} F_\odot \exp\left(-\frac{\tau}{\mu_0}\right) p_m\left(\tau, \mu, \mu_0\right), \tag{4.39}$$

$$J_{ms,m}\left(\tau, \mu\right) = \frac{\omega\left(\tau\right)}{2} \int_{-1}^{1} p_m\left(\tau, \mu, \mu'\right) L_m(\tau, \mu')d\mu', \tag{4.40}$$

$$J_{th,m}\left(\tau\right) = \delta_{m0}\left(1 - \omega\left(\tau\right)\right) B\left(T\left(\tau\right)\right). \tag{4.41}$$

Note that the zeroth harmonic refers to the azimuthally averaged radiance.

Analogously, the BRDF is expanded as a Fourier series in cosine of the azimuthal angle:

$$\rho\left(\mu, \mu', \varphi - \varphi'\right) = \sum_{m=0}^{\infty} \rho_m\left(\mu, \mu'\right) \cos m\left(\varphi - \varphi'\right). \tag{4.42}$$

The corresponding expansion coefficients ρ_m are computed from

$$\rho_m\left(\mu, \mu'\right) = \frac{1}{\pi} \frac{2 - \delta_{m0}}{2} \int_{-\pi}^{\pi} \rho\left(\mu, \mu', \varphi - \varphi'\right) \cos m\left(\varphi - \varphi'\right) d\left(\varphi - \varphi'\right). \tag{4.43}$$

For the Lambertian surface, ρ_0 is the surface albedo and depends on neither μ nor μ', while $\rho_m\left(\mu, \mu'\right) = 0$ for $m > 0$. Substituting Eq. (4.42) into Eq. (4.30) gives the expansion of the lower boundary condition (see Appendix 2 for details)

$$
\begin{aligned}
L_m\left(\tau_s, \mu > 0\right) &= \left(1 + \delta_{m0}\right) \int_0^1 L_m\left(\tau_s, -\mu', \varphi'\right) \rho_m\left(\mu, \mu'\right) \mu' d\mu' \\
&+ \frac{\mu_0 F_\odot}{\pi} \exp\left(-\frac{\tau_s}{\mu_0}\right) \rho_m\left(\mu, \mu_0\right) + \delta_{m0}\varepsilon\left(\mu\right) B\left(T_s\right).
\end{aligned}
\tag{4.44}
$$

Here we would like to make a remark. In our derivations we have implicitly assume that L is a even function of the relative azimuthal angle φ making possible expansion (4.34). In addition, the BRDF and the single-scattering phase functions depend on $\varphi - \varphi'$. Although from the mathematical point of view, such azimuthally symmetric scattering is a special case and a strong restriction, it is usually adopted when discussing most of radiative transfer problems. However, it is important to keep in mind that some problems do not exhibit azimuthal symmetry (e.g., oriented ice crystals in clouds and snowpack).

4.7   Discrete Ordinate Method

4.7.1   General Remarks

One of the most common methods to solve RTE is the discrete ordinate method, wherein the radiation field is considered along specific directions (discrete ordinates). Here it is necessary to make a reservation: there is no "discrete ordinate method" as such, but rather there is a class of methods in which the angular variables are discretized [30] (the examples of them include the eigenvalue approach, the matrix exponential approach and the matrix operator method).

In particular, the range of polar angles μ is divided into some number of discrete angular intervals, and continuous functions of μ (such as L and p) are replaced by a discrete set of direction vectors. Then the integral in Eq. (4.40) becomes a sum, while the radiative transfer equation is converted into a set of coupled linear ordinary differential equations. The method was developed by Wick [31], first, to solve the radiative transfer equation with an isotropic phase function, generalizing the original concept of the two-stream approximation of Schuster [32] and Schwarzschild [33]. Later, this technique was extensively used by Chandrasekhar [1], and sometimes was referred to as the Wick-Chandrasekhar discrete ordinate method. The proof of convergence of the discrete ordinate method is given in [34].

The main advantage of the discrete ordinate method is its generality, that is, it can be applied for the optical medium of arbitrary thickness and it solves system (4.30) with a full account for multiple light scattering effects.

4.7.2   Quadrature Rule

According to the Gauss-Legendre quadrature rule, the integral of a function $f(\mu)$ can be well approximated by a finite sum [35]:

$$\int_{-1}^{1} f(\mu)\, d\mu \approx \sum_{i=1}^{N_{do}} f(\hat{\mu}_i)\, \hat{w}_i, \tag{4.45}$$

where $\hat{\mu}_i$ and $\hat{w}_i$ for $i = 1, \ldots, N_{do}$ are the nodes and weights of the quadrature, respectively. The nodes $\hat{\mu}_i$ are given by the roots of Legendre polynomial $P_{N_{do}}(\mu)$ which occur symmetrically about 0, while the weights are given by the formula [36]

$$\hat{w}_i = \frac{2}{\left(1 - \hat{\mu}_i^2\right)\left[P_n'\left(\hat{\mu}_i\right)\right]^2}. \tag{4.46}$$

The weights must satisfy the following condition:

$$\sum_{i=1}^{N_{do}} \hat{w}_i = 2. \tag{4.47}$$

Equation (4.45) is exact for $f(\mu)$ being a polynomial of degree not higher than $2N_{do} - 1$.

In order to deal with Eq. (4.40) the integral is replaced by a symmetric quadrature rule with $2N_{do}$ nodes and weights (so that N_{do} is the number of discrete ordinates per hemisphere). If μ_i with $i = 1, \ldots, N_{do}$ is a node associated with the weight w_i, then $-\mu_i$ is also a node associated with the same weight. Usually, the quadrature is chosen to be double Gaussian [37] and, the positive nodes are given by

$$\mu_i = 0.5\left(\hat{\mu}_i + 1\right), \tag{4.48}$$

The negative nodes are given by $-\mu_i$. The weights for the double Gaussian quadrature have to be renormalized:

$$w_i = \frac{\hat{w}_i}{2}. \tag{4.49}$$

In this discrete ordinate space, the radiative transfer equation is converted into the set of $2N_{do}$ equations, one for each discrete ordinate μ_i, that is,

$$\mu_i \frac{d L_m(\tau, \mu_i)}{d\tau} = -L_m(z, \mu_i) + J_{ss,m}(\tau, \mu_i) + J_{ms,m}(\tau, \mu_i) + J_{th,m}(\tau), \tag{4.50}$$

where

$$J_{\mathrm{ss},m}(\tau, \mu_i) = (2 - \delta_{m0})\frac{\omega(\tau)}{4\pi}F_{\odot}\exp\left(-\frac{\tau}{\mu_0}\right)p(\tau, \mu_i, \mu_0) \qquad (4.51)$$

and

$$J_{\mathrm{ms},m}(\tau, \mu_i) = \frac{\omega(\tau)}{2}\sum_{j=1}^{N_{\mathrm{do}}}\left[p_m(\tau, \mu_i, -\mu_j)L_m(\tau, -\mu_j) + p_m(\tau, \mu_i, \mu_j)L_m(\tau, \mu_j)\right].$$

$$(4.52)$$

Equation (4.50) can be regarded as a balanced equation between $2N_{\mathrm{do}}$ radiant "streams". It is known [38] that a Gauss quadrature guarantees that the phase function is correctly normalized, i.e., $\int_0^{2\pi}\int_{-1}^{1}p(\mu, \mu', \varphi - \varphi')\mathrm{d}\mu'\mathrm{d}\varphi' = 4\pi$, so that the energy is conserved in the computation.

We make an additional remark. The Gauss quadrature has a factor-of-2 advantage in its efficiency as compared to equidistant quadrature methods. A method which has almost the same performances and can be implemented effortlessly by the fast Fourier transform is the Clenshaw–Curtis scheme [39]. However, as it has been shown in [40], when the number of nodes N_q increases, the error of the Clenshaw–Curtis quadrature does not decay to zero evenly but in two distinct stages; for N_q smaller than a critical value, the error decreases by the rate $O(\rho^{-2N_q})$, where $\rho > 1$, and afterwards by the rate $O(\rho^{-N_q})$. This means that initially (for small N_q), Clenshaw–Curtis quadrature converges almost as fast as the Gauss quadrature.

4.7.3 Spatial Discretization

For an inhomogeneous atmosphere, we consider a spatial discretization using N layers: $\tau_1 < \tau_2 < \cdots < \tau_{N+1}$, where $\tau_1 = 0$ and $\tau_{N+1} = \tau_s$, as shown in Fig. 4.5. A layer l is bounded above by the level τ_l and below by the level τ_{l+1}. The scattering and extinction coefficients and also the phase function are assumed to be constant within each layer. For layer l with the optical thickness $\bar{\tau}_l = \tau_{l+1} - \tau_l$, Eq. (4.50) is rewritten as the following linear system of differential equations (for simplicity we omit the azimuthal index m):

$$\frac{d}{d\tau}\begin{bmatrix}\mathbf{i}^{\uparrow}(\tau)\\\mathbf{i}^{\downarrow}(\tau)\end{bmatrix} = -\mathbf{A}_l\begin{bmatrix}\mathbf{i}^{\uparrow}(\tau)\\\mathbf{i}^{\downarrow}(\tau)\end{bmatrix} + \begin{bmatrix}\mathbf{b}_l^{\uparrow}(\tau)\\\mathbf{b}_l^{\downarrow}(\tau)\end{bmatrix},\ \tau_l \leq \tau \leq \tau_{l+1}, \qquad (4.53)$$

where $\left[\mathbf{i}^{\uparrow}(\tau)\right]_i = L_m(\tau, -\mu_i)$ and $\left[\mathbf{i}^{\downarrow}(\tau)\right]_i = L_m(\tau, \mu_i)$, $i = 1, \ldots, N_{\mathrm{do}}$. Due to the reciprocity principle for the phase function (i.e., $p_m(\mu_i, \mu_j) = p_m(-\mu_i, -\mu_j)$ and $p_m(-\mu_i, \mu_j) = p_m(\mu_i, -\mu_j)$), the layer matrix has a block structure

$$\mathbf{A}_l = \begin{bmatrix}\mathbf{A}_l^{11} & \mathbf{A}_l^{12}\\-\mathbf{A}_l^{12} & -\mathbf{A}_l^{11}\end{bmatrix}, \qquad (4.54)$$

where

$$
\begin{aligned}
\mathbf{A}_l^{11} &= \mathbf{MS}_l^\downarrow \mathbf{W} - \mathbf{M}, \\
\mathbf{A}_l^{12} &= \mathbf{MS}_l^\uparrow \mathbf{W},
\end{aligned}
$$

$\mathbf{M}$ and $\mathbf{W}$ are diagonal matrices with entries $[\mathbf{M}]_{ij} = (1/\mu_i)\,\delta_{ij}$ and $[\mathbf{W}]_{ij} = w_i \delta_{ij}$, respectively,

$$
\left[\mathbf{S}_l^\uparrow\right]_{ij} = \frac{1}{2}\omega_l p_m\left(\tau, \mu_i, -\mu_j\right) = \frac{1}{2}\omega_l \sum_{n=m}^{N_{\max}} \chi_{ln} \bar{P}_n^m\left(\mu_i\right) \bar{P}_n^m\left(-\mu_j\right), \tag{4.55}
$$

$$
\left[\mathbf{S}_l^\downarrow\right]_{ij} = \frac{1}{2}\omega_l p_m\left(\tau, \mu_i, \mu_j\right) = \frac{1}{2}\omega_l \sum_{n=m}^{N_{\max}} \chi_{ln} \bar{P}_n^m\left(\mu_i\right) \bar{P}_n^m\left(\mu_j\right), \tag{4.56}
$$

χ_{ln} is the expansion coefficient of the phase function in the lth layer, and $i, j = 1, \ldots N_{\mathrm{do}}$. It should be remarked that the symmetry relation of the associated Legendre functions, $\bar{P}_n^m\left(-\mu\right) = (-1)^{n+m}\, \bar{P}_n^m\left(\mu\right)$, implies that $\mathbf{S}^\uparrow$ and $\mathbf{S}^\downarrow$ are symmetric matrices.

The layer vector incorporates the solar single-scattering term and the thermal emission term, i.e.,

$$
\mathbf{b}_l^{\uparrow\downarrow}\left(\tau\right) = \mathbf{b}_{l,\mathrm{ss}}^{\uparrow\downarrow} \exp\left(-\tau/\mu_0\right) + \mathbf{b}_{l,\mathrm{th}}^{\uparrow\downarrow} T\left(\tau\right), \tag{4.57}
$$

where

$$
\left[\mathbf{b}_{l,\mathrm{ss}}^{\uparrow\downarrow}\right]_i = \mp \frac{1}{\mu_i}\left(2 - \delta_{m0}\right) \frac{\omega_l}{4\pi} F_\odot \sum_{n=m}^{N_{\max}} \chi_{ln} \bar{P}_n^m\left(\mu_i\right) \bar{P}_n^m\left(\mu_0\right) \tag{4.58}
$$

and

$$
\left[\mathbf{b}_{l,\mathrm{th}}^{\uparrow\downarrow}\right]_i = \mp \frac{1}{\mu_i} \delta_{m0}\omega_l. \tag{4.59}
$$

It is convenient to introduce a layer coordinate t defined by $t = \tau - \tau_l$ and $0 \leq t \leq \bar{\tau}_l$. Then the layer vector can be written as follows:

$$
\mathbf{b}_l^{\uparrow\downarrow}\left(t\right) = \mathbf{b}_{l,\mathrm{ss}}^{\uparrow\downarrow} \exp\left(-\tau_l/\mu_0 - t/\mu_0\right) + \mathbf{b}_{l,\mathrm{th}}^{\uparrow\downarrow} T\left(\tau_l + t\right). \tag{4.60}
$$

Note that in the conventional eigenvalue approach [41], the general solution of the linear system of differential equations (4.53) consists of a linear combination of all homogeneous solutions plus the particular solution for the assumed sources. For a multilayered medium, the expansion coefficients of the homogeneous solutions are the unknowns of the discretized radiative transfer problem and are computed by imposing the continuity condition for the radiances across the layer interfaces. The alternative technique employs the matrix-exponential formalism, which dates back to Waterman [42] and Flatau and Stephens [43]. We consider it in the next section.

4.7.4 Matrix Exponential Method

The linear system of differential equations can be treated as a boundary value problem. For each layer, a so-called layer equation can be derived which relates the radiance values at the layer boundaries. The discretized radiative transfer problem then reduces to a system of linear algebraic equations for the unknown values of the radiance [44].

Integrating Eq. (4.53), we obtain the equation which relates the radiances at the layer boundaries in the following integral form:

$$
\begin{bmatrix} \mathbf{i}_{l+1}^{\uparrow} \\ \mathbf{i}_{l+1}^{\downarrow} \end{bmatrix} = e^{-\mathbf{A}_l \bar{\tau}_l} \begin{bmatrix} \mathbf{i}_l^{\uparrow} \\ \mathbf{i}_l^{\downarrow} \end{bmatrix} + e^{-\mathbf{A}_l \bar{\tau}_l} \int_0^{\bar{\tau}_l} e^{\mathbf{A}_l t} \begin{bmatrix} \mathbf{b}_l^{\uparrow}(t) \\ \mathbf{b}_l^{\downarrow}(t) \end{bmatrix} dt, \tag{4.61}
$$

where $\mathbf{i}_l^{\uparrow\downarrow} = \mathbf{i}^{\uparrow\downarrow}(\tau_l)$, as shown in Fig. 4.5. Note that the solution (4.61) can be interpreted in the same way as Eqs. (4.15)–(4.16), but in the matrix form, namely the Bouguer exponents turn into matrix exponentials, while the sources and radiance values turn into corresponding column vectors.

Multiplying Eq. (4.61) by $e^{\mathbf{A}_l \bar{\tau}_l}$ gives

$$
-\begin{bmatrix} \mathbf{i}_l^{\uparrow} \\ \mathbf{i}_l^{\downarrow} \end{bmatrix} + e^{\mathbf{A}_l \bar{\tau}_l} \begin{bmatrix} \mathbf{i}_{l+1}^{\uparrow} \\ \mathbf{i}_{l+1}^{\downarrow} \end{bmatrix} = \int_0^{\bar{\tau}_l} e^{\mathbf{A}_l t} \begin{bmatrix} \mathbf{b}_l^{\uparrow}(t) \\ \mathbf{b}_l^{\downarrow}(t) \end{bmatrix} dt. \tag{4.62}
$$

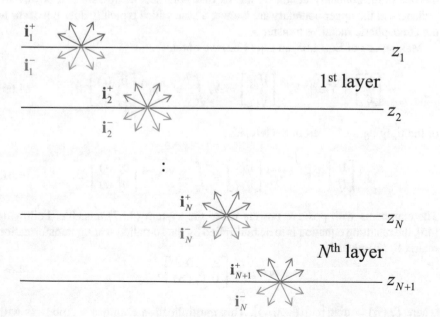

Fig. 4.5 N-layer atmosphere. The blue and green arrows correspond to the upward and downward radiances along the discrete ordinates at the layer boundaries, respectively

To evaluate $e^{\mathbf{A}_l \bar{\tau}_l}$ and $e^{\mathbf{A}_l t}$ in the framework of the discrete ordinate method with matrix exponential [44], we use the spectral decomposition of the matrix $\mathbf{A}_l$, i.e.,

$$\mathbf{A}_l = \mathbf{V}_l \Lambda_l \mathbf{V}_l^{-1}, \tag{4.63}$$

where $\mathbf{V}_l$ is the eigenvectors matrix and Λ_l is the eigenvalues matrix,

$$\Lambda_l = \begin{bmatrix} -\lambda_1 \ldots & 0 & 0 \ldots & 0 \\ \ldots & & & \\ 0 & \ldots -\lambda_{N_{do}} & 0 \ldots & 0 \\ 0 & \ldots & 0 & \lambda_1 \ldots & 0 \\ & & & \ldots \\ 0 & \ldots & 0 & 0 \ldots \lambda_{N_{do}} \end{bmatrix} \overset{\text{def}}{=} \begin{bmatrix} -\lambda_l & 0 \\ 0 & \lambda_l \end{bmatrix}, \tag{4.64}$$

and $\lambda_l = \text{diag}\left[\lambda_{l1}, \ldots, \lambda_{lN_{do}}, \right]$. Then it follows:

$$e^{\mathbf{A}_l t} = \mathbf{V}_l e^{\Lambda_l t} \mathbf{V}_l^{-1}. \tag{4.65}$$

Note that the matrix exponential $e^{-\mathbf{A}_l \bar{\tau}_l}$, reflecting the internal properties of the homogeneous medium, is called a propagator [44]. If the one-point boundary condition is given, then the solution deeper in the medium can be recovered (propagated) down from the upper boundary by applying this propagator. As we already noted, the one-point boundary condition means that both sets of upward and downward radiances at the upper boundary are known, a case which typically does not occur in the atmospheric radiative transfer.

Making use of Eq. (4.65) in Eq. (4.62) we obtain:

$$-\begin{bmatrix} \mathbf{i}_l^\uparrow \\ \mathbf{i}_l^\downarrow \end{bmatrix} + \mathbf{V}_l e^{\Lambda_l \bar{\tau}_l} \mathbf{V}_l^{-1} \begin{bmatrix} \mathbf{i}_{l+1}^\uparrow \\ \mathbf{i}_{l+1}^\downarrow \end{bmatrix} = \mathbf{V}_l \int_0^{\bar{\tau}_l} e^{\Lambda_l t} \mathbf{V}_l^{-1} \begin{bmatrix} \mathbf{b}_l^\uparrow(t) \\ \mathbf{b}_l^\downarrow(t) \end{bmatrix} dt, \tag{4.66}$$

or multiplying by $\mathbf{V}_l^{-1}$ from the left side:

$$-\mathbf{V}_l^{-1} \begin{bmatrix} \mathbf{i}_l^\uparrow \\ \mathbf{i}_l^\downarrow \end{bmatrix} + e^{\Lambda_l \bar{\tau}_l} \mathbf{V}_l^{-1} \begin{bmatrix} \mathbf{i}_{l+1}^\uparrow \\ \mathbf{i}_{l+1}^\downarrow \end{bmatrix} = \int_0^{\bar{\tau}_l} e^{\Lambda_l t} \mathbf{V}_l^{-1} \begin{bmatrix} \mathbf{b}_l^\uparrow(t) \\ \mathbf{b}_l^\downarrow(t) \end{bmatrix} dt. \tag{4.67}$$

The exponents with positive powers make the system (4.67) unstable. Following [45], the resulting equation is to be multiplied by the so-called scaling transformation matrix $\mathbf{S}_l(\bar{\tau}_l)$ given by

$$\mathbf{S}_l(\bar{\tau}_l) = \begin{bmatrix} \mathbf{E} & 0 \\ 0 & \Gamma_l(\bar{\tau}_l) \end{bmatrix}, \tag{4.68}$$

where $\Gamma_l(\bar{\tau}_l) = \text{diag}\left[\exp(-\lambda_l \bar{\tau}_l)\right]$. This multiplication eliminates exponents with positive powers, and we obtain a numerically stable layer equation, i.e.,

$$- \mathbf{S}_l \left(\bar{\tau}_l \right) \mathbf{V}_l^{-1} \begin{bmatrix} \mathbf{i}_l^\uparrow \\ \mathbf{i}_l^\downarrow \end{bmatrix} + \mathbf{S}_l \left(\bar{\tau}_l \right) e^{\Lambda_l \bar{\tau}_l} \mathbf{V}_l^{-1} \begin{bmatrix} \mathbf{i}_{l+1}^\uparrow \\ \mathbf{i}_{l+1}^\downarrow \end{bmatrix} = \mathbf{S}_l \left(\bar{\tau}_l \right) \int_0^{\bar{\tau}_l} e^{\Lambda_l t} \mathbf{V}_l^{-1} \begin{bmatrix} \mathbf{b}_l^\uparrow (t) \\ \mathbf{b}_l^\downarrow (t) \end{bmatrix} dt.$$

$$(4.69)$$

It can be written in a compact form as follows:

$$\mathbf{U}_l^1 \begin{bmatrix} \mathbf{i}_l^\uparrow \\ \mathbf{i}_l^\downarrow \end{bmatrix} + \mathbf{U}_l^2 \begin{bmatrix} \mathbf{i}_{l+1}^\uparrow \\ \mathbf{i}_{l+1}^\downarrow \end{bmatrix} = \begin{bmatrix} \mathbf{Q}_l^\uparrow \\ \mathbf{Q}_l^\downarrow \end{bmatrix}, \qquad (4.70)$$

where $\mathbf{U}_l^1 = -\mathbf{S}_l \left(\bar{\tau}_l \right) \mathbf{V}_l^{-1}$ and $\mathbf{U}_l^2 = \mathbf{D}_l \left(\bar{\tau}_l \right) \mathbf{V}_l^{-1}$ are the layer matrices, with

$$\mathbf{D}_l \left(\bar{\tau}_l \right) = \mathbf{S}_l \left(\bar{\tau}_l \right) e^{\Lambda_l \bar{\tau}_l} = \begin{bmatrix} \Gamma_l \left(\bar{\tau}_l \right) & 0 \\ 0 & \mathbf{E} \end{bmatrix} \qquad (4.71)$$

and

$$\begin{bmatrix} \mathbf{Q}_l^\uparrow \\ \mathbf{Q}_l^\downarrow \end{bmatrix} = \exp \left(-\frac{\tau_l}{\mu_0} \right) \mathbf{V}_l^{-1} \begin{bmatrix} \dfrac{e^{\left(\lambda_l - \frac{1}{\mu_0} \right) \bar{\tau}_l} - 1}{\lambda_l - \frac{1}{\mu_0}} \mathbf{b}_{l,ss}^\uparrow \\ \dfrac{e^{-\frac{1}{\mu_0} \bar{\tau}_l} - e^{-\lambda_l \bar{\tau}_l}}{\lambda_l - \frac{1}{\mu_0}} \mathbf{b}_{l,ss}^\downarrow \end{bmatrix}$$

$$+ \mathbf{S}_l \int_0^{\bar{\tau}_l} e^{\Lambda_l t} \mathbf{V}_l^{-1} \begin{bmatrix} \mathbf{b}_{l,th}^\uparrow \\ \mathbf{b}_{l,th}^\downarrow \end{bmatrix} T \left(\tau_l + t \right) dt. \qquad (4.72)$$

The azimuthal expansion coefficients of the phase functions vanish for all azimuthal harmonic numbers $m > 2$ and all layers with Rayleigh scattering only. Hence, the layer matrix $\mathbf{A}$ for such layers is diagonal, and the source vector $\mathbf{b}$ vanishes.

4.7.5 Inverse of the Eigenvector Matrix

The spectral decomposition of the layer matrix implies the solution of a $2N_{do} \times 2N_{do}$ eigenproblem (corresponding to the dimension of the layer matrix $\mathbf{A}$). As the eigenvalue problem can introduce a performance bottleneck in the radiative transfer solver, we consider this step in more detail.

Because of the special structure of the layer matrix $\mathbf{A}$ (see Eq. (4.54)), the eigenvalues occur in positive/negative pairs and the order of this algebraic eigenvalue problem may be halved. According to the Stamnes and Swanson procedure [46], the steps for computing an eigensystem of $\mathbf{A}$ consists in the calculation of

1. the eigensystem $\left\{ \alpha_k, \mathbf{w}_k^+ \right\}_{k=\overline{1,N_{do}}}$ of the asymmetric matrix $\mathbf{A}_- \mathbf{A}_+$, where $\mathbf{A}_+ = \mathbf{A}_{11} + \mathbf{A}_{12}$ and $\mathbf{A}_- = \mathbf{A}_{11} - \mathbf{A}_{12}$, where α_k and $\mathbf{w}_k^+$ are the eigenvalues and eigenvectors of $\mathbf{A}_- \mathbf{A}_+$, respectively. Note that the dimension of $\mathbf{A}_- \mathbf{A}_+$ (and

corresponding eigenvalue problem) is $N_{do} \times N_{do}$ instead of $2N_{do} \times 2N_{do}$ for the initial layer matrix $\mathbf{A}$;

2. the eigenvalues $\lambda_k = \sqrt{\alpha_k}, k = 1, \ldots, N_{do}$,
3. the eigenvectors $\mathbf{w}_k^-$ of the matrix $\mathbf{A}_+\mathbf{A}_-$, as $\mathbf{w}_k^- = (1/\lambda_k)\,\mathbf{A}_+\mathbf{w}_k^+, k = 1, \ldots, N_{do}$,
4. the linear combinations $\mathbf{v}_k^+ = \left(\mathbf{w}_k^+ + \mathbf{w}_k^-\right)/2$ and $\mathbf{v}_k^- = \left(\mathbf{w}_k^+ - \mathbf{w}_k^-\right)/2$ for $k = 1, \ldots, N_{do}$.

The eigenvector matrix $\mathbf{V}$ is a concatenation of the eigenvectors $\begin{bmatrix} \mathbf{v}_k^+ \\ \mathbf{v}_k^- \end{bmatrix}$ corresponding to λ_k, and also a concatenation of the eigenvectors $\begin{bmatrix} \mathbf{v}_k^- \\ \mathbf{v}_k^+ \end{bmatrix}$ corresponding to $-\lambda_k$. Formally we will write

$$\mathbf{V} = \left[\begin{bmatrix} \mathbf{v}_k^+ \\ \mathbf{v}_k^- \end{bmatrix} ; \begin{bmatrix} \mathbf{v}_k^- \\ \mathbf{v}_k^+ \end{bmatrix} \right]. \tag{4.73}$$

To further exploit the symmetry of the problem, we note that for any matrix $\mathbf{A}$, which has a complete set of linearly independent eigenvectors (a non-defective matrix), the inverse of the right eigenvector matrix is the transpose of the left eigenvector matrix. Indeed, let $\mathbf{A}$ be an $n \times n$ matrix with distinct eigenvalues, x_k be a right eigenvector of $\mathbf{A}$ corresponding to λ_k, i.e., $\mathbf{A}\mathbf{v}_k = \lambda_k\mathbf{v}_k$, and $\mathbf{y}_l$ be a left eigenvector of $\mathbf{A}$ corresponding to λ_l, i.e., $\mathbf{A}^T\mathbf{y}_l = \lambda_l\mathbf{y}_l$. Then, from $\mathbf{y}_l^T\mathbf{A}\mathbf{v}_k = \lambda_k\mathbf{y}_l^T\mathbf{v}_k = \lambda_l\mathbf{y}_l^T\mathbf{v}_k$, we see that for $\lambda_k \neq \lambda_l$, we have $\mathbf{y}_l^T\mathbf{v}_k = 0$. Moreover, assuming that $\mathbf{v}_k$ and $\mathbf{y}_k$ are normalized in the sense that $\mathbf{y}_k^T\mathbf{v}_k = 1, k = 1, \ldots, n$, we find that $\mathbf{V}^{-1} = \mathbf{Y}^T$, where $\mathbf{Y} = [\mathbf{y}_1, \ldots, \mathbf{y}_n]$. Thus, the spectral decomposition of $\mathbf{A}$ reads as $\mathbf{A} = \mathbf{V}\Lambda\mathbf{Y}^T$.

This procedure is effective if we are able to compute the left eigenvector matrix in an efficient way. In the so-called left eigenvector approach [47, 48], a similarity transformation [42] is applied to the matrix $\mathbf{A}$, i.e.,

$$\mathbf{A} = \mathbf{T}\hat{\mathbf{A}}\mathbf{T}^{-1}, \tag{4.74}$$

in order to obtain a scaled layer matrix $\hat{\mathbf{A}}$ with symmetric block matrices. Choosing the similarity transformation as the diagonal matrix

$$\mathbf{T} = \begin{bmatrix} \mathbf{W}^{-1/2}\mathbf{M}^{1/2} & 0 \\ 0 & \mathbf{W}^{-1/2}\mathbf{M}^{1/2} \end{bmatrix}, \tag{4.75}$$

we obtain

$$\hat{\mathbf{A}} = \begin{bmatrix} \hat{\mathbf{A}}_{11} & \hat{\mathbf{A}}_{12} \\ -\hat{\mathbf{A}}_{12} & -\hat{\mathbf{A}}_{11} \end{bmatrix} \tag{4.76}$$

with

$$\hat{\mathbf{A}}_{11} = \mathbf{M} - \mathbf{W}^{1/2}\mathbf{M}^{1/2}\mathbf{S}_+\mathbf{M}^{1/2}\mathbf{W}^{1/2}, \\ \hat{\mathbf{A}}_{12} = -\mathbf{W}^{1/2}\mathbf{M}^{1/2}\mathbf{S}_-\mathbf{M}^{1/2}\mathbf{W}^{1/2}. \tag{4.77}$$

The symmetry of the block matrices $\hat{\mathbf{A}}_{11}$ and $\hat{\mathbf{A}}_{12}$ follows from the symmetry of the matrices $\mathbf{S}_+$ and $\mathbf{S}_-$, and we have

$$\hat{\mathbf{A}}^T = \begin{bmatrix} \hat{\mathbf{A}}_{11} & -\hat{\mathbf{A}}_{12} \\ \hat{\mathbf{A}}_{12} & -\hat{\mathbf{A}}_{11} \end{bmatrix}. \tag{4.78}$$

As $\hat{\mathbf{A}}$ and $\mathbf{A}$ have the same block structure, the eigenvalues and the right eigenvectors of $\hat{\mathbf{A}}$ can be computed by using the Stamnes and Swanson procedure [46] for the asymmetric matrix $\widehat{\mathbf{A}}_-\widehat{\mathbf{A}}_+$, with $\widehat{\mathbf{A}}_+ = \widehat{\mathbf{A}}_{11} + \widehat{\mathbf{A}}_{12}$ and $\widehat{\mathbf{A}}_- = \widehat{\mathbf{A}}_{11} - \widehat{\mathbf{A}}_{12}$. Moreover, as $\hat{\mathbf{A}}$ and $\mathbf{A}$ are similar, the eigenvalues $\hat{\lambda}_k$ of $\hat{\mathbf{A}}$ coincide with the eigenvalues λ_k of $\mathbf{A}$. Then, it is not hard to see that if $\begin{bmatrix} \hat{\mathbf{v}}_k^+ & \hat{\mathbf{v}}_k^- \end{bmatrix}^T$ is the right eigenvector corresponding to λ_k, then the associated left eigenvector is $\begin{bmatrix} \hat{\mathbf{v}}_k^+ & -\hat{\mathbf{v}}_k^- \end{bmatrix}^T$. Clearly, if $\begin{bmatrix} \hat{\mathbf{v}}_k^- & \hat{\mathbf{v}}_k^+ \end{bmatrix}^T$ is the right eigenvector corresponding to $-\lambda_k$, then the associated left eigenvector is $\begin{bmatrix} -\hat{\mathbf{v}}_k^- & \hat{\mathbf{v}}_k^+ \end{bmatrix}^T$. Normalizing the right and left eigenvectors, we end up with

$$\hat{\mathbf{A}} = \hat{\mathbf{V}} \Lambda \hat{\mathbf{Y}}^T, \tag{4.79}$$

where

$$\hat{\mathbf{V}} = \begin{bmatrix} \dfrac{\mathrm{sign}\,(a_k)}{\sqrt{|a_k|}} \begin{bmatrix} \hat{\mathbf{v}}_k^+ \\ \hat{\mathbf{v}}_k^- \end{bmatrix}; & \dfrac{\mathrm{sign}\,(a_k)}{\sqrt{|a_k|}} \begin{bmatrix} \hat{\mathbf{v}}_k^- \\ \hat{\mathbf{v}}_k^+ \end{bmatrix} \end{bmatrix},$$

$$\hat{\mathbf{Y}} = \begin{bmatrix} \dfrac{1}{\sqrt{|a_k|}} \begin{bmatrix} \hat{\mathbf{v}}_k^+ \\ -\hat{\mathbf{v}}_k^- \end{bmatrix}; & \dfrac{1}{\sqrt{|a_k|}} \begin{bmatrix} -\hat{\mathbf{v}}_k^- \\ \hat{\mathbf{v}}_k^+ \end{bmatrix} \end{bmatrix},$$

and $a_k = \left\| \hat{\mathbf{v}}_k^+ \right\|^2 - \left\| \hat{\mathbf{v}}_k^- \right\|^2$. From Eqs. (4.74) and (4.79) we deduce that the right eigenvector matrix of $\mathbf{A}$ is $\mathbf{V} = \mathbf{T}\hat{\mathbf{V}}$, and that its inverse can be computed as $\mathbf{V}^{-1} = \hat{\mathbf{V}}^T \mathbf{T}^{-1}$. Obviously, the computation of $\mathbf{T}^{-1}$ is not a time-consuming process because $\mathbf{T}$ is a diagonal matrix.

The matrices $\hat{\mathbf{A}}$ and $\mathbf{A}$ have the same condition number, and therefore, the accuracies in solving the eigenvalue problems for $\hat{\mathbf{A}}$ and $\mathbf{A}$, by the Stamnes and Swanson procedure, are comparable. Alternatively, the eigenvalues and the right eigenvectors of $\hat{\mathbf{A}}$ can be computed by using the Nakajima and Tanaka procedure [49], or the Cholesky decomposition method [50]. The procedure described in [49] relies on the solution of two eigenvalue problems for the symmetric and positive-definite matrix $\widehat{\mathbf{A}}_-$ and the symmetric and positive semi-definite matrix $\widehat{\mathbf{A}}_-^{1/2}\widehat{\mathbf{A}}_+\widehat{\mathbf{A}}_-^{1/2}$. In [50], the matrix $\widehat{\mathbf{A}}_-$ is first factorized as $\widehat{\mathbf{A}}_- = \mathbf{R}^T \mathbf{R}$ by using the Cholesky decomposition method, then an eigenvalue problem for the symmetric and positive semi-definite matrix $\mathbf{R}\widehat{\mathbf{A}}_+\mathbf{R}^T$ is solved. However, as stated in [41, 50], the Stamnes and Swanson procedure with real arithmetic involves only one matrix multiplication ($\widehat{\mathbf{A}}_-\widehat{\mathbf{A}}_+$), and is more accurate and just as efficient as the other two methods.

4.7.6 Solution for the Multilayer Atmosphere

For the multilayer system, the whole-atmosphere approach can be used, in which
the layer equations are assembled into a global system of linear-algebra equations,
namely:

$$
\begin{cases}
\left[0, \mathbf{E}_{N_{do}}\right] \begin{bmatrix} \mathbf{i}_1^+ \\ \mathbf{i}_1^- \end{bmatrix} = 0, \\[2mm]
\mathbf{U}_1^1 \begin{bmatrix} \mathbf{i}_1^+ \\ \mathbf{i}_1^- \end{bmatrix} + \mathbf{U}_1^2 \begin{bmatrix} \mathbf{i}_2^+ \\ \mathbf{i}_2^- \end{bmatrix} = \begin{bmatrix} \mathbf{Q}_1^+ \\ \mathbf{Q}_1^- \end{bmatrix} \\[2mm]
\cdots \\[2mm]
\mathbf{U}_N^1 \begin{bmatrix} \mathbf{i}_N^+ \\ \mathbf{i}_N^- \end{bmatrix} + \mathbf{U}_N^2 \begin{bmatrix} \mathbf{i}_{N+1}^+ \\ \mathbf{i}_{N+1}^- \end{bmatrix} = \begin{bmatrix} \mathbf{Q}_N^+ \\ \mathbf{Q}_N^- \end{bmatrix}, \\[2mm]
\left[\mathbf{E}_{N_{do}}, \mathbf{R}_s\right] \begin{bmatrix} \mathbf{i}_{N+1}^+ \\ \mathbf{i}_{N+1}^- \end{bmatrix} = \mathbf{B}.
\end{cases}
\tag{4.80}
$$

Here $\mathbf{E}_{N_{do}}$ is the $N_{do} \times N_{do}$ identity matrix, $\mathbf{R}_s$ is the surface reflection matrix given
by

$$
[\mathbf{R}_b]_{kl} = -2w_l \left| \mu_l^- \right| \rho \left(\mu_k^+, \mu_l^- \right),
\tag{4.81}
$$

and

$$
\mathbf{B} = \delta_{m0} \varepsilon T\left(z_s\right) + (2 - \delta_{m0}) \frac{F_\odot}{\pi} |\mu_0| \exp\left(-\frac{\tau_s}{\mu_0}\right) \rho,
\tag{4.82}
$$

where $[\varepsilon]_k = \varepsilon \left(\mu_k^+\right)$ with ε being the surface emissivity, ρ the azimuthal expansion
coefficients of the BRDF, and $[\rho]_k = \rho_m \left(\mu_k^+, \mu_0\right)$. The first equation in system (4.80)
is the boundary condition at the top of the atmosphere. It states that the downwelling
diffuse radiation is absent at the top of the atmosphere. The last equation in system
(4.80) expresses the surface boundary condition. System (4.80) can be written in a
matrix form as

$$
\mathbf{UI} = \mathbf{Q}.
\tag{4.83}
$$

The entries of $\mathbf{U}$, $\mathbf{I}$ and $\mathbf{Q}$ are shown in Fig. 4.6. The resulting global matrix $\mathbf{U}$ of
system (4.80) has a sparse banded structure, and a band-compression algorithm
is used to facilitate the matrix inversion to obtain the complete solution of the
discrete-ordinate radiance field. This global matrix has dimension $2N_{do}\left(N + 1\right) \times 2N_{do}\left(N + 1\right)$, while the compression in band-storage format has dimension $(9N_{do} - 2) \times 2N_{do}(N+1)$ [51].

Once $\mathbf{I}$ is found, the azimuthal mode L_m from Eq. (4.34) can be readily computed.
At this point, thinking about the sequence $\{L_m\}_{m=0,1,2,\dots}$ as a Cauchy sequence, the
Cauchy convergence criterion can be checked. The series converges at the current
azimuthal mode m' if

$$
\max_{\mu,\varphi,\tau} \frac{L_{m'}}{\left| \sum_{m=0}^{m'} |L_m| \right|} \leq \bar{\varepsilon},
\tag{4.84}
$$

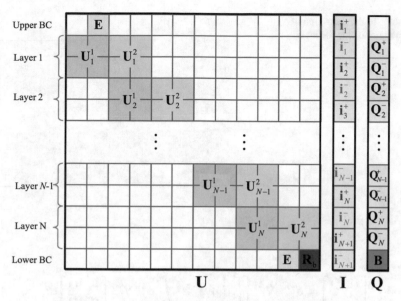

Fig. 4.6 Structure of system of layer equations equipped with boundary conditions. Each box corresponds to a $N_{do} \times N_{do}$ matrix. White boxes correspond to zero matrices

where $\bar{\varepsilon}$ is a small number corresponding to the azimuthal tolerance parameter (usually, it should be below 0.01). As recommended in [52], in order to avoid accidentally terminating the series before true convergence, azimuth summing should stop when this criterion is satisfied twice.

4.7.7 Radiance Field Inside a Layer

We have just found the radiance values at the boundaries of layers. Now our goal is to find the solution inside the lth layer surrounded by levels l and $l+1$, as shown in Fig. 4.7. Note that the direct usage of the propagator (corresponding to the one-point boundary value problem) is impossible in practice due to positive exponential terms. We split the lth layer into two sublayers with optical thicknesses ξ and $\bar{\tau}_l - \xi$, and for each of them we apply Eq. (4.67). For the upper sublayer we obtain:

Fig. 4.7 To the derivation of
the radiance inside the layer

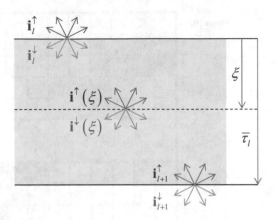

$$V_l^{-1} \begin{bmatrix} \mathbf{i}^\uparrow(\xi) \\ \mathbf{i}^\downarrow(\xi) \end{bmatrix} = e^{-\Lambda_l \xi} V_l^{-1} \begin{bmatrix} \mathbf{i}_l^\uparrow \\ \mathbf{i}_l^\downarrow \end{bmatrix} + e^{-\Lambda_l \xi} \int_0^\xi e^{\Lambda_l t} V_l^{-1} \begin{bmatrix} \mathbf{b}_l^\uparrow(t) \\ \mathbf{b}_l^\downarrow(t) \end{bmatrix} dt, \qquad (4.85)$$

while for the bottom sublayer we get

$$V_l^{-1} \begin{bmatrix} \mathbf{i}^\uparrow(\xi) \\ \mathbf{i}^\downarrow(\xi) \end{bmatrix} = e^{\Lambda_l(\bar{\tau}_l - \xi)} V_l^{-1} \begin{bmatrix} \mathbf{i}_{l+1}^\uparrow \\ \mathbf{i}_{l+1}^\downarrow \end{bmatrix} - e^{-\xi/\mu_0} \int_\xi^{\bar{\tau}_l} e^{\Lambda_l t} V_l^{-1} \begin{bmatrix} \mathbf{b}_l^\uparrow(t) \\ \mathbf{b}_l^\downarrow(t) \end{bmatrix} dt. \qquad (4.86)$$

As shown in [44, 53], to exclude the exponents with positive powers, we apply
Eq. (4.85) for the downwelling radiance and Eq. (4.86) for the upwelling radiance.
The resulting equation reads as follows:

$$V_l^{-1} \begin{bmatrix} \mathbf{i}^\uparrow(\xi) \\ \mathbf{i}^\downarrow(\xi) \end{bmatrix} = \begin{bmatrix} e^{-\lambda_l(\bar{\tau}_l - \xi)} & 0 \\ 0 & e^{-\lambda_l \xi} \end{bmatrix} \begin{bmatrix} \mathbf{V}^{11}\mathbf{i}_{l+1}^\uparrow + \mathbf{V}^{12}\mathbf{i}_{l+1}^\downarrow \\ \mathbf{V}^{21}\mathbf{i}_l^\uparrow + \mathbf{V}^{22}\mathbf{i}_l^\downarrow \end{bmatrix}$$
$$+ \begin{bmatrix} -e^{-\xi/\mu_0} \int_\xi^{\bar{\tau}_l} e^{\Lambda_l t} (\mathbf{V}^{11}\mathbf{b}_l^\uparrow(t) + \mathbf{V}^{12}\mathbf{b}_l^\downarrow(t)) dt \\ e^{-\Lambda_l \xi} \int_0^\xi e^{\Lambda_l t} (\mathbf{V}^{21}\mathbf{b}_l^\uparrow(t) + \mathbf{V}^{22}\mathbf{b}_l^\downarrow(t)) dt \end{bmatrix} \qquad (4.87)$$

where

$$V_l^{-1} = \begin{bmatrix} \mathbf{V}^{11} & \mathbf{V}^{12} \\ \mathbf{V}^{21} & \mathbf{V}^{22} \end{bmatrix}. \qquad (4.88)$$

Substituting Eq. (4.57) into Eq. (4.87) gives a continuous function $\mathbf{i}^{\uparrow\downarrow}(\xi)$ after integration.

An alternative way to find the solution inside the layer is to split it into two sublayers and solve the initial problem with an additional boundary. This approach makes
the solution more time-consuming, especially when the continuous dependence of
the radiance on the vertical coordinate is required (in this case a fine spatial layer
discretization is necessary).

4.7.8 Solution at a Given Viewing Zenith Angle: Source Function Integration and the Method of False Discrete Ordinates

So far we have found the solution at the quadrature points $\{\mu_k\}$. It is probable-that the actual cosine of the viewing zenith angle does not coincide with them. It is known that a simple interpolation does not provide required accuracy in many cases.

One of the accurate techniques is the source function integration method. We consider the integral form of the radiative transfer equation for the upwelling radiance propagating in the direction of the line of sight (LOS) $\{\mu_{LOS}, \varphi_{LOS}\}$ at the top of the atmosphere (see Eq. (4.16)):

$$L\left(0, \mu_{LOS}\varphi_{LOS}\right) = L\left(\tau_s, \mu_{LOS}\varphi_{LOS}\right) e^{-\frac{1}{\mu_{LOS}}\tau(z_s)}$$
$$+ \frac{1}{|\mu_{LOS}|} \int_0^{z_s} J\left(\tau', \mu_{LOS}\varphi_{LOS}\right) e^{-\frac{1}{|\mu_{LOS}|}\tau(z')} d\tau'. \tag{4.89}$$

It is apparent that $L\left(\tau, \mu_{LOS}, \varphi_{LOS}\right)$ solves the radiative transfer equation for $\{\mu_{LOS}, \varphi_{LOS}\}$. We can simulate this radiance in the framework of the discrete ordinate method by introducing an additional stream in the direction μ_{LOS}. Because the integral (4.52), which appears in the expression of the multiple scattering term J_{ms}, involves only the contribution of the Gaussian quadrature points $\mu_k, k = 1, \ldots, N_{do}$, we choose $w_{LOS} = 0$; for this reason, μ_{LOS} is called a false (or dummy) discrete ordinate. Since L at discrete ordinate nodes is already known, the source function J in Eq. (4.89) can be evaluated. The integration along τ' requires L as a continuous function of τ. To obtain $L\left(\tau\right)$, a technique described in the previous section can be employed. An alternative way is to apply a polynomial interpolation for the radiances found at layer boundaries.

Yet another approach is to add an additional discrete ordinate in the Gaussian quadrature with zero weight and solve the problem with such an extended set of discrete ordinates. This is the so-called dummy (or false) discrete ordinate method. Obviously, adding an additional discrete ordinate in the Gaussian quadrature increases the size of the layer matrix $\mathbf{A}$ and, hence, the dimension of the eigenvalue problem (4.63). Adding several false discrete ordinates might make computations slow. Below we consider how to avoid computational problems.

We are concerned with the computation of the left and right eigenvectors of the layer matrix for a set of false discrete ordinates $\mu_{LOS}^k > 0$, with $k = 1, \ldots, N_{LOS}$. In this case, the "augmented" layer matrix takes the form

$$\mathbf{A}_a = \begin{bmatrix} \mathbf{A}_{11} & \mathbf{A}_{12} & 0 \\ -\mathbf{A}_{12} & -\mathbf{A}_{22} & 0 \\ \mathbf{A}_{LOS}^+ & \mathbf{A}_{LOS}^- & \Sigma \end{bmatrix} = \begin{bmatrix} \mathbf{A} & 0 \\ \mathbf{A}_{LOS} & \Sigma \end{bmatrix}, \tag{4.90}$$

where

$$\left[\mathbf{A}_{\mathrm{LOS}}^{\pm}\right]_{ki} = \frac{\omega_l}{2\mu_{\mathrm{LOS}}^k} \sum_{n=m}^{N_{\max}} \chi_{ln} \bar{P}_n^m \left(\mu_{\mathrm{LOS}}^k\right) \bar{P}_n^m \left(\mp\mu_i\right) w_i, \tag{4.91}$$

for $k = 1, \ldots, N_{\mathrm{LOS}}, l = 1, \ldots, N_{\mathrm{do}}$, and $\Sigma = \mathrm{diag}\left[\left(\mu_{\mathrm{LOS}}^1\right)^{-1}, \ldots, \left(\mu_{\mathrm{LOS}}^N\right)^{-1}\right]$. The matrix $\mathbf{A}_{\mathrm{LOS}}$ is a $N_{\mathrm{LOS}} \times 2N_{\mathrm{do}}$ matrix defined by

$$\mathbf{A}_{\mathrm{LOS}} = \left[\mathbf{A}_{\mathrm{LOS}}^+, \mathbf{A}_{\mathrm{LOS}}^-\right]. \tag{4.92}$$

We sought a factorization of the augmented matrix of the form

$$\mathbf{A}_{\mathrm{a}} = \mathbf{X}_{\mathrm{a}} \mathbf{\Lambda}_{\mathrm{a}} \mathbf{Y}_{\mathrm{a}}^T, \tag{4.93}$$

with

$$\mathbf{X}_{\mathrm{a}} = \begin{bmatrix} \mathbf{T}\hat{\mathbf{X}} & 0 \\ \hat{\mathbf{X}}_{\mathrm{LOS}} & \mathbf{E} \end{bmatrix}, \tag{4.94}$$

$$\mathbf{\Lambda}_{\mathrm{a}} = \begin{bmatrix} \mathbf{\Lambda} & 0 \\ 0 & \Sigma \end{bmatrix}, \tag{4.95}$$

and

$$\mathbf{Y}_{\mathrm{a}}^T = \begin{bmatrix} \hat{\mathbf{Y}}^T \mathbf{T}^{-1} & 0 \\ \hat{\mathbf{Y}}_{\mathrm{LOS}} & \mathbf{E} \end{bmatrix}. \tag{4.96}$$

Thus, the eigenvalues corresponding to the false discrete ordinates are $\left(\mu_{\mathrm{LOS}}^k\right)^{-1}$, and the unknown $N_{\mathrm{LOS}} \times 2N_{\mathrm{do}}$ matrices $\hat{\mathbf{X}}_{\mathrm{LOS}}$ and $\hat{\mathbf{Y}}_{\mathrm{LOS}}$ have to be determined. Imposing that $\mathbf{X}_{\mathrm{a}} \mathbf{Y}_{\mathrm{a}}^T = \mathbf{I}$, we find that

$$\hat{\mathbf{Y}}_{\mathrm{LOS}} = -\hat{\mathbf{X}}_{\mathrm{LOS}} \hat{\mathbf{Y}}^T \mathbf{T}^{-1}. \tag{4.97}$$

From Eqs. (4.93)–(4.96), we obtain

$$\mathbf{A}_{\mathrm{a}} = \begin{bmatrix} \mathbf{A} & 0 \\ \hat{\mathbf{X}}_{\mathrm{LOS}} \mathbf{\Lambda} \hat{\mathbf{Y}}^T \mathbf{T}^{-1} + \Sigma \hat{\mathbf{Y}}_{\mathrm{LOS}} & \Sigma \end{bmatrix}, \tag{4.98}$$

and Eq. (4.90) gives

$$\hat{\mathbf{X}}_{\mathrm{LOS}} \mathbf{\Lambda} \hat{\mathbf{Y}}^T \mathbf{T}^{-1} + \Sigma \hat{\mathbf{Y}}_{\mathrm{LOS}} = \mathbf{A}_{\mathrm{LOS}}. \tag{4.99}$$

Denoting by $\hat{\mathbf{x}}_{\mathrm{LOS}}^k$ and $\mathbf{a}_{\mathrm{LOS}}^k$ the kth column vectors of $\hat{\mathbf{X}}_{\mathrm{LOS}}^T$ and $\mathbf{A}_{\mathrm{LOS}}^T$, respectively, and by making use of Eq. (4.97), we are led to

$$\hat{\mathbf{x}}_{\mathrm{LOS}}^k = (\mathbf{\Lambda} - \sigma_k \mathbf{E})^{-1} \hat{\mathbf{X}}^T \mathbf{T} \mathbf{a}_{\mathrm{LOS}}^k, \, k = 1, \ldots, N_{\mathrm{LOS}}. \tag{4.100}$$

Obviously, the vector $\widehat{\mathbf{x}}_{\mathrm{LOS}}^k$ is well defined if $\lambda_l \neq \left|\left(\mu_{\mathrm{LOS}}^k\right)^{-1}\right|$, for all $l = 1, \ldots, N_{\mathrm{do}}$ and all $k = 1, \ldots, N_{\mathrm{LOS}}$.

Thus, for a set of false discrete ordinates, the computation process consists in the solution of an $N_{\mathrm{do}} \times N_{\mathrm{do}}$ eigenvalue problem, and the calculation of the matrices $\widehat{\mathbf{X}}_{LOS}$ and $\widehat{\mathbf{Y}}_{\mathrm{LOS}}$ by using Eqs. (4.100) and (4.97), respectively. Alternatively, we may introduce the conjugate discrete ordinates μ_{LOS}^k and $-\mu_{\mathrm{LOS}}^k$, and solve an $(N_{\mathrm{do}} + N_{\mathrm{LOS}}) \times (N_{\mathrm{do}} + N_{\mathrm{LOS}})$ eigenvalue problem. However, in this case, $\mathbf{W}^{-1/2}$ is singular because $w_{\mathrm{LOS}}^k = 0$. This problem can be solved in practice by setting w_{LOS}^k to a very small value ε, e.g., $\varepsilon = 10^{-7}$, so that $w_l \gg \varepsilon$ for all $l = 1, \ldots, N_{\mathrm{do}}$.

In [54] it is shown that although the source function integration technique and the method of dummy discrete ordinates provide the same results, the latter is faster. In addition we also note that dummy discrete ordinate approach does not require the integration across the layer, and thus, the radiance field inside the layer is not required.

4.8   Matrix Operator Method

One can also compute the scattered radiance by the matrix operator method. The central feature of the matrix operator method is the reflection and transmission matrices (see Sect. 3.22). Transforming the solution representations of the matrix exponential method into a form which resembles the interaction principle formulation [55], equivalent expressions for the reflection and transmission matrices can be obtained.

Considering the processes in the lth layer (as shown in Fig. 4.8), we obtain:

$$\mathbf{i}_l^\uparrow = \mathbf{R}_l^\uparrow \mathbf{i}_l^\downarrow + \mathbf{T}_l^\uparrow \mathbf{i}_{l+1}^\uparrow + \mathbf{P}_l^\uparrow \tag{4.101}$$

and

$$\mathbf{i}_{l+1}^\downarrow = \mathbf{T}_l^\downarrow \mathbf{i}_l^\downarrow + \mathbf{R}_l^\downarrow \mathbf{i}_{l+1}^\uparrow + \mathbf{P}_l^\downarrow. \tag{4.102}$$

These two equations can be written as

Fig. 4.8 Illustration of the matrix operator method

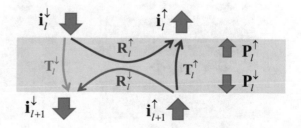

$$\begin{bmatrix} \mathbf{i}_l^\uparrow \\ \mathbf{i}_{l+1}^\downarrow \end{bmatrix} = \begin{bmatrix} \mathbf{R}_l^\uparrow & \mathbf{T}_l^\uparrow \\ \mathbf{T}_l^\downarrow & \mathbf{R}_l^\downarrow \end{bmatrix} \begin{bmatrix} \mathbf{i}_l^\downarrow \\ \mathbf{i}_{l+1}^\uparrow \end{bmatrix} + \begin{bmatrix} \mathbf{P}_l^\uparrow \\ \mathbf{P}_l^\downarrow \end{bmatrix}. \tag{4.103}$$

The reflection and transmission matrices for layer l can be obtained by first setting

$$\mathbf{V}_l^{-1} = \begin{bmatrix} \mathbf{V}_l^{11} & \mathbf{V}_l^{12} \\ \mathbf{V}_l^{22} & \mathbf{V}_l^{22} \end{bmatrix}, \tag{4.104}$$

and then, by rearranging entries in Eq. (4.70) to the form of Eq. (4.103). The result reads as follows:

$$\begin{bmatrix} \mathbf{R}_l^\uparrow & \mathbf{T}_l^\uparrow \\ \mathbf{T}_l^\downarrow & \mathbf{R}_l^\downarrow \end{bmatrix} = - \begin{bmatrix} \Gamma\mathbf{V}_l^{11} & -\mathbf{V}_l^{12} \\ \mathbf{V}_l^{21} & -\Gamma\mathbf{V}_l^{22} \end{bmatrix}^{-1} \begin{bmatrix} \Gamma\mathbf{V}_l^{12} & -\mathbf{V}_l^{11} \\ \mathbf{V}_l^{22} & -\Gamma\mathbf{V}_l^{21} \end{bmatrix} \tag{4.105}$$

and

$$\begin{bmatrix} \mathbf{P}_l^+ \\ \mathbf{P}_l^- \end{bmatrix} = \begin{bmatrix} \Gamma\mathbf{V}_l^{11} & -\mathbf{V}_l^{12} \\ \mathbf{V}_l^{21} & -\Gamma\mathbf{V}_l^{22} \end{bmatrix}^{-1} \begin{bmatrix} \mathbf{Q}_l^+ \\ \mathbf{Q}_l^- \end{bmatrix}. \tag{4.106}$$

To derive formulas for the adding algorithm, we consider the layer equation for layer $l+1$:

$$\begin{bmatrix} \mathbf{i}_{l+1}^\uparrow \\ \mathbf{i}_{l+2}^\downarrow \end{bmatrix} = \begin{bmatrix} \mathbf{R}_{l+1}^\uparrow & \mathbf{T}_{l+1}^\uparrow \\ \mathbf{T}_{l+1}^\downarrow & \mathbf{R}_{l+1}^\downarrow \end{bmatrix} \begin{bmatrix} \mathbf{i}_{l+1}^\downarrow \\ \mathbf{i}_{l+2}^\uparrow \end{bmatrix} + \begin{bmatrix} \mathbf{P}_{l+1}^\uparrow \\ \mathbf{P}_{l+1}^\downarrow \end{bmatrix}. \tag{4.107}$$

Using Eq. (4.103), we exclude $i_{l+1}^\downarrow$ and $i_{l+1}^\uparrow$ from Eq. (4.107) and obtain the equation for the larger layer encapsulating layers l and $l+1$, as shown in Fig. 4.9:

$$\begin{bmatrix} \mathbf{i}_l^+ \\ \mathbf{i}_{l+2}^- \end{bmatrix} = \begin{bmatrix} \mathbf{R}_{(l,l+1)}^+ & \mathbf{T}_{(l,l+1)}^+ \\ \mathbf{T}_{(l,l+1)}^- & \mathbf{R}_{(l,l+1)}^- \end{bmatrix} \begin{bmatrix} \mathbf{i}_l^- \\ \mathbf{i}_{l+2}^+ \end{bmatrix} + \begin{bmatrix} \mathbf{P}_{(l,l+1)}^+ \\ \mathbf{P}_{(l,l+1)}^- \end{bmatrix}, \tag{4.108}$$

where

$$\mathbf{R}_{(l,l+1)}^\uparrow = \mathbf{R}_l^\uparrow + \mathbf{T}_l^\uparrow \left(\mathbf{E} - \mathbf{R}_{l+1}^\uparrow \mathbf{R}_l^\downarrow \right)^{-1} \mathbf{R}_{l+1}^\uparrow \mathbf{T}_l^\downarrow, \tag{4.109}$$

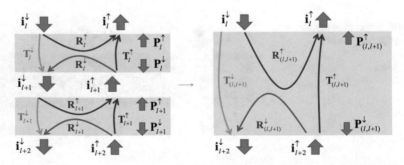

Fig. 4.9 Illustration of the adding algorithm in the framework of the matrix operator method

$$\mathbf{R}^{\downarrow}_{(l,l+1)} = \mathbf{R}^{\downarrow}_{l+1} + \mathbf{T}^{\downarrow}_{l+1} \left(\mathbf{E} - \mathbf{R}^{\downarrow}_l \mathbf{R}^{\uparrow}_{l+1}\right)^{-1} \mathbf{R}^{\downarrow}_l \mathbf{T}^{\uparrow}_{l+1}, \tag{4.110}$$

$$\mathbf{T}^{\uparrow}_{(l,l+1)} = \mathbf{T}^{\uparrow}_l \left(\mathbf{E} - \mathbf{R}^{\uparrow}_{l+1} \mathbf{R}^{\downarrow}_l\right)^{-1} \mathbf{T}^{\uparrow}_{l+1}, \tag{4.111}$$

$$\mathbf{T}^{\downarrow}_{(l,l+1)} = \mathbf{T}^{\downarrow}_{l+1} \left(\mathbf{E} - \mathbf{R}^{\downarrow}_l \mathbf{R}^{\uparrow}_{l+1}\right)^{-1} \mathbf{T}^{\downarrow}_l, \tag{4.112}$$

$$\mathbf{P}^{\uparrow}_{(l,l+1)} = \mathbf{P}^{\uparrow}_l + \mathbf{T}^{\uparrow}_l \left(\mathbf{E} - \mathbf{R}^{\uparrow}_{l+1} \mathbf{R}^{\downarrow}_l\right)^{-1} \left(\mathbf{R}^{\uparrow}_{l+1} \mathbf{P}^{\downarrow}_l + \mathbf{P}^{\uparrow}_{l+1}\right), \tag{4.113}$$

$$\mathbf{P}^{\downarrow}_{(l,l+1)} = \mathbf{P}^{\downarrow}_{l+1} + \mathbf{T}^{\downarrow}_{l+1} \left(\mathbf{E} - \mathbf{R}^{\downarrow}_l \mathbf{R}^{\uparrow}_{l+1}\right)^{-1} \left(\mathbf{R}^{\downarrow}_l \mathbf{P}^{\uparrow}_{l+1} + \mathbf{P}^{\downarrow}_l\right), \tag{4.114}$$

while lower indices in brackets refer to encapsulated layers.

Equations (4.109)–(4.114) are the essence of the adding algorithm. In the matrix operator method and for a multi layered atmosphere, the reflection matrix of the entire atmosphere is computed recursively from the reflection and transmission matrices of each layer by using the adding algorithm. Starting from the top of the atmosphere, we use the interaction principle to the first and second layers to create a thicker single layer with effective reflection and transmission matrices and source functions using Eqs. (4.109)–(4.114). Such a layer is called "a stack". The procedure is then repeated until the last layer is added to the stack. The application of the boundary conditions gives the radiance at the top of the atmosphere. The advantage of this approach is that the dimension of the matrices that have to be inverted is $N_{\text{do}} \times N_{\text{do}}$ and does not depend on the number of layers.

4.9 The Spherical Harmonics Method

Instead of using the Gaussian–Legendre quadrature and azimuthal expansion to factor out φ-dependency, it is possible to consider the spherical harmonics expansion for L:

$$L_m(\tau, \mu) = \sum_{n=m}^{N_{\max}} L_n^m(\tau) \bar{P}_n^m(\mu). \tag{4.115}$$

Substituting Eq. (4.115) in Eq. (4.40) we obtain:

$$J_{\text{ms},m}(\tau, \mu) = \frac{\omega(\tau)}{2} \int_{-1}^{1} \sum_{n=m}^{N_{\max}} \chi_n(\tau) \bar{P}_n^m(\mu) \bar{P}_n^m(\mu') \sum_{n=m}^{N_{\max}} L_n^m(\tau) \bar{P}_n^m(\mu') \, d\mu'. \tag{4.116}$$

Switching the order of summation and integration and using the orthogonality property of Legendre polynomials, we have

$$J_{\text{ms},m}(\tau, \mu) = \frac{\omega(\tau)}{2} \sum_{n=m}^{N_{\text{max}}} \chi_n(\tau) L_n^m(\tau) \bar{P}_n^m(\mu).$$ (4.117)

The single-scattering term has the same expansion kernel (Eqs. (4.37) and (4.39)). However, the problem appears with the differential term of Eq. (4.38). To expand $\mu L_m(\tau, \mu)$ in terms of Legendre polynomials we make use of the following property:

$$\mu \bar{P}_n^m(\mu) = \frac{(n+1)\bar{P}_{n+1}^m(\mu)}{\sqrt{2n+1}\sqrt{2n+3}} \sqrt{\frac{n+m+1}{n-m+1}} + \frac{n\bar{P}_{n-1}^m(\mu)}{\sqrt{2n+1}\sqrt{2n-1}} \sqrt{\frac{n-m}{n+m}}.$$ (4.118)

Thus,

$$\mu \frac{dL_m(\tau, \mu)}{d\tau} = \sum_{n=m}^{N_{\text{max}}} \left\{ \frac{(n+1)\bar{P}_{n+1}^m(\mu)}{\sqrt{2n+1}\sqrt{2n+3}} \sqrt{\frac{n+m+1}{n-m+1}} \right.$$ (4.119)
$$\left. + \frac{n\bar{P}_{n-1}^m(\mu)}{\sqrt{2n+1}\sqrt{2n-1}} \sqrt{\frac{n-m)}{n+m)}} \right\} \frac{dL_n^m(\tau)}{d\tau}.$$

Here, shifting indices of summation we obtain

$$\mu \frac{dL_m(\tau, \mu)}{d\tau} = \sum_{n=m}^{N_{\text{max}}} \left\{ \frac{n\bar{P}_n^m(\mu)}{\sqrt{2n-1}\sqrt{2n+1}} \sqrt{\frac{n+m}{n-m}} \frac{dL_{n-1}^m(\tau)}{d\tau} \right.$$ (4.120)
$$\left. + \frac{(n+1)\bar{P}_n^m(\mu)}{\sqrt{2n+1}\sqrt{2n+3}} \sqrt{\frac{n-m+1}{n+m+1}} \frac{dL_{n+1}^m(\tau)}{d\tau} \right\}.$$

Substitution of Eqs. (4.39), (4.117) and (4.120) into Eq. (4.38) leads to an infinite system of equations for $L_n^m(\tau)$. The easiest way to close the system is to assume that

$$L_n^m(\tau) = 0 \text{ for } n > N_{\text{max}}.$$ (4.121)

For instance, in the Eddington approximation, only two expansion terms are considered.

It can be shown that the spherical harmonics method and the discrete ordinate method are connected. Indeed, expanding the diffuse radiance L in terms of spherical harmonics, and integrating the multiple scattering term with respect to the azimuthal angle, we are led to an integral of the form $\int_{-1}^{1} \bar{P}_n^m(\mu') \bar{P}_{n'}^m(\mu') d\mu'$ (see Eq. (4.116)), with $n, n' = 0, \ldots, N_{\text{max}}$. For $n, n' \leq N_{\text{max}}$, $P_n^m(\mu')P_{n'}^m(\mu') = f(\mu')$ is a polynomial of degree at most $2N_{\text{max}}$. The integral reduces to $\int_{-1}^{1} f(\mu')d\mu'$, and the task is to find an exact quadrature for this integral. If this is done, the mathematical equivalence between the spherical harmonics and the discrete ordinate method is established. In general, a Gauss quadrature using N_q nodes is an exact quadrature for polynomials of degree $2N_q - 1$ or less [56]. In our case, this condition translates into $N_{max} < N_q - 1/2$, and for the choice $N_q = 2N_{do}$, it follows that $N_{max} < 2N_{do} - 1/2$. Thus, using $2N_{do}$ Gaussian nodes and weights, we need $N_{max} = 2N_{do} - 1$ expansion terms in Eq. (4.35) and $M_{max} \leq 2N_{do} - 1$ azimuthal modes in Eqs. (4.34) and (4.35).

4.10 Radiative Transfer in the Coupled Atmosphere-Underlying Surface System

Solar light as detected by the ground, airborne or satellite remote sensing instruments undergoes scattering, both in the atmosphere and the underlying surface. Therefore, the atmosphere and the underlying surface must be considered as a coupled system in the formulation of the atmospheric radiative transfer problem. This fact motivated the development of the so-called coupled atmosphere-underlying surface (e.g., ocean, ice) radiative transfer models. While in some radiative transfer models, the influence of the underlying surface can be described with the appropriate BRDF, in the full-coupled models the radiative transfer process in the atmosphere and underlying surface is treated consistently as one coupled system. In this section, we consider the coupled atmosphere—underlying surface problem, assuming that underlying surface is ocean, which covers more than 70% of the terrestrial surface.

Tanaka and Nakajima [57] applied the matrix operator method the atmosphere–ocean combined system with the flat ocean surface. Later, Nakajima and Tanaka [58] applied the matrix operator method to the atmosphere–ocean combined system with the rough surface based on the Cox-Munk model. The discrete ordinate method was applied by Jin and Stamnes [59] to solve the radiative transfer problem for a system consisting of two strata with different indices of refraction; the flat interface between the atmosphere and the ocean is considered; in [60] the same formalism was adopted for the coupled system with a rough water surface described by the Cox and Munk model. Finally, in [61] a coupled model for polarized radiative transfer was developed.

The radiative transfer model for coupled atmosphere–ocean systems is similar to that for the atmosphere. The latter can be generalized to the coupled model by adjusting the source function, the quadrature scheme and the boundary conditions. These three features are outlined below.

4.10.1 Source Function

In the case of a flat air-ocean interface, the direct solar beam is reflected and refracted by the flat surface according to the Snell's law, as shown in Fig. 4.10 (our consideration can also be applied to other problems such as radiative transfer in ice sheets

Fig. 4.10 Coupled atmosphere–ocean system with the flat interface

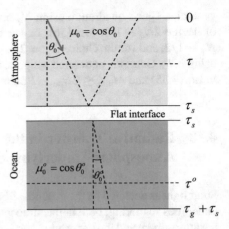

and ice floating on the surface of ocean). Consequently, in the atmosphere the single-scattering source function has two terms: one term is related to the attenuated incident beam, while the second term is due to the reflected beam ($0 < \tau < \tau_s$):

$$
\begin{aligned}
J_{ss}(\tau, \Omega) &= \frac{\omega(\tau)}{4\pi} F_\odot p(\tau, \mu_0, \mu, \varphi) \exp\left(-\frac{\tau}{\mu_0}\right) \\
&+ \frac{\omega(\tau)}{4\pi} F_\odot p(\tau, -\mu_0, \mu, \varphi) \exp\left(-\frac{\tau_s + (\tau_s - \tau)}{\mu_0}\right) R(\mu_0, n).
\end{aligned}
\tag{4.122}
$$

Here n is the refractive index of water with respect to air, and R is the Fresnel reflectance of the solar beam at the atmosphere–ocean interface. The "minus" sign at the argument μ_0 of $p(\tau, -\mu_0, \mu, \varphi)$ appears because the reflected beam propagates in the upward direction. In the ocean ($\tau_s < \tau^o < \tau_g$), the source term is the contribution of the refracted solar beam only and reads as

$$
J_{ss}(\tau^o, \Omega) = \frac{\omega(\tau^o)}{4\pi} F_\odot p(\tau, \mu_0^w, \mu, \varphi) \exp\left(-\frac{\tau_s}{\mu_0} - \frac{\tau^o}{\mu_0^w}\right) T(\mu_0, n), \tag{4.123}
$$

where μ_0^w is the solar zenith angle cosine in water related to the atmospheric solar zenith angle cosine μ_0 by Snell's law:

$$
\mu_0^w = \sqrt{1 - \frac{1 - (\mu_0)^2}{n^2}} \tag{4.124}
$$

and T is the Fresnel transmittance of the solar beam at the atmosphere–ocean interface.

The Fresnel reflectance and transmittance matrices are given by

$$R(\mu^a, n) = \frac{1}{2}\left[\left(\frac{\mu^a - n\mu^o}{\mu^a + n\mu^o}\right)^2 + \left(\frac{\mu^o - n\mu^a}{\mu^o + n\mu^a}\right)^2\right] \tag{4.125}$$

and

$$T(\mu^a, n) = \frac{1}{2}\left[\left(\frac{1}{\mu^a + n\mu^o}\right)^2 + \left(\frac{1}{\mu^o + n\mu^a}\right)^2\right], \tag{4.126}$$

respectively, where μ^a is the cosine of the zenith angle for the radiance in the atmosphere, while

$$\mu^o = \sqrt{1 - \frac{1 - (\mu^a)^2}{n^2}} \tag{4.127}$$

is the cosine of the zenith angle for the radiance in the ocean. Here we don't use negative arguments for the functions $R(\ldots)$ and $T(\ldots)$. All cosines are assumed to be positive. The R- and T-functions have the following properties of reciprocity:

$$R(\mu^a, n) = R\left(\mu^o, n^{-1}\right) \tag{4.128}$$

and

$$T(\mu^a, n) = T\left(\mu^o, n^{-1}\right). \tag{4.129}$$

If there is a roughness at the atmosphere–ocean interface, then the specular reflected light is scattered at the interface and converted to the diffuse part of the radiance. Consequently, the source function in the atmosphere is exactly the same as in the case of the system with the atmosphere only, while the source function in the ocean turns to zero.

4.10.2 Quadrature Scheme

Different numbers of discrete ordinates are used for the atmosphere and for the ocean ($2N_{do}^a$ and $2N_{do}^o$, respectively). The ocean comprises two regions of the angular domain, which are shown in Fig. 4.11. The rays from Region I interact directly with the atmosphere, while in Region II the total reflection occurs at the atmosphere–ocean interface for the rays moving in the upward direction. The demarcation between these regions occurs at the critical angle, for which the cosine is $\mu_c = \sqrt{1 - \frac{1}{n^2}}$. The discrete ordinates from Region I are related to those in the atmosphere by the Snell's law. Corresponding quadrature points and weights are given by [62]

$$\mu_{\pm(N_{do}^o - N_{do}^a + k)}^o = \sqrt{1 - \frac{1 - (\mu_{\pm k}^a)^2}{n^2}}, \quad k = 1, \ldots, N_{do}^a, \tag{4.130}$$

Fig. 4.11 Two regions in the ocean

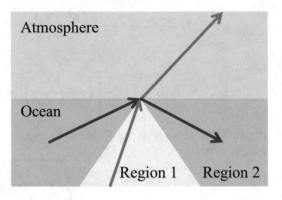

and

$$w^o_{\pm(N^o_{do}-N^a_{do}+k)} = w^a_{\pm k}\frac{\mu^a_{\pm k}}{n^2\mu^o_{\pm k}}, \quad k = 1, \dots, N^a_{do}, \tag{4.131}$$

respectively. Note that $\mu^o_k \in (\mu_c, 1)$ and $\mu^o_{-k} \in (-1, -\mu_c)$. The set $\{\mu^o_k, w^o_k\}$ for Region II is computed in two steps: in the first step, we define a temporary set $\{\hat{\mu}_{\pm k}, \hat{w}_{\pm k}\}_{k=1,\dots,N^o_{do}-N^a_{do}}$ in the angular cosine domain $[-1, 1]$, while in the second step we rescale it for the domain $[-\mu_c, \mu_c]$ by the following transformation:

$$\mu^o_{\pm k} = \hat{\mu}_{\pm k}\mu_c,$$

$$w^o_{\pm k} = \hat{w}_{\pm k}\mu_c.$$

4.10.3   Boundary Conditions for the Coupled Atmosphere–Ocean System

The set of layer equations and equation emphasizing the boundary conditions at the top of the atmosphere and at the bottom of the ocean should be accompanied with the conditions at the atmosphere–ocean interface. These conditions can be derived from the interaction principle written for the radiances at the boundaries:

$$\begin{bmatrix} L^+_{atm} \\ L^-_{ocn} \end{bmatrix} = \begin{bmatrix} R_{aa} & T_{ao} \\ T_{oa} & R_{oo} \end{bmatrix} \begin{bmatrix} L^-_{atm} \\ L^+_{ocn} \end{bmatrix}. \tag{4.132}$$

Here upper indices $+$ and $-$ correspond to upward and downward radiances, respectively, while lower indices "atm" and "ocn" correspond to the radiances in the atmosphere (at the upper boundary of the interface) and in the ocean (at the lower boundary). Note that condition (4.132) is written for the total radiances. In the flat water

surface case, reflection and transmission matrices in Eq. (4.132) are computed as
follows:

$$[R_{aa}]_{ij} = \begin{cases} R_m\,(\mu_i, n)\,, & i = j \\ 0, & i \neq j \end{cases}$$

$$[R_{oo}]_{ij} = \begin{cases} R_m\,(\mu_i, n)\,, & i = j \leq N_a \\ 1, & N_a \leq i = j \leq N_o \\ 0, & i \neq j \end{cases}$$

$$[T_{ao}]_{ij} = \begin{cases} T_m\,(\mu_i, n)\,/n^2, & i = j \\ 0, & i \neq j \end{cases}$$

$$[T_{oa}]_{ij} = \begin{cases} T_m\,(\mu_i, n)\,n^2, & i = j \\ 0, & i \neq j \end{cases}$$

Here $R_m\,(\mu_i, n)$ and $T_m\,(\mu_i, n)$ are computed using Eqs. (4.125) and (4.126) for
Fourier component $m = 0$, while for $m > 0$, $R_m = 0$ and $T_m = 1$ [63]. If the interface
is not flat (e.g., with wind-induced roughness), then the corresponding model (e.g.,
the Cox-Munk model) is used to compute corresponding reflection and transmission
matrices. Reviews and rigorous derivations of reflection and transmission matrices
from the electromagnetic scattering theory can be found in [64, 65]. The boundary
conditions at the atmosphere–ocean interface are included in the system of equations
[59, 60], similar to Eq. (4.80), or incorporated in the matrix operator method [61,
62].

4.11   Required Number of Discrete Ordinates

In principle, the radiative transfer equation can be solved by the discrete ordinate
method described above with a high accuracy. An important parameter controlling
the computational time and the accuracy of computations is the number of streams
in the polar hemisphere N_{do}. Radiative transfer models are called "multi-stream" if
$N_{do} \geq 2$. If $N_{do} = 1, 2$ or 3, we obtain the two-stream, four-stream and six-stream
models, respectively. For them, the eigenvalue decomposition (4.65) can be evaluated
analytically in a closed form [66]. Note that the implementations based on the closed
form two-stream solutions can be order of magnitude faster than the multistream
model with N_{do} set to 1 [29]. As the discrete ordinate method relies on the matrix
multiplication, matrix inversion and eigenvalue routines, the computational burden
is proportional to N_{do}^3, and the multistream models can be time-consuming.

The choice of the optimal value for N_{do} is not straight-forward. In [67] the required
N_{do} was estimated to achieve 0.1 % accuracy in radiances at the top of the atmosphere,
versus a reference calculation with 256 streams. It was shown that the optimal value
N_{do} strongly depends on the optical thickness τ of the atmosphere, namely N_{do} can
be smaller than 10 for $\tau < 0.1$. As τ increases, as many as 60 (depending on a given

observation direction and phase function) discrete ordinates might be required. The alternative way to determine N_{do} (or at least, to determine a lower constraint for N_{do}) is given in [68] and is based on the Whittaker–Nyquist–Kotelnikov–Shannon theorem [69]. In general, for a given atmospheric scenario, a convergence test is required to estimate N_{do}. In most cases, as a reference model, the multistream model with no more than $N_{do} = 128$ can be used. The accuracy also depends on the angular range of interest (e.g., rainbow and glory regions for water clouds might require more discrete ordinates).

4.12 Single-Scattering Approximation

For optically thin media, it can be assumed that the rays are scattered just once before they escape from the medium or absorbed. Following [70], it is sufficient to consider only single- and double-scattering events for the scattering optical thickness below 0.05. In fact, the accuracy of the single-scattering approximation depends on several factors, including geometry, the single-scattering albedo and the single-scattering phase function.

The solution in the single-scattering approximation can be derived from the radiative transfer equation by neglecting the multiple scattering source term. Then the radiative transfer equation turns into a differential equation:

$$\mu \frac{dL\left(\tau,\mu,\varphi\right)}{d\tau} = -L\left(\tau,\mu,\varphi\right) + \frac{\omega\left(\tau\right)}{4\pi} p\left(\tau,\mu_0,\mu,\varphi\right) e^{-\tau/\mu_0} \tag{4.133}$$

To integrate this equation, we multiply both sides by $e^{\tau/\mu}$. Then it follows:

$$e^{\tau/\mu} \mu \frac{dL\left(\tau,\mu,\varphi\right)}{d\tau} = -e^{\tau/\mu} L\left(\tau,\mu,\varphi\right) + \frac{\omega\left(\tau\right)}{4\pi} p\left(\tau,\mu_0,\mu,\varphi\right) e^{\tau\left(\frac{1}{\mu}-\frac{1}{\mu_0}\right)} \tag{4.134}$$

or

$$\frac{d\left[e^{\tau/\mu} L\left(\tau,\mu,\varphi\right)\right]}{d\left(\tau/\mu\right)} = \frac{\omega\left(\tau\right)}{4\pi} p\left(\tau,\mu_0,\mu,\varphi\right) e^{\tau\left(\frac{1}{\mu}-\frac{1}{\mu_0}\right)}. \tag{4.135}$$

Integrating Eq. (4.135) we obtain [71]:

$$L\left(\tau,\mu,\varphi\right) = \frac{1}{4\pi\mu} e^{-\tau/\mu} \int_a^\tau \omega\left(\tau'\right) p\left(\tau',\mu_0,\mu,\varphi\right) e^{\tau'\left(\frac{1}{\mu}-\frac{1}{\mu_0}\right)} d\tau'. \tag{4.136}$$

For a layer of the optical thickness τ_L with a black surface at the bottom, the boundary conditions read as follows:

$$\begin{cases} L\,(\tau = 0, \mu > 0, \varphi) = 0, \\ L\,(\tau = \tau_L, \mu < 0, \varphi) = 0. \end{cases} \tag{4.137}$$

The first boundary condition is applied to the downwelling radiance at the top while the second one is applied for the upwelling radiation at the bottom of the layer. Taking into account the boundary conditions, from Eq. (4.136) we obtain the following expression for the downwelling and upwelling radiances,

$$L\,(\tau, \mu > 0, \varphi) = \frac{1}{4\pi\mu} e^{-\tau/\mu} \int_0^\tau \omega\,(\tau')\,p\,(\tau', \mu_0, \mu, \varphi)\,e^{\tau'\left(\frac{1}{\mu} - \frac{1}{\mu_0}\right)} d\tau', \quad (4.138)$$

$$L\,(\tau, \mu < 0, \varphi) = \frac{1}{4\pi\mu} e^{-\tau/\mu} \int_{\tau_L}^\tau \omega\,(\tau')\,p\,(\tau', \mu_0, \mu, \varphi)\,e^{\tau'\left(\frac{1}{\mu} - \frac{1}{\mu_0}\right)} d\tau', \quad (4.139)$$

respectively. If the layer is homogeneous, then integration in Eqs. (4.138) and (4.139) gives

$$\begin{cases} L\,(\tau, \mu > 0, \varphi) = \dfrac{\omega F_\odot p\,(\mu_0, \mu, \varphi)}{4\pi} \dfrac{\mu_0}{\mu - \mu_0} \left[e^{-\frac{\tau}{\mu}} - e^{-\frac{\tau}{\mu_0}} \right] \\ L\,(\tau, \mu < 0, \varphi) = \dfrac{\omega F_\odot p\,(\mu_0, \mu, \varphi)}{4\pi} \dfrac{\mu_0}{\mu_0 + |\mu|} \left[e^{-\frac{\tau}{\mu_0}} - e^{-\frac{\tau}{|\mu|}} e^{-\tau_L\left(\frac{1}{\mu_0} + \frac{1}{|\mu|}\right)} \right], \end{cases} \tag{4.140}$$

The singular points $\mu = \mu_0$ can be handled by expanding the exponents in Taylor series. Indeed, expanding the exponents in Eq. (4.140) as

$$e^{-\frac{\tau}{\mu_0}} = 1 - \frac{\tau}{\mu_0} + \frac{\tau^2}{2\mu_0^2} - \frac{\tau^3}{3!\mu_0^3} + \cdots, \tag{4.141}$$

$$e^{-\frac{\tau}{\mu}} = 1 - \frac{\tau}{\mu} + \frac{\tau^2}{2\mu^2} - \frac{\tau^3}{3!\mu^3} + \cdots, \tag{4.142}$$

and using the fact that

$$\frac{1}{\mu_0^k} - \frac{1}{\mu^k} = \left(\frac{1}{\mu_0} - \frac{1}{\mu}\right)\left(\frac{1}{\mu_0^{k-1}} + \frac{1}{\mu_0^{k-2}\mu} + \cdots + \frac{1}{\mu_0\mu^{k-2}} + \frac{1}{\mu^{k-1}}\right) \tag{4.143}$$

we get

$$L\,(\tau, \mu > 0, \varphi) = \frac{\omega F_\odot p\,(\mu_0, \mu, \varphi)}{4\pi\mu} \sum_{k=1}^\infty \left(\frac{\tau^k}{k!} \sum_{i=0}^{k-1} \frac{1}{\mu^{k-1-i}\mu_0^i}\right). \tag{4.144}$$

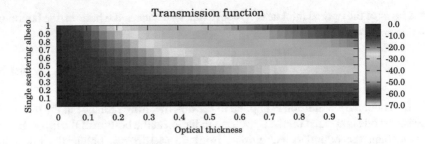

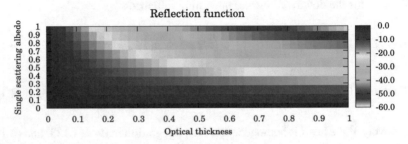

Fig. 4.12 Relative error in % of the single-scattering approximation in the case of the homogeneous layer

Recalling the definition of the transmission and reflection functions, we have

$$
\begin{cases}
T\left(\tau, \mu > 0, \varphi\right) = \dfrac{\omega p\left(\mu_0, \mu, \varphi\right)}{4} \dfrac{1}{\mu - \mu_0} \left[e^{-\frac{\tau}{\mu}} - e^{-\frac{\tau}{\mu_0}}\right], \\[2ex]
R\left(\tau, \mu < 0, \varphi\right) = \dfrac{\omega p\left(\mu_0, \mu, \varphi\right)}{4} \dfrac{1}{\mu_0 + |\mu|} \left[e^{-\frac{\tau}{\mu_0}} - e^{-\frac{\tau}{|\mu|}} e^{-\tau_L\left(\frac{1}{\mu_0} + \frac{1}{|\mu|}\right)}\right],
\end{cases}
\tag{4.145}
$$

Considering the Taylor expansion for exponents for $\tau \to 0$, Eq. (4.145) we get

$$
\begin{cases}
T\left(\tau_L, \mu > 0, \varphi\right) = \dfrac{\omega p\left(\mu_0, \mu, \varphi\right)}{4\mu\mu_0} \tau_L, \\[2ex]
R\left(0, \mu < 0, \varphi\right) = \dfrac{\omega p\left(\mu_0, \mu, \varphi\right)}{4 |\mu| \mu_0} \tau_L.
\end{cases}
\tag{4.146}
$$

As it has been pointed above, the single-scattering approximation can be used at small values of the optical thickness. Such a condition often occurs in the near infrared part of the electromagnetic spectrum for a cloudless atmosphere. The error of the single-scattering approximation increases with the optical thickness and the single-scattering albedo, as shown in Fig. 4.12.

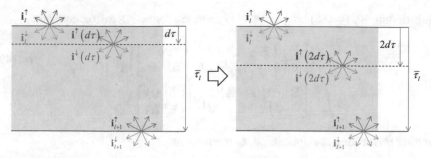

Fig. 4.13 Adding-doubling scheme

4.13 Adding-Doubling Method

Adding two identical homogeneous layers according to Eqs. (4.109)–(4.114) leads to the matrix formulation of the adding-doubling method [72, 73]. By using the adding scheme, the reflection and transmission matrices, as well as the source functions can be computed for a layer having the same composition but twice the optical thickness [74], i.e., $R_l^{\uparrow\downarrow}(2d\tau)$, $T_l^{\uparrow\downarrow}(2d\tau)$ and $P_l^{\uparrow\downarrow}(2d\tau)$, as shown in Fig. 4.13. By using the interaction principle as the recursion relations, reflection and transmission matrices for a layer with the optical thickness $d\tau$ can be transformed into those for the layer with the optical thickness $2^s d\tau$ in s iterations.

To generate starting values for the reflection and transmission matrices, we take into account that the starting optical thickness is small and thus, the computations of the matrix exponential using the eigenvalue decomposition method can be avoided. The reflection and transmission matrices for thin layers can be computed by using the infinitesimal generator initialization scheme [55], the so-called diamond initialization, Waterman initialization and others [75]. These methods are outlined in [76, 77].

The basic recursion relations of the doubling method can be obtained by using the matrix exponential formalism. The initialization matrices can be derived by approximating the matrix exponential in the layer Eq. (4.62) by the Taylor series. Indeed, in the lth layer let us choose the infinitely thin sublayer with the optical thickness $d\tau$, as shown in Fig. 4.13. Equation (4.62) for this sublayer reads as follows:

$$-\begin{bmatrix} \mathbf{i}_l^\uparrow \\ \mathbf{i}_l^\downarrow \end{bmatrix} + e^{\mathbf{A}_l d\tau}\begin{bmatrix} \mathbf{i}^\uparrow \\ \mathbf{i}^\downarrow \end{bmatrix} = \int_0^{d\tau} e^{\mathbf{A}_l t}\begin{bmatrix} \mathbf{b}_l^\uparrow(t) \\ \mathbf{b}_l^\downarrow(t) \end{bmatrix} dt. \tag{4.147}$$

Vectors $\mathbf{i}^{\uparrow\downarrow}$ refer to the radiances at the bottom of this layer. Expressing the exponent by the Taylor series up to the first order and taking into account Eq. (4.54), we obtain the following formula:

$$e^{\mathbf{A}_l d\tau} \approx \mathbf{E} + \mathbf{A}_l d\tau = \begin{bmatrix} \mathbf{E} + \mathbf{A}_l^{11} d\tau & \mathbf{A}_l^{12} d\tau \\ -\mathbf{A}_l^{12} d\tau & \mathbf{E} - \mathbf{A}_l^{11} d\tau \end{bmatrix}. \tag{4.148}$$

Substituting Eq. (4.148) into Eq. (4.147) gives the following matrix equation

$$
-\begin{bmatrix} \mathbf{i}_l^{\uparrow} \\ \mathbf{i}_l^{\downarrow} \end{bmatrix} + \begin{bmatrix} \mathbf{E} + \mathbf{A}_l^{11} d\tau & \mathbf{A}_l^{12} d\tau \\ -\mathbf{A}_l^{12} d\tau & \mathbf{E} - \mathbf{A}_l^{11} d\tau \end{bmatrix} \begin{bmatrix} \mathbf{i}^{\uparrow} \\ \mathbf{i}^{\downarrow} \end{bmatrix} =
$$

$$
\int_0^{d\tau} \begin{bmatrix} \mathbf{E} + \mathbf{A}_l^{11} t & \mathbf{A}_l^{12} t \\ -\mathbf{A}_l^{12} t & \mathbf{E} - \mathbf{A}_l^{11} t \end{bmatrix} \begin{bmatrix} \mathbf{b}_l^{\uparrow}(t) \\ \mathbf{b}_l^{\downarrow}(t) \end{bmatrix} dt,
$$

(4.149)

or after performing multiplications, two equations:

$$
-\mathbf{i}_l^{\uparrow} + \left(\mathbf{E} + \mathbf{A}_l^{11} d\tau\right) \mathbf{i}^{\uparrow} + \mathbf{A}_l^{12} d\tau \mathbf{i}^{\downarrow} = \mathbf{Y}_1
$$

(4.150)

and

$$
-\mathbf{i}_l^{\downarrow} - \mathbf{A}_l^{12} d\tau \mathbf{i}^{\uparrow} + \left(\mathbf{E} - \mathbf{A}_l^{11} d\tau\right) i^{\downarrow} = \mathbf{Y}_2,
$$

(4.151)

where

$$
\mathbf{Y}_1 = \int_0^{d\tau} \left[\left(\mathbf{E} + \mathbf{A}_l^{11} t\right) \mathbf{b}_l^{\uparrow}(t) + \mathbf{A}_l^{12} t \mathbf{b}_l^{\downarrow}(t) \right] dt
$$

(4.152)

and

$$
\mathbf{Y}_2 = \int_0^{d\tau} \left[-\mathbf{A}_l^{12} t \mathbf{b}_l^{\uparrow}(t) + \left(\mathbf{E} - \mathbf{A}_l^{11} t\right) \mathbf{b}_l^{\downarrow}(t) \right] dt.
$$

(4.153)

Then, we express $\mathbf{i}^{\downarrow}$ from Eq. (4.151):

$$
i^{\downarrow} = \left(\mathbf{E} - \mathbf{A}_l^{11} d\tau\right)^{-1} \mathbf{i}_l^{\downarrow} + \left(\mathbf{E} - \mathbf{A}_l^{11} d\tau\right)^{-1} \mathbf{A}_l^{12} d\tau \mathbf{i}^{\uparrow} + \left(\mathbf{E} - \mathbf{A}_l^{11} d\tau\right)^{-1} \mathbf{Y}_2.
$$

(4.154)

Substituting Eq. (4.154) into Eq. (4.150) leads us to the following equation:

$$
-\mathbf{i}_l^{\uparrow} + \left(\mathbf{E} + \mathbf{A}_l^{11} d\tau\right) \mathbf{i}^{\uparrow} + \mathbf{A}_l^{12} d\tau \left(\mathbf{E} - \mathbf{A}_l^{11} d\tau\right)^{-1} \mathbf{i}_l^{\downarrow}
$$
$$
+\mathbf{A}_l^{12} d\tau \left(\mathbf{E} - \mathbf{A}_l^{11} d\tau\right)^{-1} \mathbf{A}_l^{12} d\tau \mathbf{i}^{\uparrow} + \mathbf{A}_l^{12} d\tau \left(\mathbf{E} - \mathbf{A}_l^{11} d\tau\right)^{-1} \mathbf{Y}_2 = \mathbf{Y}_1.
$$

(4.155)

Comparing Eq. (4.155) with Eq. (4.101) and Eq. (4.154) with Eq. (4.102) we conclude:

$$
\mathbf{R}_l^{\uparrow}(d\tau) = \mathbf{A}_l^{12} d\tau \left(\mathbf{E} - \mathbf{A}_l^{11} d\tau\right)^{-1},
$$

(4.156)

$$
\mathbf{T}_l^{\uparrow}(d\tau) = \mathbf{E} + \mathbf{A}_l^{11} d\tau + \mathbf{A}_l^{12} d\tau \left(\mathbf{E} - \mathbf{A}_l^{11} d\tau\right)^{-1} \mathbf{A}_l^{12} d\tau,
$$

(4.157)

$$
\mathbf{P}_l^{\uparrow}(d\tau) = \mathbf{A}_l^{12} d\tau \left(\mathbf{E} - \mathbf{A}_l^{11} d\tau\right)^{-1} \mathbf{Y}_2 - \mathbf{Y}_1,
$$

(4.158)

$$
\mathbf{R}_l^{\downarrow}(d\tau) = \left(\mathbf{E} - \mathbf{A}_l^{11} d\tau\right)^{-1} \mathbf{A}_l^{12} d\tau,
$$

(4.159)

$$
\mathbf{T}_l^{\downarrow}(d\tau) = \left(\mathbf{E} - \mathbf{A}_l^{11} d\tau\right)^{-1},
$$

(4.160)

$$\mathbf{P}_l^{\downarrow}(d\tau) = \left(\mathbf{E} - \mathbf{A}_l^{11} d\tau\right)^{-1} \mathbf{Y}_2. \tag{4.161}$$

Now the key point in our derivation is to neglect quadratic and higher-order terms of $d\tau$ in Eqs. (4.156)–(4.161). We also note that

$$\left(\mathbf{E} - \mathbf{A}_l^{11} d\tau\right)^{-1} \approx \mathbf{E} + \mathbf{A}_l^{11} d\tau. \tag{4.162}$$

The integrals in Eqs. (4.152) and (4.153) are simplified by considering the Taylor expansion of $\mathbf{b}_l^{\uparrow\downarrow}(t)$ around $t = 0$ (the case of linear variation of the thermal source function was considered in [78]):

$$\mathbf{Y}_1 = \mathbf{b}_l^{\uparrow}(0)\,d\tau \tag{4.163}$$

and

$$\mathbf{Y}_2 = \mathbf{b}_l^{\downarrow}(0)\,d\tau. \tag{4.164}$$

The final result reads as follows:

$$\mathbf{R}_l^{\uparrow}(d\tau) = \mathbf{A}_l^{12} d\tau, \tag{4.165}$$

$$\mathbf{T}_l^{\uparrow}(d\tau) = \mathbf{E} + \mathbf{A}_l^{11} d\tau, \tag{4.166}$$

$$\mathbf{P}_l^{\uparrow}(d\tau) = -\mathbf{b}_l^{\uparrow}(0)\,d\tau, \tag{4.167}$$

$$\mathbf{R}_l^{\downarrow}(d\tau) = \mathbf{A}_l^{12} d\tau, \tag{4.168}$$

$$\mathbf{T}_l^{\downarrow}(d\tau) = \mathbf{E} + \mathbf{A}_l^{11} d\tau, \tag{4.169}$$

$$\mathbf{P}_l^{\downarrow}(d\tau) = b_l^{\downarrow}(0)\,d\tau. \tag{4.170}$$

Note that expressing the matrix exponential up to the second and third order, we obtain the diamond initialization scheme of Wiscombe [76] and Waterman [42], respectively.

4.14 Successive Orders of Scattering

The diffuse radiance can be expressed as follows [79]:

$$L = \sum_{s=1}^{N_s} L_s, \tag{4.171}$$

where L_s refers to the s-fold scattered radiance, while N_s is the maximum scattering order considered. The expansion (4.171) is called Neumann series, and the computational technique based on Eq. (4.171) is called the successive orders of scattering (SOS) model. Actually, SOS can be considered as an "exact" method in the sense that it converges to a "true" solution. However, if the atmosphere is transparent, the Neumann series converges slowly.

The computations of L_s can be organized iteratively. We set the multiple scattering source function

$$J_0 (\tau, \mu, \varphi) = 0 \tag{4.172}$$

and

$$L_s (\tau, \mu, \varphi) = F_\odot e^{-\tau/\mu_0} \delta (\mu_0 - \mu) \delta (\varphi). \tag{4.173}$$

The multiple scattering source function at the next iteration is given by

$$J_{s+1} = \frac{\omega (\tau)}{4\pi} \int_0^{2\pi} \int_{-1}^1 L_s (\tau, \mu', \varphi') p (\tau, \mu', \mu, \varphi - \varphi') d\mu' d\varphi'. \tag{4.174}$$

Then the radiance can be computed from (see Eq. (4.12))

$$\mu \frac{dL_s (\tau, \mu, \varphi)}{d\tau} = -L_s (\tau, \mu, \varphi) + J_s (\tau, \mu, \varphi). \tag{4.175}$$

Taking into account Eqs. (4.13) and (4.14), we have for the diffuse downwelling radiation

$$L_s (\tau, \mu > 0, \varphi) = \int_0^\tau J_s (\tau', \mu, \varphi) e^{-\frac{\tau - \tau'}{\mu}} \frac{d\tau'}{\mu} \tag{4.176}$$

and the upwelling radiation

$$L_s (\tau, \mu < 0, \varphi) = L_s (\tau_L, \mu < 0, \varphi) e^{-\frac{\tau_L - \tau}{|\mu|}} + \int_\tau^{\tau_L} J_s (\tau', \mu, \varphi) e^{-\frac{\tau' - \tau}{|\mu|}} \frac{d\tau'}{|\mu|}. \tag{4.177}$$

Note that $L_s (\tau_L, \mu < 0, \varphi)$ refers to the radiance reflected by the surface. Considering one single reflection by the surface as a single-scattering, we have

$$L_1 (\tau_L, \mu < 0, \varphi) = R (\mu, \mu_0, \varphi) F_\odot e^{-\tau_L/\mu_0} \mu_0/\pi, \tag{4.178}$$

and

$$L_{s+1} (\tau_L, \mu < 0, \varphi) = \frac{1}{\pi} \int_0^{2\pi} \int_0^1 \mu' R (\mu', \mu, \varphi - \varphi') L_{s+1} (\tau_L, \mu', \varphi') d\mu' d\varphi, \tag{4.179}$$

where R is the surface reflection matrix.

In particular, for $s = 1$, we have

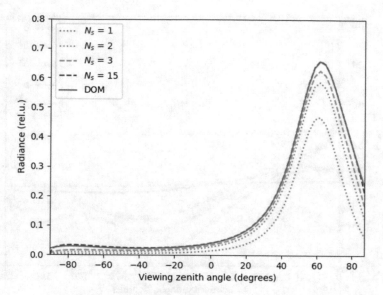

Fig. 4.14 Transmitted radiance computed by the discrete ordinate method (DOM) and successive orders of scattering taking into account 1, 2, 3 and 15 orders of scattering. The layer optical thickness is 0.8, the single-scattering albedo is 0.99, the asymmetry parameter of the Henyey–Greenstein phase function is 0.7, while the angle of incidence is 60°

$$J_1\left(\tau, \mu, \varphi\right) = \frac{\omega}{4\pi} F_\odot p\left(\mu_0, \mu, \varphi\right) e^{-\tau/\mu_0}. \tag{4.180}$$

The single scattered radiance L_1 is found using Eq. (4.180) in Eq. (4.175). Obviously, in this way we obtain the solution given by Eq. (4.140). At the second iteration, the double-scattering approximation L_2 is derived [80]. Its accuracy was analyzed in, e.g., [81–83]. The successive orders of scattering method for the Stokes parameters are described in [84–86]. Figure 4.14 shows how the successive orders of scattering technique converges to the discrete ordinate method solution.

Examining Eqs. (4.174)–(4.175), we can conclude that the angular distribution of the s-fold scattered radiance is related to the s-fold convolved phase function, i.e.,

$$L_s\left(\mu, \varphi\right) \sim \bar{p}^s\left(\mu_0, \mu, \varphi\right), \tag{4.181}$$

where

$$\bar{p}^s\left(\mu_0, \mu, \varphi\right) = \int_0^{2\pi} \int_{-1}^{1} \bar{p}^{s-1}\left(\mu_0, \mu', \varphi'\right) p\left(\mu', \mu, \varphi - \varphi'\right) d\mu' d\varphi'. \tag{4.182}$$

Let us consider

$$\bar{p}^2\left(\mu_0, \mu, \varphi\right) = \int_0^{2\pi} \int_{-1}^{1} p\left(\mu_0, \mu', \varphi'\right) p\left(\mu', \mu, \varphi - \varphi'\right) d\mu' d\varphi'. \tag{4.183}$$

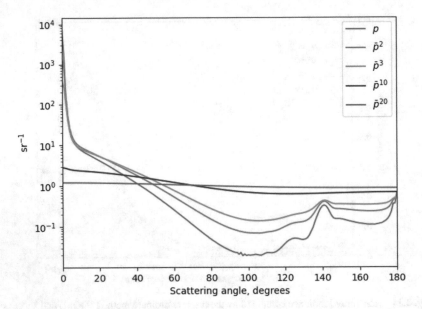

Fig. 4.15 s-fold convolved phase functions ($s = 1, 2, 3, 10$ and 20)

Expanding p in associated Legendre polynomial series (see Eq. (4.35)) and applying the orthogonality property, we obtain:

$$\bar{p}^2(\mu_0, \mu, \varphi) = \sum_{m=0}^{M_{\max}} (2 - \delta_{m0}) \sum_{n=m}^{N_{\max}} \chi_n^2 \bar{P}_n^m(\mu_0) \bar{P}_n^m(\mu) \cos(m\varphi). \tag{4.184}$$

Repeating this procedure for $\bar{p}^s$ yields

$$\bar{p}^s(\mu_0, \mu, \varphi) = \sum_{m=0}^{M_{\max}} (2 - \delta_{m0}) \sum_{n=m}^{N_{\max}} \chi_n^s \bar{P}_n^m(\mu_0) \bar{P}_n^m(\mu) \cos(m\varphi). \tag{4.185}$$

Note that $\{\chi_n^s\}_{n=0,1,2\ldots}$ decays faster as s increases, and hence, $\bar{p}^s$ becomes more isotropic, as shown in Fig. 4.15. Thus, the angular features of the scattered radiance distribution are mostly due to low-order scattering events, while the "background" is due to high-order scattered radiation.

Keeping in mind that the photons lose memory of their origin after several scattering events and the radiation approaches a diffusion equilibrium state, we can assume that the ratio of two successive orders of scattering L_s/L_{s-1} is constant [87]. Therefore Eq. (4.171) can be truncated at the $\bar{N}_s$th order of scattering, while the rest of the terms replaced by a geometric series $L_{\bar{N}_s}/(1-c)$, where $c = L_{s+1}/L_s$. Then Eq. (4.171) is replaced by

$$L = \sum_{s=1}^{\bar{N}_s-1} L_s + \frac{L_{\bar{N}_s}}{1-c}. \tag{4.186}$$

4.15 Padé and Taylor Series Approximations for Computing Matrix Exponentials

Two approximations for optically thin media can be derived in the framework of matrix exponential formalism. The Padé approximant (i.e., the "best" approximation of a function by a rational function of given order) has been suggested by Flatau and Stephens [43] for computing $e^{-\mathbf{A}\tau}$. More recently, McGarragh and Gabriel [88] used this approximation in connection with the matrix operator method. Essentially, the nth diagonal Padé approximation to the exponential of the matrix $\mathbf{A}\tau$ is defined as

$$\frac{N_n(\mathbf{A}\tau)}{D_n(\mathbf{A}\tau)}, \tag{4.187}$$

where $D_n(\mathbf{A}\tau)$ and $N_n(\mathbf{A}\tau)$ are polynomials in $\mathbf{A}\tau$ of degree n, given by

$$D_n(\mathbf{A}\tau) = \sum_{k=0}^{n} c_k \tau^k \mathbf{A}^k, \tag{4.188}$$

$$N_n(\mathbf{A}\tau) = \sum_{k=0}^{n} (-1)^k c_k \tau^k \mathbf{A}^k \tag{4.189}$$

and

$$c_k = \frac{(2n-k)!n!}{(2n)!k!(n-k)!} \tag{4.190}$$

From Eqs. (4.188) and (4.189) it is can be seen that $N_n(\mathbf{A}\tau) = D_n(-\mathbf{A}\tau)$. To compute $D_n(\mathbf{A}\tau)$ and $N_n(\mathbf{A}\tau)$ we have to compute powers of $\mathbf{A}$. By taking advantage of the block symmetries within $\mathbf{A}$ (cf. Eq. (4.54)), we find

$$\mathbf{A}^k = \begin{bmatrix} \mathbf{X}_k & \mathbf{Y}_k \\ (-1)^k \mathbf{Y}_k & (-1)^k \mathbf{X}_k \end{bmatrix}, \tag{4.191}$$

where the matrices $\mathbf{X}_k$ and $\mathbf{Y}_k$ are computed recursively at $k \geq 2$ as follows:

$$\mathbf{X}_k = \mathbf{X}_{k-1}\mathbf{A}_{11} - \mathbf{Y}_{k-1}A_{12}, \tag{4.192}$$

$$\mathbf{Y}_k = \mathbf{X}_{k-1}\mathbf{A}_{12} - \mathbf{Y}_{k-1}A_{11}, \tag{4.193}$$

with initial values $\mathbf{X}_1 = \mathbf{A}_{11}$ and $\mathbf{Y}_1 = \mathbf{A}_{12}$. The coefficients c_k are also computed recursively by means of

$$c_k = \frac{n - k + 1}{k \, (2n - k + 1)} c_{k-1} \tag{4.194}$$

with the initial value $c_1 = 0.5$.

The second approximation in the framework of the matrix exponential formalism is the Taylor series approximation, which uses the definition of the matrix exponential, namely

$$e^{-\mathbf{A}\tau} \approx \mathbf{E} + \sum_{k=1}^{n} (-1)^k \frac{\tau^k}{k!} \mathbf{A}^k. \tag{4.195}$$

Accounting of Eq. (4.191), we obtain

$$e^{-\mathbf{A}\tau} \approx \mathbf{E} + \sum_{k=1}^{n} (-1)^k \frac{\tau^k}{k!} \begin{bmatrix} \mathbf{X}_k & \mathbf{Y}_k \\ (-1)^k \mathbf{Y}_k & (-1)^k \mathbf{X}_k \end{bmatrix}. \tag{4.196}$$

In general, both the Padé approximation and Taylor series approximation can be applied as long as the condition $\|\mathbf{A}\tau\| < 1$ is fulfilled.

4.16   Small-Angle Approximation

For sighting angles close to incident angles (i.e., $\mu \approx \mu_0$), the small-angle approximation can be used to compute the transmitted radiance for the case of forward-peaked phase functions. This leads to the following boundary-value problem [89]:

$$\begin{cases} \mu_0 \dfrac{dL}{d\tau} (\tau, \mu, \varphi) = \\ -L (\tau, \mu, \varphi) + \dfrac{\omega(\tau)}{4\pi} \int_0^{2\pi} \int_{-1}^{1} p \left(\tau, \mu, \mu', \varphi' \right) L \left(\tau, \mu', \varphi' \right) d\mu' d\varphi', \\ L (0, \mu, \varphi) = F_\odot \delta (\mu - \mu_0) \, \delta (\varphi) . \end{cases} \tag{4.197}$$

Note that μ is substituted by μ_0 at the derivative term. The advantage of the small-angle radiative transfer equation is that it can be solved analytically, and several interesting results can be obtained, e.g., including studies of optical transfer and coherence functions [90]. Equation (4.197) can be solved by using the spherical harmonics method, i.e., expanding the radiance into Fourier cosine series (see Eq. (4.34)), and then Legendre series (see Eq. (4.115)). Note that on the left-hand side of Eq. (4.197) we have $\mu_0 dL/d\tau$ instead of $\mu dL/d\tau$, and thus, all terms in Eq.(4.197) have the same expansion kernels. The solution reads as follows:

$$L_{SAA}(\tau, \mu, \varphi) = \tag{4.198}$$

$$\frac{F_\odot}{4\pi} \sum_{m=0}^{M_{max}} \sum_{n=m}^{N_{max}} (2 - \delta_{m0}) \, e^{-\frac{1}{\mu_0}(1 - \omega\chi_n)\tau} \, \bar{P}_n^m(\mu) \, \bar{P}_n^m(\mu_0) \cos(m\varphi).$$

The small-angle solution gives a Dirac delta function at the upper boundary and a peaked function inside the scattering medium. The amplitude of the peaked function decreases with the increasing optical depth, while its width increases with the increasing optical depth.

We note that the small-angle solution just as the direct solar beam solves the boundary-value problem for a semi-infinite medium since no reflection from the lower boundary is allowed. Note that the solution (4.198) contains a singularity term due to contribution of the direct solar beam. For numerical computations the direct solar beam should be removed. The final expression reads

$$L_{SAA}(\tau, \mu, \varphi) =$$

$$\frac{F_\odot}{4\pi} \sum_{n,m} (2 - \delta_{m0}) \left[e^{-\frac{1}{\mu_0}(1 - \omega\chi_n)\tau} - e^{-\frac{1}{\mu_0}\tau} \right] \bar{P}_n^m(\mu) \, \bar{P}_n^m(\mu_0) \cos(m\varphi). \tag{4.199}$$

Expanding exponents in Taylor series we obtain

$$L_{SAA}(\tau, \mu, \varphi) =$$

$$\frac{F_\odot}{4\pi} e^{-\frac{1}{\mu_0}\tau} \sum_{n,m} (2 - \delta_{m0}) \sum_{s=0}^{\infty} \frac{1}{s!} \left(\frac{1}{\mu_0} \omega\chi_n\tau \right)^s \bar{P}_n^m(\mu) \, \bar{P}_n^m(\mu_0) \cos(m\varphi). \tag{4.200}$$

Switching the order of summation and taking into account Eq. (4.185), we express the solution as a scattering orders series:

$$L_{SAA}(\tau, \mu, \varphi) = \frac{F_\odot}{4\pi} e^{-\frac{1}{\mu_0}\tau} \sum_{s=0}^{\infty} \frac{1}{s!} \left(\frac{\omega\tau}{\mu_0} \right)^s \bar{p}^s(\mu_0, \mu, \varphi). \tag{4.201}$$

As we can see, for $\omega \to 1$, the s-fold convolved phase functions are weighted by the Poisson distribution coefficients. Thus, Eq. (4.201) can be regarded as a solution of the radiative transfer equation in the small-angle approximation in the framework of successive orders of scattering.

For the case of the normal incidence we have $\mu_0 = 1$ and all azimuthal harmonics $m > 0$ vanish. Consequently, Eq. (4.199) is simplified:

$$L_{SAA}(\tau, \mu, \varphi) = \frac{F_\odot}{4\pi} \sum_{n=0}^{\infty} (2n + 1) \left[e^{-\frac{1}{\mu_0}(1 - \omega\chi_n)\tau} - e^{-\frac{1}{\mu_0}\tau} \right] P_n(\mu). \tag{4.202}$$

Equation (4.202) can be rewritten in the integral form [71, 91]. We recall the formula for the expansion coefficient χ_n (see Eq. (3.137)):

$$\chi_n = \frac{1}{2} \int_{-1}^{1} P_n (\mu) \, p (\mu) \, d\mu. \tag{4.203}$$

Then, substituting Eq. (4.203) into Eq. (4.202) and making use of the asymptotic formula [92]

$$\lim_{\theta \to 0} P_n (\mu) = J_0 \left(\theta \left(n + \frac{1}{2} \right) \right), \tag{4.204}$$

where $\mu = \cos \theta$, we obtain

$$L_{\mathrm{SAA}} (\tau, \mu, \varphi) =$$
$$\frac{F_\odot}{4\pi} \sum_{n=0}^{\infty} 2 \left(n + \frac{1}{2} \right) \left[e^{-\frac{1}{\mu_0} \left(1 - \omega \frac{1}{2} \int_{-1}^{1} J_0 \left(\theta \left(n + \frac{1}{2} \right) \right) p(\mu) d\mu \right) \tau} - e^{-\frac{1}{\mu_0} \tau} \right].$$
$$\times J_0 \left(\theta \left(n + \frac{1}{2} \right) \right) \tag{4.205}$$

Introducing a new function

$$f \left(n + \frac{1}{2} \right) = \tag{4.206}$$
$$\left(n + \frac{1}{2} \right) \left[e^{-\frac{1}{\mu_0} \left(1 - \omega \frac{1}{2} \int_{-1}^{1} J_0 \left(\theta \left(n + \frac{1}{2} \right) \right) p(\mu) d\mu \right) \tau} - e^{-\frac{1}{\mu_0} \tau} \right] J_0 \left(\theta \left(n + \frac{1}{2} \right) \right),$$

Eq. (4.205) can be rewritten as follows:

$$L_{\mathrm{SAA}} (\tau, \mu, \varphi) = \frac{F_\odot}{2\pi} \sum_{n=0}^{\infty} f \left(n + \frac{1}{2} \right). \tag{4.207}$$

Then, approximating the sum by the integral, i.e.,

$$\sum_{n=0}^{\infty} f \left(n + \frac{1}{2} \right) = \int_{0}^{\infty} f (\xi) \, d\xi, \tag{4.208}$$

the small-angle solution can be represented in the integral form:

$$L_{\mathrm{SAA}} (\tau, \mu, \varphi) = \frac{F_\odot}{2\pi} \int_{0}^{\infty} f (\xi) \, d\xi. \tag{4.209}$$

The advantage of the integral solution (4.209) is that it can be analytically integrated for special types of phase functions [93].

Figure 4.16 illustrates the small-angle solution. As we can see, it agrees well with the discrete ordinate solution for viewing angles close to the incident angle.

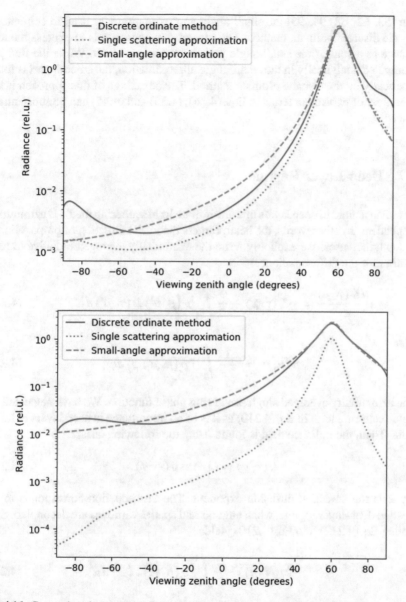

Fig. 4.16 Comparison between the discrete ordinate solution, the small-angle approximation and the single-scattering approximation. The computations are performed for the Henyey–Greenstein phase function with the asymmetry parameter 0.9. The single-scattering albedo is 0.99. The angle of incidence is 60°. (Top) The optical thickness is 0.3; (bottom) the optical thickness is 2.0

In [53, 62, 68, 94, 95] the small-angle approximation was used in conjunction with the discrete ordinate method. The idea of this approach is to express the total radiance as a sum of the small-angle part and the smooth part. While the first part is computed analytically in the small-angle approximation, the smooth part is found numerically by the discrete ordinate method. The advantage of this approach is that the number of expansion terms in Eqs. (4.34), (4.35) and (4.45) can be substantially reduced.

4.17 Deep-Layer Regime

The radiance field in deep layers of semi-infinite light scattering media is azimuthally independent. In other words, the beam forgets the initial direction of propagation as the depth increases. We explicitly write the azimuthally averaged radiative transfer equation for the diffuse radiance:

$$\mu \frac{dL\left(\tau, \mu\right)}{d\tau} = -L\left(\tau, \mu\right) + \frac{\omega}{2} \int_{-1}^{1} p\left(\mu, \mu'\right) L\left(\tau, \mu'\right) d\mu', \qquad (4.210)$$

where

$$p\left(\mu, \mu'\right) = \frac{1}{2\pi} \int_{0}^{2\pi} p\left(\mu, \mu', \varphi\right) d\varphi \qquad (4.211)$$

is the azimuthally averaged single-scattering phase function. We have neglected the single-scattering term in Eq. (4.210) as it is of no importance in deep layers of turbid media. Then, the radiance field is found using the following ansatz:

$$L\left(\tau, \mu\right) = i\left(\mu\right) \exp\left(-k\tau\right). \qquad (4.212)$$

Here k is the so-called diffusion exponent. The last equation corresponds to the so-called deep-layer regime, when angular and spatial variables are decoupled. Substituting Eq. (4.212) into Eq. (4.210) yields

$$\left(1 - k\mu\right) i\left(\mu\right) = \frac{\omega}{2} \int_{-1}^{1} p\left(\mu, \mu'\right) i\left(\mu'\right) d\mu'. \qquad (4.213)$$

This integral equation can be solved numerically by taking into account the normalization condition

$$\frac{\omega}{2} \int_{-1}^{1} i\left(\mu'\right) d\mu' = 1. \qquad (4.214)$$

Also, it follows for isotropic ($p = 1$) scattering from Eq. (4.213):

$$i(\mu) = \frac{1}{1 - k\mu}. \tag{4.215}$$

The diffusion exponent k can be derived by substituting this solution to Eq. (4.213). Then it follows:

$$\frac{\omega_0}{2k} \ln \frac{1+k}{1-k} = 1. \tag{4.216}$$

This transcendent equation can be used to derive k for a given single-scattering albedo in the case of isotropic scattering. The parameter k is equal to zero, and the internal light field is equal in all directions for non-absorbing media. For weakly absorbing media ($k \to 0$), it follows:

$$\ln(1+k) \approx k - \frac{k^2}{2} + \frac{k^3}{3} \tag{4.217}$$

and, therefore,

$$\frac{1}{2k} \ln \frac{1+k}{1-k} \approx 1 + \frac{k^2}{3}. \tag{4.218}$$

Finally, one derives

$$k = \sqrt{3(1 - \omega)} \tag{4.219}$$

for weakly absorbing media with isotropic scattering. This equation can be extended for the case of weakly absorbing media with anisotropic scattering ($p \neq 1$) using the notion of average cosine of the scattering angle g [96, 97]:

$$k = \sqrt{3(1 - g)(1 - \omega)}. \tag{4.220}$$

One can see that the diffusion exponent decreases with the increase of g. It means that the diffuse light flux is attenuated with a smaller rate in semi-infinite media with larger values of g.

The reflection function of semi-infinite light scattering media can be derived by solving the radiative transfer equation for a given large value of optical thickness. In particular, one can assume that $\tau = 5000$ in the case of non-absorbing semi-infinite media. This value is much smaller for absorbing semi-infinite turbid media.

Interestingly, the radiative transfer equation can be formulated in such a way (for semi-infinite media) that τ does not appear in the equation at all [98]. This can be done using the invariant imbedding method proposed by V. A. Ambartsumian in the first half of the twentieth century [99]. This method is based on the fact that the addition to the semi-infinite medium of a layer with an arbitrary thickness and the same optical properties does not change the reflected light intensity. As the layer can be of arbitrary thickness, we add an infinitely thin layer so that in this layer the multiple scattering collisions can be neglected. The different processes contributing to the reflection function due to adding a thin layer are shown in Fig. 4.17.

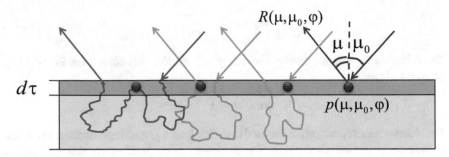

Fig. 4.17 Processes in the infinitely thin layer added above a semi-infinite medium

By setting the change of the reflection function to zero, the Ambartsumian's non-linear integral equation can be derived. This equation can be written in the following form:

$$R = R_{ss} + R_{ms}, \tag{4.221}$$

where

$$R_{ss} = \frac{\omega p (\theta)}{4 (\mu_0 + \eta)} \tag{4.222}$$

is the single scattered light contribution and

$$R_{ms} = \frac{\omega (\mu_0 L(\eta) + \eta L(\mu_0) + \mu_0 \eta M (\mu_0, \eta))}{\mu_0 + \eta} \tag{4.223}$$

is the multiple scattered light contribution and $\eta = |\mu|$. This expression satisfies the reciprocity principle (the invariance of reflectance due to interchange of the cosines μ_0 and η). Also it follows [99]:

$$L(\eta) = \frac{1}{4\pi} \int_0^1 \int_0^{2\pi} p(\eta, \varphi, \eta', \varphi') R(\eta', \varphi', \mu_0, \varphi_0) d\eta' d\varphi', \tag{4.224}$$

$$L(\mu_0) = \frac{1}{4\pi} \int_0^1 \int_0^{2\pi} p(\mu_0, \varphi_0, \eta', \varphi') R(\eta', \varphi', \eta, \varphi) d\eta' d\varphi' \tag{4.225}$$

and

$$M(\mu_0) = \frac{1}{4\pi^2} \int_0^{2\pi} d\varphi' \int_0^1 d\eta' R(\eta', \varphi', \eta, \varphi) \int_0^{2\pi} d\varphi'' \tag{4.226}$$

$$\times \int_0^1 d\eta'' p(-\eta', \varphi', \eta'', \varphi'') R(\eta'', \varphi'', \mu_0 \varphi_0).$$

This equation can be solved by various methods, including the simple iteration technique [100], the small-angle approximation [101] and Riccati equation solvers [102].

It follows for the case of isotropic scattering:

$$R(\eta, \mu_0) = \frac{\omega}{4(\eta + \mu_0)} + \frac{\omega \mu_0}{2(\eta + \mu_0)} \int_0^1 R(\eta', \mu_0)\, d\eta'$$
$$+ \frac{\omega \eta}{2(\eta + \mu_0)} \int_0^1 R(\eta', \eta)\, d\eta' + \frac{\omega \mu_0 \eta}{\eta + \mu_0} \int_0^1 \int_0^1 R(\eta, \eta')\, R(\eta'', \mu_0)\, d\eta' d\eta''. \tag{4.227}$$

Let us define the so-called H-function:

$$H(\eta) = 1 + 2\eta \int_0^1 R(\mu_0, \eta)\, d\mu_0. \tag{4.228}$$

Then the Ambartsumian equation can be written in a simpler form:

$$R(\eta, \mu_0) = \frac{\omega H(\eta) H(\mu_0)}{4(\eta + \mu_0)}. \tag{4.229}$$

One can see that the calculation of the reflection function is reduced to the determination of the function of one variable $H(\eta)$. The integral equation for this function can be derived substituting Eq. (4.229) in Eq. (4.228). Then it follows:

$$H(\eta) = 1 + \frac{\omega}{2}\eta H(\eta) \int_0^1 \frac{H(\mu_0)}{\mu_0 + \eta}\, d\mu_0. \tag{4.230}$$

One can derive from this equation that $H(0) = 1$. Also the following analytical solution of this integral equation is possible [103]:

$$H(\eta) = \exp(-\alpha \eta), \tag{4.231}$$

where

$$\alpha = \frac{1}{\pi} \int_0^\infty \frac{\ln z(x)\, dx}{1 + \eta^2 x^2} \tag{4.232}$$

and

$$z(x) = 1 - \omega \frac{\arctan x}{x}. \tag{4.233}$$

Various analytical approximations for this function are available. In particular, it has been proposed by Hapke [104]:

$$H(\eta) = \frac{1 + 2\eta}{1 + 2\eta\sqrt{1 - \omega}}. \tag{4.234}$$

with an error smaller than 4%. In Fig. 4.18 the reflection function for a homogeneous layer with isotropic scattering computed for different values of optical thickness is shown. The solution for a semi-infinite medium is computed by using Eqs. (4.229) and (4.234). One can see that the discrete ordinate solutions converge to the solution

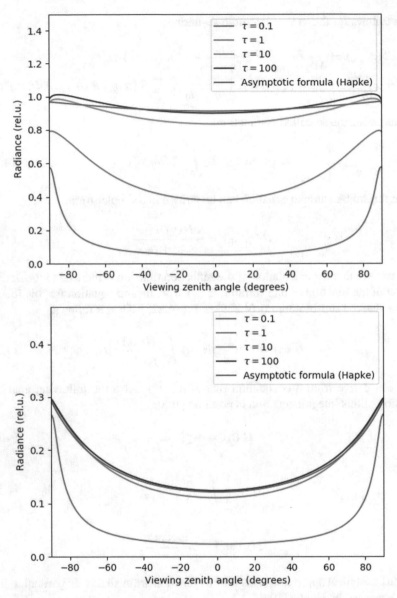

Fig. 4.18 Reflection function for a layer with isotropic scattering. As the optical thickness increases, the discrete ordinate solution converges to the asymptotic solution. The latter is computed by using Eq. (4.229), in which the Hapke approximation (Eq. (4.234)) is used. The angle of incidence is 60°, while the single-scattering albedo is 0.999 and 0.5 in top and bottom plots, respectively

for the semi-infinite medium as the layer thickness increases. In addition, the convergence occurs at lower values of the optical thickness τ when the single-scattering albedo is small.

From Eq. (4.234) it follows that for non-absorbing media $H(1) = 3$, while the exact solution gives: $H(1) = 2.9078$ for this case. The H-function can be used to derive the reflection function of a semi-infinite medium with isotropic scattering as discussed above. In particular, it follows from Eq. (4.229) under assumption that

$$R(\eta, \mu_0) = \frac{(1 + 2\eta)(1 + 2\mu_0)}{4(\eta + \mu_0)} \tag{4.235}$$

in case of non-absorbing semi-infinite media with isotropic scattering. One can see that $R(1, 1) = 9/8$ (more exactly, 1.06, if the numerical solution of the integral equation is used) at nadir illumination and observation conditions. This result only slightly differs from that for the ideal Lambertian reflector ($R = 1$) signifying the fact that the turbid media can have values of the reflection function larger than that for the absolutely white Lambertian reflector in some particular directions.

The approximation given by Eq. (4.235) can be generalized to the case of non-absorbing turbid media with arbitrary phase functions. First of all, we note that Eq. (4.227) can be written in the following form:

$$R = R_{ss} + R_{ms}, \tag{4.236}$$

where

$$R_{ss} = \frac{1}{4(\mu_0 + \eta)}, \tag{4.237}$$

$$R_{ms} = \frac{A + B(\eta + \mu_0) + C\eta\mu_0}{4(\eta + \mu_0)}, \tag{4.238}$$

$A = 1, B = 4, C = 2$. We assume that Eq. (4.238) remains valid for non-absorbing semi-infinite media with arbitrary phase functions. However, R_{ss} is given by Eq. (4.222) at $\omega = 1$, and coefficients A, B and C in Eq. (4.238) can be derived from numerical calculations of the function $F = 2\eta R_{ms}(\eta, \eta)$ by means of the exact model (e.g., the discrete ordinate method) and the asymptotic representation based on Eq. (4.238), which is a quadratic function with respect to η. In particular, Kokhanovsky [105, 106] advised to compute the reflection function as

$$\frac{A + B(\eta + \mu_0) + C\eta\mu_0 + (p(\eta, \mu_0, \varphi) - p_0(\eta, \mu_0))}{4(\eta + \mu_0)} \tag{4.239}$$

and to use the following sets of parameters for the case of typical phase functions of terrestrial clouds:

$$A = 3.944, \quad B = -2.5, \quad C = 10.664 \tag{4.240}$$

for water clouds and

$$A = 1.247, \quad B = 1.186, \quad C = 5.157 \tag{4.241}$$

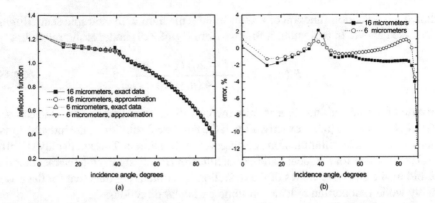

Fig. 4.19 **a** Comparison of calculations of the reflection function for cloudy media with effective droplet radii of 6 and 16 μm, using Equation Eqs. (4.238) and (4.240) and exact radiative transfer calculations; **b** Difference between calculations using Eqs. (4.238) and (4.240) and exact calculations [from [107]]

for ice clouds. The example of computations according to Eqs. (4.239)–(4.240) is shown in Fig. 4.19.

4.18   Optically Thick Layers

When the optical thickness is sufficiently large, the reflection and transmission functions can be expressed by simple analytical expressions known as the asymptotic theory of thick layers. This analytical model is much faster and more convenient for theoretical considerations than numerical models based on discrete ordinate schemes. In practice, equations shown in this section can be applied when the optical thickness of the medium is greater than 10 (and in some cases even at optical thickness larger than 5). This case covers most cloudy situations occurring in the terrestrial atmosphere.

In the classical asymptotic theory, the reflection and transmission functions for optically thick layers of optical thickness τ_0 are given by the following equations:

$$R\left(\mu, \mu_0, \varphi, \tau_0\right) = R_\infty\left(\mu, \mu_0, \varphi\right) - \frac{ml \exp\left(-2k\tau_0\right)}{1 - l^2 \exp\left(-2k\tau_0\right)} K\left(\mu\right) K\left(\mu_0\right) \quad (4.242)$$

and

$$T\left(\mu, \mu_0, \tau_0\right) = \frac{m \exp\left(-k\tau_0\right)}{1 - l^2 \exp\left(-2k\tau_0\right)} K\left(\mu\right) K\left(\mu_0\right), \quad (4.243)$$

respectively. To simplify equations we use μ instead of η in this section.

The escape function $K(\mu)$ is given by the relation

$$mK(\mu) = i(\mu) - 2\int_0^1 R_\infty(\mu, \mu') i(-\mu') \mu' d\mu \qquad (4.244)$$

and satisfies the normalization condition

$$2\int_0^1 K(\mu) i(\mu) \mu d\mu = 1, \qquad (4.245)$$

where the constant m is

$$m = 2\int_{-1}^1 i^2(\mu) \mu d\mu. \qquad (4.246)$$

The normalization condition for the escape function of non-absorbing turbid media $K_0(\mu)$ is:

$$2\int_0^1 K_0(\mu) \mu d\mu = 1. \qquad (4.247)$$

The constant l is computed as

$$l = 2\int_0^1 K(\mu) i(-\mu) \mu d\mu. \qquad (4.248)$$

The value of k is the diffusion exponent describing the attenuation of the radiation in the diffusion domain and being computed as the smallest positive eigenvalue of Eq. (4.213). The function $i(\mu)$ can be regarded as the corresponding eigenfunction, or diffusion pattern, satisfying the Sobolev–van de Hulst relation

$$i(-\mu) = 2\int_0^1 R_\infty(\mu, \mu') i(\mu') \mu' d\mu' \qquad (4.249)$$

and the normalization condition (4.214). Relations (4.242) and (4.243) show that the reflection function R depends on the azimuthal angle through the reflection function of a semi-infinite atmosphere R_∞, while the transmission function T is azimuthally independent.

For a layer with an underlying Lambertian surface of albedo ρ, the reflection and transmission functions can be presented as

$$R_\rho(\mu, \mu', \varphi, \tau_0) = R(\mu, \mu', \varphi, \tau_0) + \frac{\rho}{1 - \rho r_s} t(\mu, \tau_0) t(\mu', \tau_0) \qquad (4.250)$$

and

$$T_\rho(\mu, \mu', \tau_0) = T(\mu, \mu', \tau_0) + \frac{\rho}{1 - \rho r_s} r(\mu, \tau_0) t(\mu', \tau_0), \qquad (4.251)$$

respectively, with

$$r\left(\mu, \tau_0\right) = r_\infty\left(\mu\right) - \frac{mnl\exp\left(-2k\tau_0\right)}{1 - l^2\exp\left(-2k\tau_0\right)}K\left(\mu\right), \tag{4.252}$$

$$t\left(\mu, \tau_0\right) = r_\infty\left(\mu\right) - \frac{mn\exp\left(-k\tau_0\right)}{1 - l^2\exp\left(-2k\tau_0\right)}K\left(\mu\right), \tag{4.253}$$

$$r_s\left(\tau_0\right) = r_{s\infty}\left(\mu\right) - \frac{mn^2l\exp\left(-2k\tau_0\right)}{1 - l^2\exp\left(-2k\tau_0\right)}. \tag{4.254}$$

Here, r_∞ and $r_{s\infty}$ are the plane albedo and spherical albedo of a semi-infinite atmosphere, respectively, and n is the μ-weighted mean of the escape function

$$n = 2\int_0^1 K\left(\mu\right)\mu d\mu. \tag{4.255}$$

The reflection function of a semi-infinite atmosphere can be obtained as a solution to Eq. (4.221) by simple iteration [100], while the diffusion pattern and the diffusion exponent can be obtained by solving the integral equation (4.213) in conjunction with Eqs. (4.214) [108]. The constant m is then computed from Eq. (4.246), while $K\left(\mu\right)$, l, and n follow from Eqs. (4.244), (4.248) and (4.255), respectively. Note that all relations of the asymptotic theory can also be derived in the framework of the matrix exponential formalism [77].

The equations given above can be simplified in the case of weakly absorbing optically thick media ($\omega \to 1$). Then it follows [71]:

$$k = \sqrt{3(1 - g)(1 - \omega)}, \tag{4.256}$$

$$m = 1 - l^2, \tag{4.257}$$

$$l = \exp(-y), \tag{4.258}$$

$$K(\eta) \approx K_0(\eta), \tag{4.259}$$

where

$$y = 4\sqrt{\frac{\beta}{3(1 - g)}} \tag{4.260}$$

is the similarity parameter and

$$\beta = 1 - \omega \tag{4.261}$$

is the probability of photon absorption. Then it follows:

$$R(\mu, \mu_0, \varphi, \tau_0) = R_\infty(\mu, \mu_0, \varphi)$$
$$-\frac{(1 - \exp(-2y)) \exp(-2k\tau_0 - y)}{1 - \exp(-2k\tau_0 - 2y)} K_0(\mu) K_0(\mu_0), \tag{4.262}$$

$$T(\mu, \mu_0, \tau_0) = \frac{(1 - \exp(-2y)) \exp(-k\tau_0)}{1 - \exp(-2k\tau_0 - 2y)} K_0(\mu) K_0(\mu_0). \tag{4.263}$$

The escape function $K_0(\eta)$ can be approximated by the following linear dependence:

$$K_0(\eta) = a + b\eta, \tag{4.264}$$

where $a = 3/7$, $b = 6/7$. One can conclude that the transmittance $T(\mu, \mu_0, \tau_0)$ is determined by two parameters of a turbid layer: $x = k\tau_0$ and y. Therefore, different turbid layers with the same set of parameters (x, y) will be characterized by the similar transmission functions. The reflection function of a semi-infinite weakly absorbing layer can be related to the reflection function $R_{0\infty}(\mu, \mu_0, \varphi)$ of a non-absorbing layer having the same phase function [109]:

$$R_\infty(\mu, \mu_0, \varphi) = R_{0\infty}(\mu, \mu_0, \varphi) e^{-fy}, \tag{4.265}$$

where

$$f = \frac{K_0(\mu_0) K_0(\eta)}{R_{0\infty}(\mu_0, \eta, \varphi)}. \tag{4.266}$$

Therefore, the equations given above can be written as

$$R(\mu, \mu_0, \varphi, \tau_0) = R_{0\infty}(\mu, \mu_0, \varphi) e^{-fy} - T(\mu, \mu_0, \tau_0) e^{-x-y} \tag{4.267}$$

and

$$T(\mu, \mu_0, \tau_0) = t_s K_0(\mu) K_0(\mu_0), \tag{4.268}$$

where

$$t_s = \frac{\sinh(y)}{\sinh(x + y)}. \tag{4.269}$$

One can see that the calculation of the transmission function can be done using a closed analytical form. The calculation of $R(\mu, \mu_0, \varphi, \tau_0)$ is reduced to the calculation of the reflection function of a semi-infinite non-absorbing layer, which can be done using the radiative transfer codes, look-up-tables or approximations as discussed above. The obtained equations can be used to find approximate equations for plane/spherical albedo and transmittance. Equations given in Table 4.1 can be used to calculate spherical/plane absorptances:

$$a_{s,p} = 1 - r_{s,p} - t_{s,p}. \tag{4.270}$$

Table 4.1 Approximations for spherical/plane albedo and transmittance of weakly absorbing optically thick turbid layers ($\varsigma = \frac{3}{4}(1-g)$). The results for a non-absorbing layer are given in the last column

Parameter	Symbol	Equation	$\omega = 1$
Spherical albedo	r_s	$\dfrac{\sinh(x)}{\sinh(x+y)}$	$\dfrac{\varsigma\tau}{1+\varsigma\tau}$
Plane albedo	r_p	$\dfrac{\exp(-yK_0(\mu_0))}{-t_s K_0(\mu_0)\exp(-x-y)}$	$1 - \dfrac{K_0(\mu_0)}{1+\varsigma\tau}$
Spherical transmittance	t_s	$\dfrac{\sinh(y)}{\sinh(x+y)}$	$\dfrac{1}{1+\varsigma\tau}$
Plane transmittance	t_p	$t_s K_0(\mu_0)$	$\dfrac{1}{1+\varsigma\tau}K_0(\mu_0)$

All the results become especially simple in the case of non-absorbing media. Then it follows taking corresponding limits:

$$R(\mu, \mu_0, \varphi, \tau_0) = R_{0\infty}(\mu, \mu_0, \varphi) - T(\mu, \mu_0, \tau_0), \qquad (4.271)$$

$$T(\mu, \mu_0, \tau_0) = \frac{1}{1+\varsigma\tau_0}K_0(\mu)K_0(\mu_0), \qquad (4.272)$$

where

$$\varsigma = \frac{3}{4}(1-g). \qquad (4.273)$$

The equations for spherical/plane reflectances and transmittances of non-absorbing layers are given in Table 4.1. It should be pointed out that the classical asymptotic theory for non-absorbing optically thick layers leads to the same result as discussed above (see Eq. (4.272)) except 1 in the denominator is substituted by

$$\delta = 3\int_0^1 K_0(\mu)\mu^2 d\mu, \qquad (4.274)$$

which is equal to $45/42 \approx 1.071$ in the approximation given by Eq.(4.264). The exact calculations for the Henyey–Greenstein phase function with g in the range 0–0.9 give the value of this parameter in the range 1.066–1.072, which is close to the result reported above.

4.19 Phase Function Truncation Methods

The discrete ordinate method uses the Legendre polynomial expansion of the phase function (see Eq. (4.35)). The phase function associated with the scattering by aerosol and cloud particles have a pronounced peak in the forward scattering direction. In the case of strongly forward-peaked phase functions, the number of expansion coefficients can range up to several thousands. That leads to time-consuming computations. In this regard, several techniques have been proposed to approximate the phase function by a smoother function. The idea of truncation algorithms is to substitute the original phase function p having a lot of Legendre expansion coefficients by another phase function $\tilde{p}$ with fewer expansion coefficients, which though is close in a certain sense to the original phase function p.

In [110] it was suggested to replace the forward diffraction peak of the phase function by the delta function:

$$p\left(\cos\Theta\right) \approx f\delta\left(1 - \cos\Theta\right) + \tilde{p}\left(\cos\Theta\right), \tag{4.275}$$

where $f = \int_{-1}^{1}(p(\cos\Theta) - \tilde{p}(\cos\Theta))\,d\left(\cos\Theta\right)$. The truncated phase function outside a forward cone $\cos\Theta \in \left[-1, \tilde{\mu}\right]$ coincides with the original phase function, while for $\cos\Theta \in \left[\tilde{\mu}, 1\right]$ the phase function is extrapolated preserving the slope of p at $\cos\Theta = \tilde{\mu}$. Essentially, the region of the forward cone is defined ad hoc. For instance, in [111] $\tilde{\mu} \approx 0.94$. Based on this idea, the so-called delta-S method [112] was developed, in which the phase function in the forward code is assumed to be constant.

In the delta-M method [113] of Wiscombe, the forward peak is approximated by the delta function, i.e.,

$$p\left(\cos\Theta\right) = 2f\delta\left(1 - \cos\Theta\right) + (1 - f)\,\tilde{p}\left(\cos\Theta\right), \tag{4.276}$$

where f is the truncation factor and $\tilde{p}$ is the smoothed (truncated) phase function expressed by N_{tr} Legendre expansion coefficients. Considering the Legendre series expansion of Eq. (4.276), in which

$$2f\delta\left(1 - \cos\Theta\right) = \sum_{k=0}^{\infty}(2k + 1)\,fP_k\left(\cos\Theta\right) \tag{4.277}$$

and

$$\tilde{p}\left(\cos\Theta\right) = \sum_{k=0}^{N_{\text{tr}}}(2k + 1)\,\tilde{\chi}_k P_k\left(\cos\Theta\right), \tag{4.278}$$

and equating the first N_{tr} moments of p and $\tilde{p}$, we obtain:

$$\tilde{\chi}_k = \frac{\chi_k - f}{1 - f}, \, k = 0, \ldots, N_{\text{tr}}, \tag{4.279}$$

where $\tilde{\chi}_k$ is the expansion coefficient of $\tilde{p}$ (cf. Eq. (4.278)). Then, the substitution of Eq. (4.276) into the radiative transfer equation leads to the radiative transfer equation with the single-scattering phase function $\tilde{p}$ and also with other scaled parameters, i.e.,

$$\tilde{\sigma}_{\text{ext}} = \sigma_{\text{ext}} - \sigma_{\text{sca}} f, \tag{4.280}$$

$$\tilde{\sigma}_{\text{sca}} = \sigma_{\text{sca}} \frac{1 - f}{1 - f \frac{\sigma_{\text{sca}}}{\sigma_{\text{ext}}}}, \tag{4.281}$$

$$\tau' = \tau (1 - \omega f) \tag{4.282}$$

and

$$\omega' = \omega \frac{1 - f}{1 - \omega f}. \tag{4.283}$$

The truncation factor f is set to $\chi_{N_{\text{tr}}}$ where N_{tr} is the number of expansion coefficients of the truncated phase function. Usually, it follows: $N_{\text{tr}} = 2N_{\text{do}}$. The RTE with scaled parameters can be easily solved numerically or even (under certain assumptions) analytically.

One drawback of the delta-M method is that it leads to errors in the phase function at large scattering angles (i.e., in backscattering directions). To reduce the truncation error, the delta-fit method has been designed [114]. The idea of this method is to minimize the residual between $\tilde{p}$ and p:

$$r_{\delta-\text{fit}} = \sum_{i=1}^{q} \left(\frac{\tilde{p}(\mu_i) - p(\mu_i)}{p(\mu_i)} \right)^2 w_i \tag{4.284}$$

where w_i is the weight for each scattering angle cosine μ_i (for the sake of numerical efficiency, μ_i are taken outside a forward cone, i.e., $\mu_i < \tilde{\mu}$). In practice, μ_i and w_i are taken as the nodes and weights of the Legendre quadrature rule of the qth order. The Legendre expansion coefficients $\tilde{\chi}_k$ of $\tilde{p}$ are founded as a solution of the optimization problem

$$\tilde{\chi}_k = \arg \min_{\tilde{\chi}_k} (r_{\delta-\text{fit}}) \tag{4.285}$$

through the least-squares fitting problem

$$\frac{\partial r_{\delta-\text{fit}}}{\partial \tilde{\chi}_k} = 0, \text{ for } k = 0, \ldots, N_{\text{tr}}. \tag{4.286}$$

After substituting Eq. (4.278) into Eq. (4.284), the minimization problem (4.285)–(4.286) leads to the following system of linear equations:

$$Ax = b, \tag{4.287}$$

where

$$A_{kl} = \sum_{i=0}^{q} w_i \frac{P_l(\mu_i) P_k(\mu_i)}{(p(\mu_i))^2}, \tag{4.288}$$

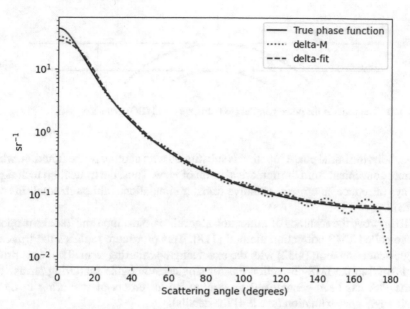

Fig. 4.20 Illustration of the truncation procedure: the original Henyey–Greenstein phase function with the asymmetry parameter 0.8 and those truncated according to the delta-M and delta-fit methods preserving 16 Legendre expansion coefficients

$$x_l = \tilde{\chi}_l \tag{4.289}$$

and

$$b_k = \sum_{i=0}^{q} w_i \frac{P_k(\mu_i)}{p(\mu_i)}. \tag{4.290}$$

Note that the resulting $\tilde{\chi}_l$ have to be normalized so that $\tilde{\chi}_0 = 1$. To use the delta-fit method, the optical thickness and the single-scattering albedo for the layer have to be adjusted similarly as the delta-M method (see Eqs. (4.282) and (4.283)).

The delta-M method has been further improved in [115] by approximating the delta function in Eqs. (4.276) and (4.277) by a forward-peaked function, yet with a finite width as follows:

$$2f\delta(1 - \cos\Theta) \approx \sum_{k=0}^{\infty} (2k + 1)\,c\exp\left(-\frac{k^2}{2\sigma^2}\right) f P_k(\cos\Theta). \tag{4.291}$$

Here coefficients c and σ are found by matching higher-order moments (see [115] for computational details). A comprehensive analysis of the truncation techniques is given in [116]. Adaptation of the delta-M and delta-fit truncation methods to phase matrices (i.e., for the vector radiative transfer equation) is outlined in [86, 117]. As shown in Fig. 4.20, truncation methods lead to artifacts in the phase functions, and

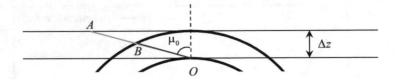

Fig. 4.21 The paths in the plane-parallel (AO) and spherical (BO) geometry

essentially, the initial phase function is substituted with another phase function, which is more convenient from the numerical point of view. Thus, the truncation techniques always introduce an error in radiative transfer simulations (although small in some cases).

To improve the accuracy of truncation algorithms, Nakajima and Tanaka proposed the so-called TMS-correction method [118]. This procedure replaces the truncated single-scattering term (4.32) with the exact single-scattering term. The same principle is used in the IMS-correction but for single- and double-scattering terms. This framework has been generalized by computing all orders of scattering using the small-angle approximation (see [94] for details).

4.20 Pseudo-spherical Model

Throughout the chapter, we have been focused on one-dimensional models, in which the atmosphere is treated as a collection of plane-parallel slabs. It seemed to be reasonable because the Earth radius, which is around 6400 km, is much larger than the geometrical thickness of the atmosphere and the height of the low Earth orbit (below 1000 km). However, the assumption of a plane-parallel atmosphere breaks down when the solar zenith angle or the viewing zenith angle approaches 90° or limb observations of terrestrial atmosphere are of interest. In this case, the sphericity of the atmosphere should be taken into account. The difference between plane-parallel and spherical models consists of two aspects. First, there is a difference in optical paths computed in plane-parallel and spherical geometries, as shown in Fig. 4.21. Second, the solar-view angles estimated with respect to the tangent to the sphere are altitude-dependent in the spherical geometry, as illustrated in Fig. 4.22. Apparently, the viewing zenith angle at the top of the atmosphere is not equal to that at the bottom of the atmosphere. The difference can reach 2° in the case of Sun-synchronous orbit satellites.

The radiative transfer equation in spherical geometry is similar to that in slab geometry but the derivative of the radiance consists of three terms instead of one (which is just $\mu \dfrac{d}{dz}$):

$$
\mu \frac{\partial}{\partial r} + \frac{1 - \mu^2}{r} \frac{\partial}{\partial \mu} + \frac{1}{r} \sqrt{1 - \mu^2} \sqrt{1 - \mu_0^2}
$$

$$
\times \left[\cos (\varphi - \varphi_0) \frac{\partial}{\partial \mu_0} + \frac{\mu_0}{1 - \mu_0^2} \sin (\varphi - \varphi_0) \frac{\partial}{\partial (\varphi - \varphi_0)} \right], \quad (4.292)
$$

where r is the absolute value of the radius vector going from the sphere center. The last two terms take into account the change in the radiance associated with changes in the polar angle, the azimuthal angle and the solar zenith angle while moving across the spherical shell. Following [119, 120], the attenuation of the direct solar beam as it goes through a homogeneous layer j of the geometrical thickness Δz_j is computed as $\exp \left(-\beta \Delta z_j \right)$, where β is a so-called Chapman factor $\varXi$ [121], which is equal to the path in the spherical layer divided by Δz_j. Apparently, $\beta < \mu_0^{-1}$ in the spherical model while in the plane-parallel atmosphere we have $\beta = \mu_0^{-1}$. Thus, the single-scattering source function is given as follows:

$$
J_{ss} (r, \mu, \varphi) = \frac{\sigma_{sca} (r)}{4\pi} F_\odot p (r, \mu_0, \mu, \varphi) \exp \left(-\tau \left(r_{TOA}, r \right) \varXi (r, \mu_0) \right), \quad (4.293)
$$

where $r_{TOA,}$ is the coordinate of the top of the atmosphere (TOA) level. The full spherical radiative transfer model is computationally expensive compared to the one-dimensional models and is usually a subject to Monte-Carlo methods (see, e.g., MC++ model [122] and libRadtran's Monte CarloRT solver MYSTIC [123]). To take the sphericity into account at the same computational costs as of the plane-parallel model, the pseudo-spherical approximation is used, in which the single-scattering source term is computed in the spherical geometry according to Eq. (4.293) while the multiple scattering source term and the derivative term are computed with a plane-parallel model. In some implementations, the multiple scattering term can be then adjusted empirically to account for the spherical effects. Several improvements of the pseudo-spherical model have been proposed (e.g., [124]). More advanced pseudo-spherical models include corrections of the solar zenith angle, the viewing zenith angle and the relative azimuthal angle at each boundary of atmospheric layers. A comprehensive analysis of the accuracy of the pseudo-spherical model can be found in [125] and numerous references therein.

4.21 Polarized Radiative Transfer Equation

The one-dimensional vector radiative transfer equation is formulated in terms of the Stokes vector $\mathbf{L} = [I, Q, U, V]^T$, as defined in Chap. 3, in a manner similar to the scalar one. The scalar radiance L turns into the Stokes vector, the single-scattering phase function turns into the phase matrix $\mathbf{Z}$, while the extinction coefficient turns into the extinction matrix [64, 126, 127]. For the diffuse component of the Stokes vector the radiative transfer equation reads as follows:

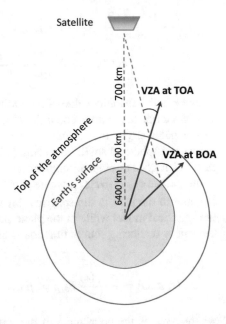

$$\mu\frac{d\mathbf{L}\left(\tau,\mu,\varphi\right)}{d\tau} = -\mathbf{L}\left(\tau,\mu,\varphi\right) + \delta_{m0}\left(1 - \omega\left(\tau\right)\right)\mathbf{B}\left(\tau\right)$$
$$+\frac{\omega\left(\tau\right)}{4\pi}\int_0^{2\pi}\int_{-1}^{1}\mathbf{Z}\left(\tau,\mu',\mu,\varphi-\varphi'\right)\mathbf{L}\left(\tau,\mu',\varphi'\right)d\mu'd\varphi' \qquad (4.294)$$
$$+\frac{\omega\left(\tau\right)}{4\pi}\mathbf{F}_\odot\mathbf{Z}\left(\tau,\mu_0,\mu,\varphi\right)\exp\left(-\frac{\tau}{\mu_0}\right),$$

where $\mathbf{B} = [B, 0, 0, 0]^T$ and $\mathbf{F}_\odot = \left[F_\odot, 0, 0, 0\right]$ are the unpolarized thermal and solar terms.

The boundary conditions for the diffuse component of the Stokes vector are formulated similar to those for the scalar radiance:

$$\begin{cases} \mathbf{L}\left(0,\mu > 0,\varphi\right) = 0, \\ \mathbf{L}\left(z_s,\mu < 0,\varphi\right) = \frac{1}{\pi}\int_0^{2\pi}\int_{-1}^{0}\mathbf{L}\left(z_s,-\mu,\varphi\right)\boldsymbol{\rho}\left(\mu,\mu',\varphi-\varphi'\right)\mu'd\mu'd\varphi' \\ +\frac{\mu_0 F_\odot}{\pi}\exp\left(-\frac{\tau(z_s)}{\mu_0}\right)\boldsymbol{\rho}\left(\mu,\mu_0,\varphi\right) \end{cases}$$

$$(4.295)$$

where $\boldsymbol{\rho}$ is the bidirectional reflectance distribution matrix (BRDM).

There are various approaches to solve the vector radiative transfer equation. All methods developed for the scalar equation can be extended to the vector one, including Monte-Carlo technique [128], the adding-doubling technique [129], successive orders of scattering [84], the matrix operator method [61], the small-angle approximation [130], the invariance principle [131] and the asymptotic theory [132]. A discrete ordinate method, which converts the radiative transfer equation into a linear system of differential equations by discretizing the angular variation of the phase matrix and Stokes vector, has been developed by Siewert [133].

Recalling the description of polarization in Chap. 3, the phase matrix can be expanded in Fourier series of the azimuthal angle as follows:

$$\mathbf{Z}\left(\tau, \mu', \mu, \varphi - \varphi'\right) = \sum_{m=0}^{M} \Phi_1^m \left(\varphi - \varphi'\right) \mathbf{W}^m(\mu', \mu)\mathbf{D}_1 + \Phi_2^m \left(\varphi - \varphi'\right) \mathbf{W}^m(\mu', \mu)\mathbf{D}_2,$$

(4.296)

where

$$\Phi_1^m \left(\xi\right) = (2 - \delta_{m0}) \begin{bmatrix} \cos\xi & 0 & 0 & 0 \\ 0 & \cos\xi & 0 & 0 \\ 0 & 0 & \sin\xi & 0 \\ 0 & 0 & 0 & \sin\xi \end{bmatrix},$$

(4.297)

$$\Phi_2^m \left(\xi\right) = (2 - \delta_{m0}) \begin{bmatrix} -\sin\xi & 0 & 0 & 0 \\ 0 & -\sin\xi & 0 & 0 \\ 0 & 0 & \cos\xi & 0 \\ 0 & 0 & 0 & \cos\xi \end{bmatrix},$$

(4.298)

$$\mathbf{D}_1 = \begin{bmatrix} 1 & 0 & 0 & 0 \\ 0 & 1 & 0 & 0 \\ 0 & 0 & 0 & 0 \\ 0 & 0 & 0 & 0 \end{bmatrix}$$

(4.299)

and

$$\mathbf{D}_2 = \begin{bmatrix} 0 & 0 & 0 & 0 \\ 0 & 0 & 0 & 0 \\ 0 & 0 & 1 & 0 \\ 0 & 0 & 0 & 1 \end{bmatrix}.$$

(4.300)

Note that unlike the scalar case, the azimuthal expansion of the Stokes vector involves also sine terms.

$$\mathbf{L}\left(\tau, \mu, \varphi\right) = \sum_{m=0}^{M} \sum_{k=1}^{2} \Phi_k^m \left(\varphi\right) \mathbf{L}_k^m \left(\tau, \mu\right).$$

(4.301)

The azimuthal expansion of the phase matrix follows from the addition theorem for generalized spherical functions and the result is

$$\mathbf{W}^m(\mu, \mu') = \sum_{l=m}^{N_{\max}} \Pi_l^m (\mu)\mathbf{B}_l \left(\tau\right) \Pi_l^m (\mu').$$

(4.302)

Substituting Eqs. (4.296), (4.301) and (4.302) into Eq. (4.294) we obtain the equation for the azimuthal components $L_k^m(\tau, \mu)$:

$$\mu \frac{d\mathbf{L}_k^m(\tau, \mu)}{d\tau} = -\mathbf{L}_k^m(\tau, \mu) + \mathbf{J}_{\mathrm{ss},k}^m(\tau, \mu) + \mathbf{J}_{\mathrm{ms},k}^m(\tau, \mu) + \mathbf{J}_{\mathrm{th},k}^m(\tau) \qquad (4.303)$$

with

$$\mathbf{J}_{\mathrm{ss},m}^k(\tau, \mu) = (2 - \delta_{m0}) \frac{\omega(\tau)}{4\pi} F_\odot \mathbf{D}_k \exp\left(-\frac{\tau}{\mu_0}\right) \sum_{l=m}^{N_{\max}} \Pi_l^m(\mu) \mathbf{B}_l(\tau) \Pi_l^m(\mu_0),$$
$$(4.304)$$

$$\mathbf{J}_{\mathrm{ms},k}^m(\tau, \mu) = \frac{\omega(\tau)}{2} \int_{-1}^{1} \sum_{n=m}^{N_{\max}} \Pi_l^m(\mu) \tilde{\mathbf{B}}_l(z) \Pi_l^m(\mu') \mathbf{L}_k^m(\tau, \mu') d\mu', \qquad (4.305)$$

and

$$\mathbf{J}_{\mathrm{th},m}(\tau) = \delta_{m0}(1 - \omega(\tau)) \mathbf{B}(\tau). \qquad (4.306)$$

Note that Eq. (4.303) can be solved independently for $k = 1$ and $k = 2$.

Defining the $8N_{\mathrm{do}}$-dimensional Stokes vector function in the discrete ordinate space by $\left[\mathbf{i}_m^\uparrow \ \mathbf{i}_m^\downarrow\right]^T$, where

$$\left[\mathbf{i}_m^{\uparrow\downarrow}(\tau)\right]_i = \mathbf{L}(\tau, \mp\mu_i), \quad i = 1, \dots, N_{\mathrm{do}}, \qquad (4.307)$$

we are led to a system of differential equations, similar to Eq. (4.53)

$$\frac{d}{d\tau} \begin{bmatrix} \mathbf{i}^\uparrow(\tau) \\ \mathbf{i}^\downarrow(\tau) \end{bmatrix} = -\mathbf{A}_l \begin{bmatrix} \mathbf{i}^\uparrow(\tau) \\ \mathbf{i}^\downarrow(\tau) \end{bmatrix} + \begin{bmatrix} \mathbf{b}_l^\uparrow(\tau) \\ \mathbf{b}_l^\downarrow(\tau) \end{bmatrix}, \ \tau_l \le \tau \le \tau_{l+1}, \qquad (4.308)$$

where $\mathbf{A}$ is the layer matrix given by

$$\mathbf{A} = \begin{bmatrix} \mathbf{A}_{11} & \mathbf{A}_{12} \\ -\mathbf{D}\mathbf{A}_{12}\mathbf{D} & -\mathbf{D}\mathbf{A}_{11}\mathbf{D} \end{bmatrix}. \qquad (4.309)$$

Here the $4N_{\mathrm{do}} \times 4N_{\mathrm{do}}$ block matrices are expressed as

$$\mathbf{A}_l^{11} = \mathbf{M} - \mathbf{M}\mathbf{S}_l^\uparrow \mathbf{W},$$
$$\mathbf{A}_l^{12} = -\mathbf{M}\mathbf{S}_l^\downarrow \mathbf{W},$$

$\mathbf{M}$ and $\mathbf{W}$ are diagonal matrices with entries $[\mathbf{M}]_{ij} = (1/\mu_i) \mathbf{I}\delta_{ij}$ and $[\mathbf{W}]_{ij} = w_i \mathbf{I}\delta_{ij}$ and $\mathbf{I}$ being the 4-by-4 identity matrix, respectively,

$$\left[\mathbf{S}_l^{\uparrow\downarrow}\right]_{ij} = \frac{1}{2}\omega_l \sum_{n=m}^{N_{\max}} \Pi_n^m(\mu_i) \mathbf{B}_{ln} \Pi_n^m(\mp\mu_j), \quad i, j = 1, \dots N_{\mathrm{do}}, \qquad (4.310)$$

$\mathbf{B}_{ln}$ is the expansion coefficient of the phase matrix in the lth layer, while $\mathbf{D} = \mathrm{diag}$ $[\mathbf{d},\ldots,\mathbf{d}]$, $\mathbf{d} = \mathrm{diag}[1, 1, -1, -1]$.

The layer vector incorporates the solar single-scattering term and the thermal emission term, i.e.,

$$\mathbf{b}_l^{\uparrow\downarrow}(\tau) = \mathbf{b}_{l,ss}^{\uparrow\downarrow} \exp\left(-\tau/\mu_0\right) + \mathbf{b}_{l,th}^{\uparrow\downarrow}(\tau), \tag{4.311}$$

where

$$\left[\mathbf{b}_{l,ss}^{\uparrow\downarrow}\right]_i = \mp\frac{1}{\mu_i}\left(2 - \delta_{m0}\right)\frac{\omega_l}{4\pi}F_\odot \sum_{n=m}^{N_{max}} \Pi_n^m\left(\mp\mu_i\right)\mathbf{B}_{ln}\Pi_n^m\left(\mu_0\right) \tag{4.312}$$

and

$$\left[\mathbf{b}_{l,th}^{\uparrow\downarrow}\right]_i = \mp\frac{1}{\mu_i}\delta_{m0}\left(1 - \omega_l\right)B(T(\tau)). \tag{4.313}$$

Further computations can be performed in the framework of the matrix exponential formalism [95, 134] or by using the eigenvalue approach. In addition, all approximate models, such as the Padé approximation, can be applied.

In the spherical harmonics method [135], $\mathbf{L}_k^m(\tau, \mu)$ is expanded similar to Eq. (4.302), i.e.,

$$\mathbf{L}_k^m(\tau, \mu) = \sum_{l=m}^{L+m} \Pi_l^m(\mu)\,\mathbf{L}_{lk}^m(\tau), \tag{4.314}$$

in Eqs. (4.303)–(4.306). As in the scalar case, to get rid of summation, the orthogonality property can be used, namely,

$$\int_{-1}^{1} \Pi_l^m(\mu)\,\Pi_l^n(\mu)\,d\mu = \frac{2}{2l+1}\Delta_l\delta_{kl}\delta_{mn} \tag{4.315}$$

with

$$\Delta_l = \mathrm{diag}\left\{1, (1 - \delta_{0l})(1 - \delta_{1l}), (1 - \delta_{0l})(1 - \delta_{1l}), 1\right\}. \tag{4.316}$$

4.22 Radiative Transfer Models for Horizontally Inhomogeneous Media

In this section we briefly outline the radiative transfer models which can be used to compute the radiance field in the case of horizontally inhomogeneous media, such as the atmosphere with the inhomogeneous cloud and the surface with the surface albedo varying in space. In this case, the radiance field is a function of three spatial coordinates and two angles, i.e., $L(x, y, z, \mu, \varphi)$, where x and y are two horizontal spatial dimensions. Essentially, this is a subject of three-dimensional radiative transfer models. Note that such models are much more time-consuming than previously considered one-dimensional models, and hence cannot be used in near-real-time analysis of the satellite data. Nevertheless, they can be used for assessing the effects of spatial inhomogenieties on the result of data interpretation. As noted in [136], the differences between one- and three-dimensional radiative transfer models are more pronounced in the cases involving clouds at visible wavelengths, where multiple scattering is dominant and cloud extinction varies on spatial scales smaller than the radiation mean free path. More information about the three-dimensional effects can be found in [137]. An excellent review of the three-dimensional models has been done by Nikolaeva et al. [138]. A common technique to treat a multidimensional problem is the Monte-Carlo method. However, this topic is out of the scope of this book, and readers are encouraged to refer to the classical book of Marchuk et al. [139]. Below we consider the so-called deterministic and stochastic models.

4.22.1   Deterministic Models

In the so-called characteristics method, three-dimensional radiative transfer equation is solved numerically on a certain spatial grid, e.g., 3D Cartesian grid. Thus, the space is split into a set of cells, as shown in Fig. 4.23. Characteristic schemes are based on the known streaming properties of the transport equation and the use of the integral form of the radiative transfer equation to calculate the radiance at each grid point. If the integration is performed along the whole characteristic of the differential operator, we speak about the long characteristic method, and if the integration is performed along a segment of the characteristic located inside a cell, we speak about the short characteristic method.

Computations can be organized iteratively. Dividing the atmosphere into a set of columns, we solve a one-dimensional problem for each column and compute corresponding source functions in each node. Then, given that the incident radiances

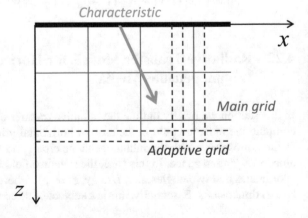

Fig. 4.23 Illustration of the characteristics method: bold black line is the upper boundary, at which the incident radiation is known. To find the radiation at the nodes, the radiative transfer equation is integrated along the discrete ordinate (green arrow). The additional splitting due to the adaptive grid technique is shown with red lines

at the domain boundaries are known, we integrate the RTE in a similar manner as it is done in the one-dimensional case (see Eqs. (4.13) and (4.14)) to compute radiance values at the cell nodes. To implement this idea numerically, it is convenient to use the discrete ordinate representation of the radiance field (i.e., the integration is performed along the discrete ordinate, while to compute the source function, the quadrature rule can be applied). In the 3D code SHDOM [140], DOM is used in conjunction with the spherical harmonics method to compute the source function. An important feature of SHDOM is the adaptive grid technique. This technique improves the solution accuracy by increasing the spatial resolution in regions where the source function is changing more rapidly.

In order to perform integration along the characteristic, some assumptions have to be made regarding the behavior of the radiance field and the source function. An extensive analysis of several characteristics schemes can be found in [141]. Apart from characteristics schemes, integro-interpolation schemes which deal with the balanced equation for a cell can be used (this method is implemented in the RADUGA code [142]).

4.22.2 Stochastic Radiative Transfer Models

Real clouds are an inhomogeneous three-dimensional scattering medium. Moreover, the precise structure of clouds remains unknown so that there is not sufficient information to formulate a fully three-dimensional problem. Instead, a cloud can be characterized by a small number of statistical properties. In this regard, two computational strategies can be proposed for simulating the radiance field and obtaining unknown statistical characteristics of the radiance field. The first one is based on multidimensional deterministic models and consists in the following steps:

1. Simulation of sampling random realizations of an inhomogeneous cloudy field;
2. Solution of radiative transfer equation for each realization, and
3. Averaging of the obtained solutions over the ensemble of cloudiness realizations.

For each realization of a cloud field, a three-dimensional radiative transfer problem has to be solved by using either the Monte-Carlo method or by deterministic multidimensional solvers.

In the second strategy, cloud fields are regarded as stochastic scattering media due to their internal inhomogeneity and stochastic geometry. The radiative transfer through these media is described by stochastic radiative transfer models (SRTM), in which new transport equations, relating the statistical parameters of the clouds to those of the radiance field, are derived. The goal of this approach is to obtain a relatively simple relationship between the statistical parameters of clouds and radiation that can be numerically evaluated in practice. Within the second strategy, the computations of the radiance field can be organized as follows:

1. Simulation of random realizations of an inhomogeneous cloudy field,
2. Obtaining the relevant statistical structure information from a set of realizations of a cloudy field, and
3. Solution of the stochastic radiative transfer problem using as input the cloud statistics parameters obtained at the previous step.

In this way, time-consuming three-dimensional computations are avoided, but the SRTM is less accurate than 3D models. However, the SRTM seems to be a good compromise between accuracy and computational costs.

Techniques of computing the mean outgoing radiation field from any type of stochastic ensemble of cloud structures can be broadly categorized into two major groups. The first group is based on the Monte-Carlo algorithm for the scattered medium in which some optical parameters are random functions. To minimize computational costs, the random cloud field is modeled together with the photon trajectories. The details can be found in [143, 144]. This technique proved to be very efficient for forward simulations, in particular, for computing the radiance field in broken clouds.

In the second group of methods, the analytical procedure of statistical averaging is applied directly to the radiative transfer equation. A corresponding SRTM for analyzing the statistical structure of the radiance field in media with random spatial fluctuations of the optical properties was studied in, e.g., [71, 145–147].

The idea of statistical averaging consists in representing the cloud extinction field $\sigma(r)$ as

$$\sigma(r) = f(r)\,\bar{\sigma}(r), \qquad (4.317)$$

where $\bar{\sigma}(r)$ is a deterministic function, while f is the random geometric function, which reflects the statistics of the cloud field. Note that for broken clouds, described by binary statistical mixtures, we have $f = 1$ inside the cloud, and $f = 0$ outside the cloud. The function f is known as the random indicator function. Obviously, the fluctuations in f lead to the corresponding fluctuations the radiance L. Then, let us consider the Reynolds decomposition for f and L, namely,

$$f = \bar{f} + f', \quad L = \bar{L} + L', \qquad (4.318)$$

where $\bar{f}$ and $\bar{L}$ are the mean values, and f' and L' are the fluctuations of the indicator function and radiance, respectively (as shown in Fig. 4.24). Substituting Eqs. (4.317) and (4.318) in the original radiative transfer equation and taking the ensemble average we derive a new radiative transfer equation with respect to $\bar{L}$. It also contains an unknown covariance term $\overline{L'f'}$. Repeating the steps for this equation, we obtain an infinite system (a hierarchy) of stochastic equations involving higher-order covariance terms. For broken clouds, there is a simple closure relation (namely, $f = f^2$) and only first two equations remain. The resulting system of equations can be solved by using the discrete ordinate method [148]. A review of stochastic radiative transfer models can be found in [149].

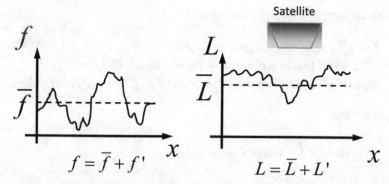

Fig. 4.24 Fluctuations of the indicator function f and the corresponding fluctuations of the radiance in the case of stochastic cloud inhomogeneity

Appendix 1: Azimuthal Expansion of the Radiative Transfer Equation

Consider the multiple scattering term (Eq. (4.18)):

$$J_{ms}(z, \mu, \varphi) = \frac{\sigma_{sca}(z)}{4\pi} \int_0^{2\pi} \int_{-1}^1 p(z, \mu', \mu, \varphi - \varphi') L(z, \mu', \varphi') d\mu' d\varphi'.$$

(4.319)

Substituting expansions of L (Eq. (4.34)) and p (Eq. (4.35)) into Eq. (4.18) gives

$$J_{ms}(z, \mu, \varphi) =$$
$$\frac{\sigma_{sca}(z)}{4\pi} \int_0^{2\pi} \int_{-1}^1 \left\{ \left[\sum_{m'=0}^n (2 - \delta_{m'0}) \sum_{n=m'}^{\infty} \chi_n \bar{P}_n^{m'}(\mu) \bar{P}_n^{m'}(\mu') \cos[m'(\varphi - \varphi')] \right] \right.$$
$$\times \left. \left[\sum_{m=0}^{\infty} L_m(z, \mu') \cos m\varphi' \right] \right\} d\mu' d\varphi'.$$

Note that to avoid confusion, the summation in the second sum is performed over m' rather than m. Interchanging integration and summation we have

$$\frac{\sigma_{sca}(z)}{4\pi} \sum_{m=0}^M \sum_{m'=0}^n (2 - \delta_{m'0}) \int_0^{2\pi} \cos[m'(\varphi - \varphi')] \cos m\varphi' d\varphi' \qquad (4.320)$$

$$\times \int_{-1}^1 \sum_{n=m'}^{\infty} L_m(z, \mu') \chi_n \bar{P}_n^{m'}(\mu) \bar{P}_n^{m'}(\mu') d\mu'. \qquad (4.321)$$

Consider the integral over $d\varphi'$ and denote it as I:

$$I = \int_0^{2\pi} \cos m\varphi' \cos m' (\varphi - \varphi') d\varphi' =$$
$$\int_0^{2\pi} \cos m\varphi' \left[\cos m'\varphi \cos m'\varphi' + \sin m'\varphi \sin m'\varphi' \right] d\varphi' =$$
$$\cos m'\varphi \int_0^{2\pi} \cos m\varphi' \cos m'\varphi' d\varphi' + \sin m'\varphi \int_0^{2\pi} \cos m\varphi' \sin m'\varphi' d\varphi'.$$

The second term is

$$\sin m'\varphi \int_0^{2\pi} \cos m\varphi' \sin m'\varphi' d\varphi' =$$
$$\frac{1}{2} \sin m'\varphi \int_0^{2\pi} \left[\sin \left(m'\varphi' - m\varphi' \right) + \sin \left(m'\varphi' + m\varphi' \right) \right] d\varphi'$$

and equals to zero for any m and m'. The first term is

$$\cos m'\varphi \int_0^{2\pi} \cos m\varphi' \cos m'\varphi' d\varphi' =$$
$$\frac{1}{2} \cos m'\varphi \int_0^{2\pi} \left[\cos \varphi' \left(m + m' \right) + \cos \varphi' \left(m - m' \right) \right] d\varphi'$$
$$= \frac{1}{2} \cos m'\varphi \int_0^{2\pi} \cos \varphi' \left(m + m' \right) d\varphi' + \frac{1}{2} \cos m'\varphi \int_0^{2\pi} \cos \varphi' \left(m - m' \right) d\varphi'.$$

Obviously, the first integral here is not zero for $m = m' = 0$, while the second is not zero if $m = m'$. Thus, their sum equals

$$I = (1 + \delta_{m0}) \pi \cos m\varphi. \tag{4.322}$$

Substituting this result into Eq. (4.320) gives

$$J_{\text{ms}}(z, \mu, \varphi) = \frac{\sigma_{\text{sca}}(z)}{4\pi} \sum_{m=0}^{M} (2 - \delta_{m0}) (1 + \delta_{m0}) \pi \cos m\varphi$$

$$\times \int_{-1}^{1} \sum_{n=m'}^{\infty} L_m (z, \mu') \chi_n \bar{P}_n^{m'} (\mu) \bar{P}_n^{m'} (\mu') d\mu'.$$

Finally, we note that $(2 - \delta_{m0}) (1 + \delta_{m0}) = 2$ for any $m \geq 0$, and that proves Eq. (4.40).

Appendix 2: Azimuthal Expansion of the Boundary Condition at The lower Boundary

Consider the boundary condition at the ground (cf. Eq. (4.30)):

$$
L\left(z_s, \mu > 0, \varphi\right) = \frac{1}{\pi} \int_0^{2\pi} \int_0^1 L\left(z_s, -\mu', \varphi'\right) \rho\left(\mu, \mu', \varphi - \varphi'\right) \mu' d\mu' d\varphi'
$$
$$
+ \mu_0 \frac{F_\odot}{\pi} \exp\left(-\frac{\tau(z_s, z_{TOA})}{\mu_0}\right) \rho\left(\mu, \mu_0, \varphi\right). \tag{4.323}
$$

Following Eq. (4.34) we have

$$
L\left(z_s, \mu > 0, \varphi\right) = \sum_{m=0}^M L_m\left(z_s, \mu > 0\right) \cos m\varphi \tag{4.324}
$$

and

$$
L\left(z_s, \mu < 0, \varphi'\right) = \sum_{m=0}^M L_m\left(z_s, \mu < 0\right) \cos m\varphi'. \tag{4.325}
$$

The BRDF is expanded as follows:

$$
\rho\left(\mu, \mu', \varphi - \varphi'\right) = \sum_{m=0}^M \rho_m\left(\mu, \mu'\right) \cos m\left(\varphi - \varphi'\right). \tag{4.326}
$$

Substitution of Eqs. (4.325) and (4.326) into integral term of Eq. (4.323) gives:

$$
\frac{1}{\pi} \int_0^{2\pi} \int_0^1 L\left(z_s, -\mu', \varphi'\right) \rho\left(\mu, \mu', \varphi - \varphi'\right) \mu' d\mu' d\varphi' =
$$
$$
= \frac{1}{\pi} \int_0^{2\pi} \int_0^1 \left[\sum_{m=0}^M L_m\left(z_s, \mu < 0\right) \cos m\varphi'\right]
$$
$$
\times \left[\sum_{m=0}^M \rho_m\left(\mu, \mu'\right) \cos m\left(\varphi - \varphi'\right)\right] \mu' d\mu' d\varphi'.
$$

Note that to avoid confusion, the summation in the second sum is performed over m' rather than m. Interchanging integration and summation we have

$$
\frac{1}{\pi} \sum_{m=0}^M \sum_{m'=0}^M \int_0^{2\pi} \int_0^1 \left[L_m\left(z_s, \mu < 0\right) \cos m\varphi'\right]
$$
$$
\times \left[\rho_{m'}\left(\mu, \mu'\right) \cos m'\left(\varphi - \varphi'\right)\right] \mu' d\mu' d\varphi'.
$$

The integration over φ' and μ' can be performed independently

$$\frac{1}{\pi} \sum_{m=0}^{M} \sum_{m'=0}^{M} \int_0^1 \left[L_m \left(z_s, \mu < 0 \right) \rho_{m'} \left(\mu, \mu' \right) \right] \mu' d\mu'$$

$$\times \int_0^{2\pi} \cos m\varphi' \cos m' \left(\varphi - \varphi' \right) d\varphi'.$$

The integral over φ' was already considered in Appendix 1. Making use of Eq. (4.322) we have

$$\frac{1}{\pi} \int_0^{2\pi} \int_0^1 L \left(z_s, -\mu', \varphi' \right) \rho \left(\mu, \mu', \varphi - \varphi' \right) \mu' d\mu' d\varphi' =$$
$$= \sum_{m=0}^M \int_0^1 \left[L_m \left(z_s, \mu < 0 \right) \rho_m \left(\mu, \mu' \right) \right] \mu' d\mu' \left(1 + \delta_{m0} \right) \cos m\varphi. \qquad (4.327)$$

The expansion is trivial for the second term in Eq.(4.323):

$$\frac{\mu_0 F_\odot}{\pi} \exp \left(-\frac{\tau(z_s, z_{TOA})}{\mu_0} \right) \rho \left(\mu, \mu_0, \varphi \right)$$
$$= \frac{\mu_0 F_\odot}{\pi} \exp \left(-\frac{\tau(z_s, z_{TOA})}{\mu_0} \right) \sum_{m=0}^M \rho_0 \left(\mu, \mu_0 \right) \cos m\varphi. \qquad (4.328)$$

Finally, it follows:

$$L \left(z_s, \mu > 0, \varphi \right) = \sum_{m=0}^M L_m \left(z_s, \mu > 0 \right) \cos m\varphi. \qquad (4.329)$$

Substitution of Eqs. (4.327)–(4.329) into Eq. (4.323) proves Eq. (4.44).

References

1. S. Chandrasekhar, *Radiative transfer*. Dover publications, inc. New York (1950)
2. Y.L. Pan, S. Hill, R.G. Pinnick, H. Huang, J. Bottiger, R. Chang, Fluorescence spectra of atmospheric aerosol particles measured using one or two excitation wavelengths: Comparison of classification schemes employing different emission and scattering results. Opt. Express **18**(12), 12436 (2010). https://doi.org/10.1364/oe.18.012436
3. J.F. Grainger, J. Ring, Anomalous Fraunhofer line profiles. Nature **193**(4817), 762–762 (1962). https://doi.org/10.1038/193762a0
4. C. McLinden, (1999). https://www.ess.uci.edu/~cmclinden/link/xx/. [Online; accessed 14 May 2019]
5. N. Shefov, Intensities of some emissions of the twilight and night sky. Spektr. Elektrofotom. Radiolokatsionnye Issled. Polyarn. Siyanii Svecheniya Nochnogo Neba **1**, 25–29 (1959). [in Russian]
6. J. Noxon, R. Goody, Noncoherent scattering of skylight. Atmos. Ocean Phys. **1**, 163–166 (1965)
7. A.W. Harrison, Diurnal variation of the ring effect. Can. J. Physics **54**(10), 1000–1005 (1976). https://doi.org/10.1139/p76-117

8. R.T. Brinkmann, Rotational raman scattering in planetary atmospheres. Astro. J. **154**, 1087 (1968). https://doi.org/10.1086/149827

9. A. Smekal, Zur Quantentheorie der Dispersion. Die Naturwissenschaften **11**(43), 873–875 (1923). https://doi.org/10.1007/bf01576902. [in German]

10. C. Raman, A new radiation. Indian J. Phys. **2**, 387–398 (1927)

11. A. Mohammed, Theoretical studies of Raman scattering (PhD dissertation). KTH, Stockholm. http://urn.kb.se/resolve?urn=urn:nbn:se:kth:diva-28332 (2011). [Online; accessed 1 June 2020]

12. G.W. Kattawar, A.T. Young, T.J. Humphreys, Inelastic scattering in planetary atmospheres. I - The Ring effect, without aerosols. The Astrophysical Journal **243**, 1049 (1981). https://doi.org/10.1086/158669

13. M. Conde, P. Greet, F. Jacka, The ring effect in the sodium d2 fraunhofer line of day skylight over mawson, antarctica. Journal of Geophysical Research **97**(D11), 11561 (1992). https://doi.org/10.1029/92jd00768

14. D.J. Fish, R.L. Jones, Rotational raman scattering and the ring effect in zenith-sky spectra. Geophysical Research Letters **22**(7), 811–814 (1995). https://doi.org/10.1029/95gl00392

15. K.V. Chance, R.J.D. Spurr, Ring effect studies: Rayleigh scattering, including molecular parameters for rotational raman scattering, and the fraunhofer spectrum. Applied Optics **36**(21), 5224 (1997). https://doi.org/10.1364/ao.36.005224

16. L. Lelli, V. Rozanov, M. Vountas, J. Burrows, Polarized radiative transfer through terrestrial atmosphere accounting for rotational raman scattering. J Quant Spectrosc Radiat Transfer **200**, 70–89 (2017). https://doi.org/10.1016/j.jqsrt.2017.05.027

17. R. Spurr, J. de Haan, R. van Oss, A. Vasilkov, Discrete-ordinate radiative transfer in a stratified medium with first-order rotational Raman scattering. J Quant Spectrosc Radiat Transfer **109**(3), 404–425 (2008). https://doi.org/10.1016/j.jqsrt.2007.08.011

18. V. Rozanov, M. Vountas, Radiative transfer equation accounting for rotational raman scattering and its solution by the discrete-ordinates method. J Quant Spectrosc Radiat Transfer **133**, 603–618 (2014). https://doi.org/10.1016/j.jqsrt.2013.09.024

19. J. Landgraf, O. Hasekamp, R. van Deelen, I. Aben, Rotational raman scattering of polarized light in the earth atmosphere: a vector radiative transfer model using the radiative transfer perturbation theory approach. J Quant Spectrosc Radiat Transfer **87**(3-4), 399–433 (2004). https://doi.org/10.1016/j.jqsrt.2004.03.013

20. R. van Deelen, J. Landgraf, I. Aben, Multiple elastic and inelastic light scattering in the earth's atmosphere: a doubling-adding method to include rotational raman scattering by air. J Quant Spectrosc Radiat Transfer **95**(3), 309–330 (2005). https://doi.org/10.1016/j.jqsrt.2004.11.002

21. W. Cochran, Raman scattering as a probe of planetary atmospheres. Advances in Space Research **1**(9), 143–153 (1981). https://doi.org/10.1016/0273-1177(81)90228-3

22. A. Oklopčić, C.M. Hirata, K. Heng, Raman scattering by molecular hydrogen and nitrogen in exoplanetary atmospheres. The Astrophysical Journal **832**(1), 30 (2016). https://doi.org/10.3847/0004-637x/832/1/30

23. E.A. Milne, The reflection effect in eclipsing binaries. Monthly Notices of the Royal Astronomical Society **87**(1), 43–55 (1926). https://doi.org/10.1093/mnras/87.1.43

24. E.A. Milne, Radiative equilibrium in the outer layers of a star: the temperature distribution and the law of darkening. Monthly Notices of the Royal Astronomical Society **81**(5), 361–375 (1921). https://doi.org/10.1093/mnras/81.5.361

25. M.M.R. Williams, The Milne problem in radiative transfer. Monthly Notices of the Royal Astronomical Society **140**(3), 403–407 (1968). https://doi.org/10.1093/mnras/140.3.403

26. V. Ambartsumyan, The scattering of light by planetary atmospheres. Aston. Zh. **19**, 30 (1942). [in Russian]

27. V.S. Remizovich, Unconventional application of the two-flux approximation for the calculation of the ambartsumyan-chandrasekhar function and the angular spectrum of the backward-scattered radiation for a semi-infinite isotropically scattering medium. Laser Physics **20**(6), 1368–1389 (2010). https://doi.org/10.1134/s1054660x10120030

28. V.V. Grigoryev, D.I. Nagirner, S.I. Grachev, H-functions in radiative transfer theory: Calculation of Voigt functions and justification of a model for formation of cyclotron lines in the spectra of neutron stars. Astrophysics **62**(1), 129–146 (2019). https://doi.org/10.1007/s10511-019-09569-4

29. R. Spurr, V. Natraj, A linearized two-stream radiative transfer code for fast approximation of multiple-scatter fields. J Quant Spectrosc Radiat Transfer **112**(16), 2630–2637 (2011). https://doi.org/10.1016/j.jqsrt.2011.06.014

30. C. Dullemond, What makes radiative transfer hard, and how to solve it - an introduction. in *Radiative Transfer in Astrophysics Theory, Numerical Methods and Applications*, chap. 4. University of Heidelberg (2012). http://www.ita.uniheidelberg.de/~dullemond/lectures/radtrans_2012/Chapter_4.pdf

31. G.C. Wick, Über ebene Diffusionsprobleme. Zeitschrift für Physik **121**(11-12), 702–718 (1943). https://doi.org/10.1007/bf01339167. [in German]

32. A. Schuster, Radiation through a foggy atmosphere. Astro. J **21**, 1 (1905). https://doi.org/10.1086/141186

33. K. Schwarzschild, Über das Gleichgewicht der Sonnenatmosphäre. Nachr. Konig. Gesel. der Wiss., Gottingen **1**, 41–53 (1906). [in German]

34. P.M. Anselone, Convergence of the Wick-Chandrasekhar approximation technique in radiative transfer. The Astrophysical Journal **128**, 124 (1958). https://doi.org/10.1086/146523

35. C. Gauss, Methodus nova integralium valores per approximationem inveniendi. Comment. Soc. Reg. Scient. Gotting. Recent. pp. 39–76 (1814)

36. M. Abramowitz, I. Stegun, *Handbook of Mathematical Functions: with Formulas, Graphs, and Mathematical Tables*. Dover Publications; 0009-Revised edition (1965)

37. J.B. Sykes, Approximate integration of the equation of transger. Monthly Notices of the Royal Astronomical Society **111**(4), 377–386 (1951). https://doi.org/10.1093/mnras/111.4.377

38. K. Stamnes, G. Thomas, J. Stamnes, *Radiative Transfer in the Atmosphere and Ocean*. Cambridge University Press (2017). https://doi.org/10.1017/9781316148549

39. C. Clenshaw, A. Curtis, A method for numerical integration on an automatic computer. Numerische Mathematik **2**(1), 197–205 (1960). https://doi.org/10.1007/bf01386223

40. J. Weideman, L. Trefethen, The kink phenomenon in fejér and clenshaw-curtis quadrature. Numerische Mathematik **107**(4), 707–727 (2007). https://doi.org/10.1007/s00211-007-0101-2

41. K. Stamnes, S. Tsay, W. Wiscombe, K. Jayaweera, Numerically stable algorithm for discrete-ordinate-method radiative transfer in multiple scattering and emitting layered media. Appl. Opt. **12**, 2502–2509 (1988). https://doi.org/10.1364/AO.27.002502

42. P. Waterman, Matrix-exponential description of radiative transfer. J Opt Soc Am **71**(4), 410–22 (1981). https://doi.org/10.1364/JOSA.71.000410

43. P. Flatau, G. Stephens, On the fundamental solution of the radiative transfer equation. J. Geophys. Res. **93**(D9), 11037 (1988). https://doi.org/10.1029/jd093id09p11037

44. A. Doicu, T. Trautmann, Discrete-ordinate method with matrix exponential for a pseudo-spherical atmosphere: Scalar case. J Quant Spectrosc Radiat Transfer **110**(1-2), 146–158 (2009). https://doi.org/10.1016/j.jqsrt.2008.09.014

45. A.H. Karp, J. Greenstadt, J.A. Fillmore, Radiative transfer through an arbitrarily thick, scattering atmosphere. J Quant Spectrosc Radiat Transfer **24**(5), 391–406 (1980). https://doi.org/10.1016/0022-4073(80)90074-6

46. K. Stamnes, R. Swanson, A new look at the discrete ordinate method for radiative transfer calculations in anisotropically scattering atmospheres. J Atmos Sci **38**(2), 387–389 (1981). https://doi.org/10.1175/1520-0469(1981)038<0387:ANLATD>2.0.CO;2

47. D. Efremenko, A. Doicu, D. Loyola, T. Trautmann, Acceleration techniques for the discrete ordinate method. J Quant Spectrosc Radiat Transfer **114**, 73–81 (2013). https://doi.org/10.1016/j.jqsrt.2012.08.014

48. S. Korkin, A. Lyapustin, Matrix exponential in C/C++ version of vector radiative transfer code IPOL. J Quant Spectrosc and Radiat Transfer **227**, 106–110 (2019). https://doi.org/10.1016/j.jqsrt.2019.02.009

49. T. Nakajima, M. Tanaka, Matrix formulations for the transfer of solar radiation in a plane-parallel scattering atmosphere. J Quant Spectrosc Radiat Transfer **35**(1), 13–21 (1986). https://doi.org/10.1016/0022-4073(86)90088-9

50. K. Stamnes, S. Tsay, T. Nakajima, Computation of eigenvalues and eigenvectors for the discrete ordinate and matrix operator methods in radiative transfer. J Quant Spectrosc Radiat Transfer **39**(5), 415–419 (1988). https://doi.org/10.1016/0022-4073(88)90107-0

51. R. Spurr, LIDORT and VLIDORT: Linearized pseudo-spherical scalar and vector discrete ordinate radiative transfer models for use in remote sensing retrieval problems. in A.A. Kokhanovsky (ed.) *Light Scattering Reviews* 3, vol. 3, pp. 229–275. Springer (2008). https://doi.org/10.1007/978-3-540-48546-9_7

52. K. Stamnes, S.C. Tsay, W. Wiscombe, I. Laszlo, DISORT - A general-purpose FORTRAN program for Discrete-Ordinate Method radiative transfer in scattering and emitting layered media: documentation of methodology, version 1.1 (2000). https://www.meteo.physik.uni-muenchen.de/~emde/lib/exe/fetch.php?media=teaching:radiative_transfer:disortreport1.1.pdf

53. V. Budak, G. Kaloshin, O. Shagalov, V. Zheltov, Numerical modeling of the radiative transfer in a turbid medium using the synthetic iteration. Optics Express **23**(15), A829 (2015). https://doi.org/10.1364/oe.23.00a829

54. E. Chalhoub, R. Garcia, The equivalence between two techniques of angular interpolation for the discrete-ordinates method. J Quant Spectrosc Radiat Transfer **64**(5), 517–535 (2000). https://doi.org/10.1016/s0022-4073(99)00134-x

55. I. Grant, G. Hunt, Discrete space theory of radiative transfer i. fundamentals. Proceedings of the Royal Society of London. A. Mathematical and Physical Sciences **313**(1513), 183–197 (1969). https://doi.org/10.1098/rspa.1969.0187

56. L.M. Delves, J.L. Mohamed, *Computational Methods for Integral Equations*. Cambridge University Press. Cambridge (1985)

57. M. Tanaka, T. Nakajima, Effects of oceanic turbidity and index of refraction of hydrosols on the flux of solar radiation in the atmosphere-ocean system. J Quant Spectrosc Radiat Transfer **18**(1), 93 – 111 (1977). https://doi.org/10.1016/0022-4073(77)90130-3

58. M. Tanaka, T. Nakajima, Effect of wind-generated waves on the transfer of solar radiation in the atmosphere-ocean system. J Quant Spectrosc Radiat Transfer **29**(6), 521 – 537 (1983). https://doi.org/10.1016/0022-4073(83)90129-2

59. Z. Jin, K. Stamnes, Radiative transfer in nonuniformly refracting layered media: atmosphere–ocean system. Appl Opt **33**(3), 431–442 (1994). https://doi.org/10.1364/AO.33.000431

60. Z. Jin, T. Charlock, K. Rutledge, K. Stamnes, Y. Wang, Analytical solution of radiative transfer in the coupled atmosphere-ocean system with a rough surface. Appl Opt **45**(28), 7443–7455 (2006). https://doi.org/10.1364/AO.45.007443

61. Y. Ota, A. Higurashi, T. Nakajima, T. Yokota, Matrix formulations of radiative transfer including the polarization effect in a coupled atmosphere-ocean system. J Quant Spectrosc Radiat Transfer **111**(6), 878–894 (2010). https://doi.org/10.1016/j.jqsrt.2009.11.021

62. V. Budak, D. Klyuykov, S. Korkin, Convergence acceleration of radiative transfer equation solution at strongly anisotropic scattering. In A. Kokhanovsky (ed.) *Light scattering reviews*, vol. 5, pp. 147–203. Springer Berlin Heidelberg (2010). https://doi.org/10.1007/978-3-642-10336-0_5

63. R. Spurr, K. Stamnes, H. Eide, W. Li, K. Zhang, J. Stamnes, Simultaneous retrieval of aerosols and ocean properties: A classic inverse modeling approach. i. analytic jacobians from the linearized cao-disort model. J Quant Spectrosc Radiat Transfer **104**(3), 428–449 (2007). https://doi.org/10.1016/j.jqsrt.2006.09.009

64. L. Tsang, J.A. Kong, R.T. Shin, *Theory of microwave remote sensing*. Wiley-Interscience; 1st edition (1985)

65. A. Doicu, M. Mishchenko, Radiative transfer in a discrete random medium adjacent to a half-space with a rough interface. J Quant Spectrosc Radiat Transfer **218**, 194–202 (2018). https://doi.org/10.1016/j.jqsrt.2018.07.016

66. R.F. van Oss, R. Spurr, Fast and accurate 4 and 6 stream linearized discrete ordinate radiative transfer models for ozone profile retrieval. J Quant Spectrosc Radiat Transfer **75**(2), 177–220 (2002). https://doi.org/10.1016/s0022-4073(01)00246-1

67. C. O'Dell, Acceleration of multiple-scattering, hyperspectral radiative transfer calculations via low-streams interpolation. Journal of Geophysical Research **115**(D10) (2010). https://doi.org/10.1029/2009jd012803

68. V. Afanas'ev, A.Y. Basov, V. Budak, D. Efremenko, A. Kokhanovsky, Analysis of the discrete theory of radiative transfer in the coupled "ocean–atmosphere" system: Current status, problems and development prospects. Journal of Marine Science and Engineering **8**(3), 202 (2020). https://doi.org/10.3390/jmse8030202

69. C. Shannon, Communication in the presence of noise. Proceedings of the IRE **37**(1), 10–21 (1949). https://doi.org/10.1109/jrproc.1949.232969

70. J. Dave, Importance of higher order scattering in a molecular atmosphere. J Opt Soc Am **54**(3), 307–315 (1964). https://doi.org/10.1364/JOSA.54.000307

71. A. Kokhanovsky, *Cloud Optics*. Springer Netherlands (2006). https://doi.org/10.1007/1-4020-4020-2

72. J. Fischer, H. Grassl, Radiative transfer in an atmosphere-ocean system: an azimuthally dependent matrix-operator approach. Applied Optics **23**(7), 1032 (1984). https://doi.org/10.1364/ao.23.001032

73. S. Prahl, M. van Gemert, A. Welch, Determining the optical properties of turbid media by using the adding-doubling method. Applied Optics **32**(4), 559 (1993). https://doi.org/10.1364/ao.32.000559

74. J.E. Hansen, Radiative transfer by doubling very thin layers. The Astrophysical Journal **155**, 565 (1969). https://doi.org/10.1086/149892

75. G.N. Plass, G.W. Kattawar, F.E. Catchings, Matrix operator theory of radiative transfer 1: Rayleigh scattering. Applied Optics **12**(2), 314 (1973). https://doi.org/10.1364/ao.12.000314

76. W. Wiscombe, On initialization, error and flux conservation in the doubling method. J Quant Spectrosc Radiat Transfer **16**(8), 637–658 (1976). https://doi.org/10.1016/0022-4073(76)90056-x

77. D. Efremenko, V. Molina García, S. Gimeno García, A. Doicu, A review of the matrix-exponential formalism in radiative transfer. J Quant Spectrosc Radiat Transfer **196**, 17–45 (2017). https://doi.org/10.1016/j.jqsrt.2017.02.015

78. Q. Liu, F. Weng, Advanced doubling-adding method for radiative transfer in planetary atmospheres. Journal of the Atmospheric Sciences **63**(12), 3459–3465 (2006). https://doi.org/10.1175/jas3808.1

79. J. Lenoble, *Radiative transfer in scattering and absorbing atmospheres: Standard computational procedures*. Hampton, VA, A. Deepak Publishing (1985)

80. A. Hammad, S. Chapman, VII. The primary and secondary scattering of sunlight in a plane-stratified atmosphere of uniform composition. The London, Edinburgh, and Dublin Philosophical Magazine and Journal of Science **28**(186), 99–110 (1939). https://doi.org/10.1080/14786443908521165

81. J. Hovenier, Multiple scattering of polarized light in planetary atmospheres. Astronomy and Astrophysics **13**, 7 (1971)

82. V. Natraj, R. Spurr, A fast linearized pseudo-spherical two orders of scattering model to account for polarization in vertically inhomogeneous scattering–absorbing media. J Quant Spectrosc Radiat Transfer **107**(2), 263–293 (2007). https://doi.org/10.1016/j.jqsrt.2007.02.011

83. S. Korkin, A. Lyapustin, A. Marshak, On the accuracy of double scattering approximation for atmospheric polarization computations. J Quant Spectrosc Radiat Transfer **113**(2), 172–181 (2012). https://doi.org/10.1016/j.jqsrt.2011.10.008

84. J. Lenoble, M. Herman, J. Deuzé, B. Lafrance, R. Santer, D. Tanré, A successive order of scattering code for solving the vector equation of transfer in the earth's atmosphere with aerosols. J Quant Spectrosc Radiat Transfer **107**(3), 479–507 (2007). https://doi.org/10.1016/j.jqsrt.2007.03.010

85. Q. Min, M. Duan, A successive order of scattering model for solving vector radiative transfer in the atmosphere. J Quant Spectrosc Radiat Transfer **87**(3-4), 243–259 (2004). https://doi.org/10.1016/j.jqsrt.2003.12.019

86. P.W. Zhai, Y. Hu, C.R. Trepte, P.L. Lucker, A vector radiative transfer model for coupled atmosphere and ocean systems based on successive order of scattering method. Optics Express **17**(4), 2057 (2009). https://doi.org/10.1364/oe.17.002057

87. M. Duan, Q. Min, D. Lü, A polarized radiative transfer model based on successive order of scattering. Advances in Atmospheric Sciences **27**(4), 891–900 (2010). https://doi.org/10.1007/s00376-009-9049-8

88. G. McGarragh, P. Gabriel, Efficient computation of radiances for optically thin media by padé approximants. J Quant Spectrosc Radiat Transfer **111**(12-13), 1885–1899 (2010). https://doi.org/10.1016/j.jqsrt.2010.03.011

89. S. Goudsmit, J. Saunderson, Multiple scattering of electrons. Phys Rev **57**, 24–9 (1940)

90. L. Dolin, Propagation of a narrow light beam in a random medium. Izv. VUZov, Radiofiz **7**, 380–391 (1964). In Russian

91. A. Kokhanovsky, Small-angle approximations of the radiative transfer theory. Journal of Physics D: Applied Physics **30**(20), 2837–2840 (1997). https://doi.org/10.1088/0022-3727/30/20/009

92. G. Szegö, *Orthogonal Polynomials*. American Mathematical Society, Providence, Rhode Island (1992)

93. E. Zege, A. Kokhanovsky, Analytical solution to the optical transfer function of a scattering medium with large particles. Applied Optics **33**(27), 6547 (1994). https://doi.org/10.1364/ao.33.006547

94. V. Budak, D. Klyuykov, S. Korkin, Complete matrix solution of radiative transfer equation for pile of horizontally homogeneous slabs. J Quant Spectrosc Radiat Transfer **112**(7), 1141–1148 (2011). https://doi.org/10.1016/j.jqsrt.2010.08.028

95. V. Budak, D. Efremenko, O. Shagalov, Efficiency of algorithm for solution of vector radiative transfer equation in turbid medium slab. Journal of Physics: Conference Series **369**, 012021 (2012). https://doi.org/10.1088/1742-6596/369/1/012021

96. V.V. Sobolev, *Light Scattering in Planetary Atmospheres*. Pergamon Press (1972)

97. E. Yanovitskij, *Light Scattering in Inhomogeneous Atmospheres*. Springer Berlin Heidelberg (1997). https://doi.org/10.1007/978-3-642-60465-2

98. W.M. Irvine, Multiple scattering in planetary atmospheres. Icarus **25**(2), 175–204 (1975). https://doi.org/10.1016/0019-1035(75)90019-6

99. V. Ambartsumyan, On the problem of diffuse light reflection by a turbid medium. Dokl. Akad. Nauk SSSR **38**, 257 (1943). [in Russian]

100. M. Mishchenko, J. Dlugach, E. Yanovitskij, N. Zakharova, Bidirectional reflectance of flat, optically thick particulate layers: an efficient radiative transfer solution and applications to snow and soil surfaces. J Quant Spectrosc Radiat Transfer **63**(2-6), 409–432 (1999). https://doi.org/10.1016/s0022-4073(99)00028-x

101. V. Afanas'ev, D. Efremenko, A. Lubenchenko, On the application of the invariant embedding method and the radiative transfer equation codes for surface state analysis. In *Light Scattering Reviews* 8, pp. 363–423. Springer Berlin Heidelberg (2013). https://doi.org/10.1007/978-3-642-32106-1_8

102. V. Afanas'ev, D. Efremenko, P. Kaplya, Analytical and numerical methods for computing electron partial intensities in the case of multilayer systems. Journal of Electron Spectroscopy and Related Phenomena **210**, 16–29 (2016). https://doi.org/10.1016/j.elspec.2016.04.006. http://dx.doi.org/10.1016/j.elspec.2016.04.006

103. V. Fok, On some integral equations of mathematical physics. Mathematical Collection **14**(56), 3–50 (1944). [in Russian]

104. B. Hapke, *Theory of Reflectance and Emittance Spectroscopy*. Cambridge University Press, New York (1993)

105. A. Kokhanovsky, Reflection of light from nonasbsorbing semi-infinite cloudy media: a simple approximation. J Quant Spectrosc Radiat Transfer **85**(1), 25–33 (2004). https://doi.org/10.1016/s0022-4073(03)00192-4

106. A. Kokhanovsky, Reflection of light from particulate media with irregularly shaped particles. J Quant Spectrosc Radiat Transfer **96**(1), 1–10 (2005). https://doi.org/10.1016/j.jqsrt.2004.12.008

107. A. Kokhanovsky, Simple approximate formula for the reflection function of a homogeneous, semi-infinite turbid medium. Journal of the Optical Society of America A **19**(5), 957 (2002). https://doi.org/10.1364/josaa.19.000957

108. X.Sun, Y. Han, X. Shi, Application of asymptotic theory for computing the reflection of optically thick clouds. Journal of Optics A: Pure and Applied Optics **8**(12), 1074–1079 (2006). https://doi.org/10.1088/1464-4258/8/12/007

109. E. Zege, A. Ivanov, I. Katsev, *Image Transfer Through a Scattering Medium*. Springer Berlin Heidelberg (1991). https://doi.org/10.1007/978-3-642-75286-5

110. J. Potter, The delta function approximation in radiative transfer theory. Journal of the Atmospheric Sciences **27**(6), 943–949 (1970). https://doi.org/10.1175/1520-0469(1970)027<0943:tdfair>2.0.co;2

111. J.E. Hansen, Exact and approximate solutions for multiple scattering by cloudy and hazy planetary atmospheres. Journal of the Atmospheric Sciences **26**(3), 478–487 (1969). https://doi.org/10.1175/1520-0469(1969)026<0478:eaasfm>2.0.co;2

112. C. Mitrescu, G. Stephens, On similarity and scaling of the radiative transfer equation. J Quant Spectrosc Radiat Transfer **86**(4), 387–394 (2004). https://doi.org/10.1016/j.jqsrt.2003.12.028

113. W. Wiscombe, The delta-m method: Rapid yet accurate radiative flux calculations for strongly asymmetric phase functions. J Atmos Sci **34**(9), 1408–1422 (1977). https://doi.org/10.1175/1520-0469(1977)034<1408:TDMRYA>2.0.CO;2

114. Y.X. Hu, B. Wielicki, B. Lin, G. Gibson, S.C. Tsay, K. Stamnes, T. Wong, δ-fit: A fast and accurate treatment of particle scattering phase functions with weighted singular-value decomposition least-squares fitting. J Quant Spectrosc Radiat Transfer **65**(4), 681–690 (2000). https://doi.org/10.1016/s0022-4073(99)00147-8

115. Z. Lin, N. Chen, Y. Fan, W. Li, K. Stamnes, S. Stamnes, New treatment of strongly anisotropic scattering phase functions: The delta-m+ method. Journal of the Atmospheric Sciences **75**(1), 327–336 (2018). https://doi.org/10.1175/jas-d-17-0233.1

116. V. Rozanov, A. Lyapustin, Similarity of radiative transfer equation: Error analysis of phase function truncation techniques. J Quant Spectrosc Radiat Transfer **111**(12-13), 1964–1979 (2010). https://doi.org/10.1016/j.jqsrt.2010.03.018

117. S. Sanghavi, G. Stephens, Adaptation of the delta-m and delta-fit truncation methods to vector radiative transfer: Effect of truncation on radiative transfer accuracy. J Quant Spectrosc Radiat Transfer **159**, 53–68 (2015). https://doi.org/10.1016/j.jqsrt.2015.03.007

118. T. Nakajima, M. Tanaka, Algorithms for radiative intensity calculations in moderately thick atmos using a truncation approximation. J Quant Spectrosc Radiat Transfer **40**(1), 51–69 (1988). https://doi.org/10.1016/0022-4073(88)90031-3

119. A. Dahlback, K. Stamnes, A new spherical model for computing the radiation field available for photolysis and heating at twilight. Planetary and Space Science **39**(5), 671–683 (1991). https://doi.org/10.1016/0032-0633(91)90061-e

120. R. Spurr, Simultaneous derivation of intensities and weighting functions in a general pseudo-spherical discrete ordinate radiative transfer treatment. J Quant Spectrosc Radiat Transfer **75**(2), 129–175 (2002). https://doi.org/10.1016/S0022-4073(01)00245-X

121. R. Spurr, VLIDORT: A linearized pseudo-spherical vector discrete ordinate radiative transfer code for forward model and retrieval studies in multilayer multiple scattering media. J Quant Spectrosc Radiat Transfer **102**(2), 316–342 (2006). https://doi.org/10.1016/j.jqsrt.2006.05.005

122. O. Postylyakov, Linearized vector radiative transfer model MCC++ for a spherical atmosphere. J Quant Spectrosc Radiat Transfer **88**(1-3), 297–317 (2004). https://doi.org/10.1016/j.jqsrt.2004.01.009

123. C. Emde, B. Mayer, Simulation of solar radiation during a total eclipse: a challenge for radiative transfer. Atmospheric Chemistry and Physics **7**(9), 2259–2270 (2007). https://doi.org/10.5194/acp-7-2259-2007

124. A. Rozanov, V. Rozanov, J. Burrows, Combined differential-integral approach for the radiation field computation in a spherical shell atmosphere: Nonlimb geometry. Journal of Geophysical Research: Atmospheres **105**(D18), 22937–22942 (2000). https://doi.org/10.1029/2000jd900378

125. S. Korkin, E.S. Yang, R. Spurr, C. Emde, N. Krotkov, A. Vasilkov, D. Haffner, J. Mok, A. Lyapustin, Revised and extended benchmark results for rayleigh scattering of sunlight in spherical atmospheres. J Quant Spectrosc Radiat Transfer p. 107181 (2020). https://doi.org/10.1016/j.jqsrt.2020.107181

126. G. Rozenberg, Vector-parameter of stokes (matrix methods for accounting for radiation polarization in the ray optics approximation). Uspekhi Fizicheskih Nauk **56**(5), 77–110 (1955). https://doi.org/10.3367/ufnr.0056.195505c.0077. [in Russian]

127. M. Mishchenko, Vector radiative transfer equation for arbitrarily shaped and arbitrarily oriented particles: a microphysical derivation from statistical electromagnetics. Applied Optics **41**(33), 7114 (2002). https://doi.org/10.1364/ao.41.007114

128. H. Tynes, G. Kattawar, E. Zege, I. Katsev, A. Prikhach, L. Chaikovskaya, Monte Carlo and multicomponent approximation methods for vector radiative transfer by use of effective Mueller matrix calculations. Applied Optics **40**(3), 400 (2001). https://doi.org/10.1364/ao.40.000400

129. J.F. de Haan, P.B. Bosma, J.W. Hovenier, The adding method for multiple scattering calculations of polarized light. Astronomy and Astrophysics **183**(2), 371–391 (1987)

130. V. Boudak, B. Veklenko, Light-field polarization of the point monodirected source in small-angle approximation. In *Seventh International Symposium on Laser Metrology Applied to Science, Industry, and Everyday Life*. SPIE (2002). https://doi.org/10.1117/12.484464

131. H. Domke, E. Yanovitskij, Principles of invariance applied to the computation of internal polarized radiation in multilayered atmospheres. J Quant Spectrosc Radiat Transfer **36**(3), 175–186 (1986). https://doi.org/10.1016/0022-4073(86)90067-1

132. H. Domke, Linear Fredholm integral equations for radiative transfer problems in finite plane parallel media. I. Imbedding in an infinite medium. Astronomische Nachrichten **299**(2), 87–93 (1978). https://doi.org/10.1002/asna.19782990205

133. C. Siewert, A discrete-ordinates solution for radiative-transfer models that include polarization effects. J Quant Spectrosc Radiat Transfer **64**(3), 227–254 (2000). https://doi.org/10.1016/s0022-4073(99)00006-0

134. A. Doicu, T. Trautmann, Discrete-ordinate method with matrix exponential for a pseudo-spherical atmosphere: Vector case. J Quant Spectrosc Radiat Transfer **110**(1-2), 159–172 (2009). https://doi.org/10.1016/j.jqsrt.2008.09.013

135. R. Tapimo, H. Kamdem, D. Yemele, A discrete spherical harmonics method for radiative transfer analysis in inhomogeneous polarized planar atmosphere. Astrophysics and Space Science **363**(3) (2018). https://doi.org/10.1007/s10509-018-3266-5

136. R. Pincus, F. Evans, Computational cost and accuracy in calculating three-dimensional radiative transfer: Results for new implementations of monte carlo and SHDOM. Journal of the Atmospheric Sciences **66**(10), 3131–3146 (2009). https://doi.org/10.1175/2009jas3137.1

137. A. Marshak, A. Davis (eds.), *3D Radiative Transfer in Cloudy Atmospheres* (Springer, Berlin, 2005). https://doi.org/10.1007/3-540-28519-9

138. O.V. Nikolaeva, L.P. Bass, T.A. Germogenova, V.S. Kuznetsov, A.A. Kokhanovsky, Radiative transfer in horizontally and vertically inhomogeneous turbid media. In *Springer Praxis Books*, pp. 295–347. Springer Berlin Heidelberg (2007). https://doi.org/10.1007/978-3-540-68435-0_8

139. G. Marchuk, G. Mikhailov, M. Nazaraliev, R.A. Darbinjan, B. Kargin, B. Elepov, *The Monte Carlo Methods in Atmospheric Optics*, *Springer Series in Optical Sciences*, vol. 12. Springer Berlin Heidelberg (1980). https://doi.org/10.1007/978-3-540-35237-2

140. F. Evans, The spherical harmonics discrete ordinate method for three-dimensional atmospheric radiative transfer. Journal of the Atmospheric Sciences **55**(3), 429–446 (1998). https://doi.org/10.1175/1520-0469(1998)055<0429:tshdom>2.0.co;2

141. A. Doicu, D. Efremenko, T. Trautmann, An analysis of the short-characteristic method for the spherical harmonic discrete ordinate method (SHDOM). J Quant Spectrosc Radiat Transfer **119**, 114–127 (2013). https://doi.org/10.1016/j.jqsrt.2012.12.024

142. O. Nikolaeva, L. Bass, T. Germogenova, A. Kokhanovsky, V. Kuznetsov, B. Mayer, The influence of neighbouring clouds on the clear sky reflectance studied with the 3-d transport code RADUGA. J Quant Spectrosc Radiat Transfer **94**(3-4), 405–424 (2005). https://doi.org/10.1016/j.jqsrt.2004.09.037

143. S. Prigarin, B. Kargin, U. Oppel, Random fields of broken clouds and their associated direct solar radiation, scattered transmission and albedo. Pure Appl. Opt. **7**(6), 1389–1402 (1998). https://doi.org/10.1088/0963-9659/7/6/017

144. B. Kargin Statistical modeling of stochastic problems of the atmosphere and ocean optics. In: G. Matvienko, M. Panchenko (eds.) *Seventh International Symposium on Atmospheric and Ocean Optics*. SPIE-Intl Soc Optical Eng (2000). https://doi.org/10.1117/12.411950

145. O. Anisimov, L. Fukshansky, Stochastic radiation in macroheterogeneous random optical media. J Quant Spectrosc Radiat Transfer **48**(2), 169–186 (1992). https://doi.org/10.1016/0022-4073(92)90087-K

146. G. Titov, T. Zhuravleva, V. Zuev, Mean radiation fluxes in the near-IR spectral range: Algorithms for calculation. Journal of Geophysical Research: Atmospheres **102**(D2), 1819–1832 (1997). https://doi.org/10.1029/96jd02218

147. G. Titov, Statistical description of radiative transfer in clouds. J Atmos Sci **47**(1), 24–38 (1990). https://doi.org/10.1175/1520-0469(1990)047<0024:SDORTI>2.0.CO;2

148. A. Doicu, D., Efremenko, D. Loyola, T. Trautmann, Discrete ordinate method with matrix exponential for stochastic radiative transfer in broken clouds. J Quant Spectrosc Radiat Transfer **138**, 1–16 (2014). https://doi.org/10.1016/j.jqsrt.2014.01.011

149. D. Efremenko, A. Doicu, D. Loyola, T. Trautmann, Fast stochastic radiative transfer models for trace gas and cloud property retrievals under cloudy conditions. In *Springer Series in Light Scattering*, pp. 231–277. Springer International Publishing (2018). https://doi.org/10.1007/978-3-319-70796-9_3

Chapter 5
Inverse Problems

5.1 Direct and Inverse Radiative Transfer Problems

The radiative transfer models considered in previous chapters belong to the so-called *direct problems*. Given *the model parameters* comprising the information about atmospheric constituents and the underlying surface, the *response of the model* (i.e., the radiance field) can be found. By definition, an *inverse problem* is a process of extracting from a set of observations the causal factors that produced them. Thus, the model parameters are retrieved from the response of the model. Inverse problems arise in various fields of science and engineering, including astrophysics, geophysics, quantum mechanics, material science, etc. Moreover, we are confronted with the inverse problems in our daily lives when interpreting observations or making conclusions about parameters that we cannot directly observe. For instance, customers solve inverse problems assessing the quality of food by visually examining it; doctors solve inverse problems setting a diagnosis by interpreting the blood test results. In their treatise, Bohren and Huffman [1] defined metaphorically the concepts of direct and inverse problems in the following way: direct problem is to describe the tracks of a given dragon, while the inverse problem is aimed at describing a dragon from an examination of its tracks. An impressive example of the inverse problem solution is the discovery of Neptune by examining the perturbed trajectory of Uranus [2, 3].

Satellite-based remote sensing aims at deriving the properties of atmosphere and underlying surfaces from the analysis of satellite measurements performed in a broad spectral range from ultraviolet to microwaves. The atmospheric parameters of interest are often continuous along a vertical coordinate (e.g., a trace gas profile). However, as we can only perform a finite number of measurements and calculations, continuous parameters should be represented in a discrete form by using the state vector:

$$\mathbf{x} = (x_1, ..., x_n)^T. \tag{5.1}$$

Here n is the number of unknown parameters which characterize the atmosphere and which we want to retrieve. Thus, it follows: $\mathbf{x} \in \mathbb{R}^n$. For example, for trace

© Springer Nature Switzerland AG 2021
D. Efremenko and A. Kokhanovsky, *Foundations of Atmospheric Remote Sensing*,
https://doi.org/10.1007/978-3-030-66745-0_5

gas vertical profile retrieval, the model state vector consists of number densities of atmospheric constituents defined at discrete altitude levels. Note that the elements of $\mathbf{x}$ do not have to be of the same type (e.g., in addition to number densities, $\mathbf{x}$ can contain the surface albedo, aerosol optical thickness, etc.).

The measurement vector $\mathbf{y}$ incorporates a finite number m of observations

$$\mathbf{y} = (y_1, ..., y_m)^T , \qquad (5.2)$$

which are often the spectral radiances in corresponding channels measured by the sensor onboard a satellite. We may write $\mathbf{y} \in \mathbb{R}^m$. The measurements can be taken at different wavelengths, different geometries and even at different moments of time. The complete set of measurements must include the spectral Stokes vector of reflected light $\mathbf{L}(\lambda, \mathbf{\Omega}_0, \mathbf{\Omega})$. Here $\mathbf{\Omega}_0$ and $\mathbf{\Omega}$ are unit vectors characterizing directions of incident radiation and observation. Some optical instruments have the capability to observe the same ground scene not only at different wavelengths but also from different directions $\mathbf{\Omega}$. This makes it possible to increase the accuracy of retrievals. Future advanced remote sensing spaceborne platforms will hold different instrumentation from ultraviolet to microwaves and one will be able to retrieve all essential atmospheric (trace gases, atmospheric aerosol, cloudiness, temperature/atmospheric pressure profiles) and the underlying surface parameters by analyzing the data coming from a single platform. This will make it possible to pose and solve the inverse problems in its most general form, which includes the determination of all essential atmospheric and the underlying surface parameters. Currently, the number of instruments on satellite platforms is limited and only a partial solution of the inverse atmospheric optics problem is possible for a given combination of instruments or for a single instrument (say, spaceborne spectrometer, polarimeter or imager). Recent advances in spaceborne remote sensing are summarized by Burrows et al. [4] and Liang et al. [5].

Using the terminology of algebra, we can say that the forward model (or the forward operator) F maps from state space of dimension n onto measurement space of dimension m:

$$F : \mathbb{R}^n \to \mathbb{R}^m, \ F(\mathbf{x}) = \mathbf{y}. \qquad (5.3)$$

In our case, the forward model F simulates $\mathbf{y}$ for a given vector $\mathbf{x}$ and incorporates the radiative transfer model accompanied by an appropriate instrument model. To estimate $\mathbf{x}$ from a given set of data $\mathbf{y}$, the forward model should be inverted, i.e.,

$$F^{-1} : \mathbb{R}^m \to \mathbb{R}^n, \ \mathbf{x} = F^{-1}(\mathbf{y}), \qquad (5.4)$$

where F^{-1} is the inverse model (or the inverse operator). The schematic representation of the forward and inverse models is shown in Fig. 5.1. Note that the radiative transfer model considered in the previous chapter is concealed in F. In addition to that, F comprises the instrument-related features, such as the influence of the slit-function, sensor degradation [6], instrument noise etc.

Fig. 5.1 Schematic representation of the forward and inverse models. If we deal with polarimetric measurements, then the Stokes parameters are used instead of spectral radiance

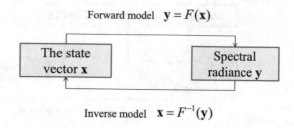

Despite apparent simplicity of Eq. (5.4), there are many difficulties associated with the inversion procedure. The first problem we are confronted to is that F is a nonlinear function, and F^{-1} cannot be estimated directly (i.e., the forward model is not explicitly invertible). Special retrieval algorithms have to be developed in order to establish an inverse mapping from $\mathbf{y}$-space into $\mathbf{x}$-space. The second problem (and in fact, the most challenging) is that the inverse problems of remote sensing are *severely ill-posed*. Actually, the concept of ill-posedness is the cornerstone of the inversion theory [7, 8]. In practice, the measured data always contains noise, i.e.,

$$\mathbf{y}^{\delta} = \mathbf{y} + \boldsymbol{\delta}. \tag{5.5}$$

In addition to that, there is a systematic forward model error caused by some adopted assumptions in the physical model. The systematic error is repeatable in different measurements and its mean value (so-called bias) cannot be zero. These two types of errors affect the retrieval results. Consequently, instead of the true state vector $\mathbf{x}$, a noisy state vector $\mathbf{x}^{\delta}$ is retrieved. For ill-posed problems the relative error in retrieved $\mathbf{x}$ is comparable to (or even exceeds) the relative error of the measurements. In addition to that, it can happen that the inverse problem does not have a unique solution in the sense that there may exist several state vectors (sometimes infinite numbers) leading to the same observations [9] (using the metaphor of Bohren and Huffman, we may say that several animals have similar traces). Also, one might get a physically meaningless result in the retrieval procedure. In practice, the ill-posed problem cannot be solved directly and must be converted into a well-posed problem. In this regard, it is common to think about forward modeling as a craft and inversion as an art.

Despite a wide range of applications in which the inverse problems arise, the basic approaches to solve them are application-independent to some extent. Most of the inversion techniques reduce the initial inverse problem to an exercise in optimization, i.e., minimizing the difference between the measured data $\mathbf{y}^{\delta}$ and the prediction of the forward model which uses the retrieved quantity as input. However, there are several methods which were specifically designed for solving inverse problems in atmospheric remote sensing. Moreover, if there are several parameters to be retrieved, they can be retrieved simultaneously or in turn [10]. In the latter case the algorithm is expected to be more stable (as other parameters are assumed to be constant) although at the cost of accuracy.

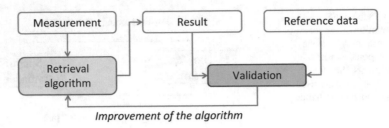

Fig. 5.2 Retrieval algorithm and validation procedure

Note that the retrieval algorithms and the data provided by them have to be validated. Validation is mandatory to quantify the reliability of remote sensing products. In general, validation can be defined as the process of assessing, by independent means, the quality of the data products derived from the system outputs [11]. The accuracy of the extracted information from spaceborne measurements should be independently assessed with the reference data. In practice, it can be done via comparison with ground-based measurements or comparison with other alternative techniques. Validation is essential in various remote sensing applications. In this regard, several validation sites and networks have been established (see, e.g., [12] for general review). Depending on how we are confident in the reference data, the retrieval algorithm can be updated and improved, as shown in Fig. 5.2.

5.2   Example of Trace Gas Vertical Column Density Retrieval

In this section we consider an example of trace gas vertical column retrieval. This method was used to make first retrievals of atmospheric ozone vertical total column by using the Dobson spectrophotometer [13].

Let us assume that we measure the transmitted solar radiation (see Fig. 5.3) for the cloudless sky at two wavelengths (namely $\lambda_1 = 305$ nm and $\lambda_2 = 325$ nm) looking directly at the Sun. Note that the radiation corresponding to λ_1 is absorbed stronger than that to λ_2. It follows from the Bouguer's attenuation law:

$$y_1 = E(\lambda_1)\exp(-\tau(\lambda_1)/\mu), \quad y_2 = E(\lambda_2)\exp(-\tau(\lambda_2)/\mu), \qquad (5.6)$$

where y_1 and y_2 are the corresponding irradiances measured on the ground, E refers to the incident solar irradiance at the top of the atmosphere, τ is the atmospheric optical thickness, while μ is the incident angle cosine. It follows for the atmospheric optical thickness:

$$\tau(\lambda) = \tau_a(\lambda) + \tau_{sca}^{mol}(\lambda) + \tau_{abs}^{mol}(\lambda). \qquad (5.7)$$

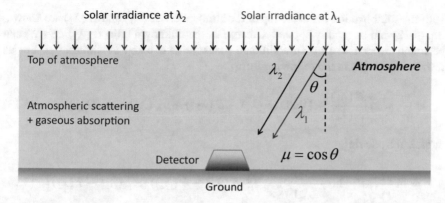

Fig. 5.3 Measurement of the irradiance due to the direct solar beam at two wavelengths

Here $\tau_a(\lambda)$ is the aerosol optical thickness, $\tau_{sca}^{mol}(\lambda)$ is the molecular scattering optical thickness, and $\tau_{abs}^{mol}(\lambda)$ is the molecular absorption optical thickness.

We shall neglect the contribution of all atmospheric gases except for ozone at the wavelengths λ_1, λ_2. Then it follows for the molecular absorption optical thickness:

$$\tau_{abs}^{mol} \equiv \tau_{abs}^{O_3} = \int_{z_1}^{z_2} n_{O_3}(z)\bar{\sigma}_{abs}^{O_3}(z)dz, \tag{5.8}$$

where z_1 is the ground level and z_2 is the top-of-atmosphere height, n_{O_3} is the concentration of ozone molecules, while $\bar{\sigma}_{abs}^{O_3}$ is the ozone absorption cross-section. Using the mean value theorem, one derives:

$$\tau_{abs}^{O_3} = \bar{\sigma}\xi, \tag{5.9}$$

where

$$\xi = \int_{z_1}^{z_2} n_{O_3}(z)dz \tag{5.10}$$

is the total ozone column and $\bar{\sigma}$ is the ozone absorption cross-section at a certain atmospheric level $z = z^*$. Note that quantities E, τ_{sca}^{mol}, τ_a and $\bar{\sigma}$ are wavelength-dependent, and

$$\bar{\sigma}(\lambda_1) > \bar{\sigma}(\lambda_2). \tag{5.11}$$

One derives from Eqs. (5.6)–(5.10):

$$\xi = \frac{1}{\bar{\sigma}(\lambda_2) - \bar{\sigma}(\lambda_1)}\left[\mu \ln\left(\frac{y_1\, E(\lambda_2)}{y_2\, E(\lambda_1)}\right) + \tilde{\tau}(\lambda_1) - \tilde{\tau}(\lambda_2)\right], \tag{5.12}$$

where $\tilde{\tau}(\lambda) = \tau_{sca}^{mol}(\lambda) + \tau_a(\lambda)$.

If the solar irradiances at the top of the atmosphere $E\left(\lambda_1\right)$ and $E\left(\lambda_2\right)$ are known, Eq. (5.12) can be readily applied. Otherwise, the unknown ratio $E\left(\lambda_2\right)/E\left(\lambda_1\right)$ can be found by considering an additional measurement taken at different μ. Let us rewrite Eq. (5.12) in the following form:

$$\ln\frac{y_2}{y_1} = \ln\frac{E\left(\lambda_2\right)}{E\left(\lambda_1\right)} + \left[\text{VCD}\left(\bar{\sigma}\left(\lambda_1\right) - \bar{\sigma}\left(\lambda_2\right)\right) + \tau_R\left(\lambda_1\right) - \tau_R\left(\lambda_2\right)\right]\frac{1}{\mu}, \quad (5.13)$$

or after introducing

$$Y = \ln\frac{y_2}{y_1}, \quad a = \ln\frac{E\left(\lambda_2\right)}{E\left(\lambda_1\right)}, \quad b = \left[\xi\left(\bar{\sigma}\left(\lambda_1\right) - \bar{\sigma}\left(\lambda_2\right)\right) + \tilde{\tau}\left(\lambda_1\right) - \tilde{\tau}\left(\lambda_2\right)\right] \quad (5.14)$$

we have:

$$Y = a + \frac{b}{\mu}. \quad (5.15)$$

Note that Y, a and b are independent of μ, while Y is known from measurements. Then, taking measurements at two different incidence angles with cosines μ_1 and μ_2, we obtain a system of two equations in the form of Eq. (5.15) with two unknowns, namely a and b. By solving this system, the irradiance ratio can be found and then Eq. (5.12) can be applied to retrieve the vertical column density.

Note that this method has its drawbacks. In particular, the contribution from multiple scattered light is not taken into account (as a matter of fact, we can only hope that the division y_2/y_1 somewhat cancels the influence of multiple scattering). Nevertheless, this technique can be used to calibrate and validate data obtained from satellite measurements (in this case, the additional information regarding aerosol and cloud parameters must be used).

This method can be modified to get information about the vertical distribution of ozone—this is the so-called Umkehr method [14] ("Umkehr" means reversal in German). Instead of direct solar light ratios, the instrument measures the ratio of the zenith sky light scattered intensities at wavelengths λ_1 and λ_2. The measurements are performed at different solar zenith angles, including the cases when the Sun is near the horizon. The value of

$$N\left(\mu\right) = \log\frac{L\left(\lambda_2, \mu\right)}{L\left(\lambda_1, \mu\right)} \quad (5.16)$$

plotted against the solar zenith angle cosine μ is analyzed. Assuming that

$$L\left(\lambda_1, \mu\right) \sim e^{-\frac{\bar{\sigma}\left(\lambda_1\right)\xi}{\mu}}, \quad L\left(\lambda_2, \mu\right) \sim e^{-\frac{\bar{\sigma}\left(\lambda_2\right)\xi}{\mu}}. \quad (5.17)$$

and taking into account Eq. (5.11) one can expect that the value of $N\left(\mu\right)$ increases with θ. Interestingly, this curve has a maximum at angles when the Sun is located close to the horizon and then decreases, as shown in Fig. 5.4. The qualitative description of

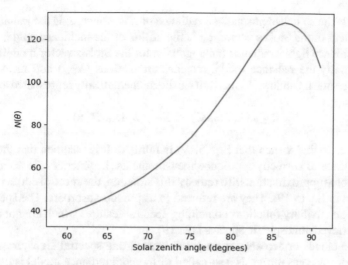

Fig. 5.4 Example of the $N(\theta)$-curve

this phenomenon was given by Götz [14], although it was later criticized by Pekeris [15].

When the Sun is close to the horizon, the direct light is strongly attenuated. Explanation of Götz is based on the assumption that there is a Rayleigh scattering process above the stratospheric ozone. Then the scattered light has more chances to pass through the ozone moving at angles close to normal, as the path through the ozone becomes shorter. The maximum in the $N(\theta)$-curve implies the existence of maximum of the ozone concentration at a certain level in the atmosphere [16]. Note that the resulting ozone profile retrieved by using the Umkehr method is strongly dependent on the algorithm used (see, e.g., [17]).

In these simple examples we have seen that the measurements carried out at different wavelengths and geometries carry information about atmospheric constituencies, including information about their vertical distribution. In the next sections we will discuss how to retrieve the information from such measurements in a more efficient manner.

5.3 Radiance and Differential Radiance Models

The retrieval of cloud parameters, aerosol parameters and surface properties uses the so-called radiance model for R, defined as

$$R = \frac{\pi L}{\mu_0 E_0},$$ (5.18)

where R is the so-called normalized radiance or reflectance, L is the radiance or the first element of the Stokes vector, μ_0 is the cosine of the incidence angle, while E_0 is the incidence light irradiance (analogously, for the Stokes vector model).

Essentially the radiance model matches simulations (R_{sim}) and measurements (R_{meas}) in a given window. Formally, it can be mathematically represented as follows:

$$R_{meas} (\lambda, \boldsymbol{\Omega}_0, \boldsymbol{\Omega}) \approx R_{sim} (\lambda, \boldsymbol{\Omega}_0, \boldsymbol{\Omega}, \mathbf{x}) . \tag{5.19}$$

The goal is to find $\mathbf{x}$ such that Eq. (5.19) is fulfilled. It is assumed that the forward model is able to correctly reproduce measurements. In practice, the computations involve some approximations. To remedy this situation, the spectral correction terms are added in Eq. (5.19). They are referred to as pseudo-absorbers. The spectral corrections are auxiliary functions containing spectral features which are not attributed to the retrieved atmospheric species [18, 19].

Unlike clouds or aerosols, trace gases have strong spectral signatures. For that reason, for trace gas retrievals a so-called differential radiance model is used:

$$\ln \bar{R}_{meas} (\lambda) \approx \ln \bar{R}_{sim} (\lambda, x) \tag{5.20}$$

with

$$\ln \bar{R}_{sim} (\lambda) = \ln R_{sim} (\lambda) - P_{sim} (\lambda, \mathbf{p}_{sim} (x)) \tag{5.21}$$

and

$$\ln \bar{R}_{meas} (\lambda) = \ln R_{meas} (\lambda) - P_{meas} (\lambda, \mathbf{p}_{meas}) . \tag{5.22}$$

Here, P_{sim} and P_{meas} are polynomials of low order with coefficients $\mathbf{p}_{sim}$ and $\mathbf{p}_{meas}$, respectively. For a state vector $\mathbf{x}$, the coefficients of the smoothing polynomials P_{sim} and P_{meas} are computed as

$$\mathbf{p}_{sim} (x) = \arg \min_{\mathbf{p}} \|\log R_{sim} (\mathbf{x}) - P_{sim}(\mathbf{p})\|^2 , \tag{5.23}$$

and

$$\mathbf{p}_{meas} = \arg \min_{\mathbf{p}} \|\log R_{meas} - P_{meas}(\mathbf{p})\|^2 , \tag{5.24}$$

respectively. In general, a smoothing polynomial is assumed to account for the low-order frequency structure due to scattering mechanisms (clouds, aerosols, and surface reflectance) so that $\ln \bar{R}$ will mainly reflect the absorption process due to gas molecules [20]. Note that the polynomial subtraction is somewhat similar to the background removal procedure [21, 22] used in optical and electron spectroscopy. This procedure could be avoided if we were able to take into account all factors which have an impact on the spectrum.

5.4 Basic Retrieval Techniques

In principle, the accurate solution of an inverse problem requires simultaneous determination gas, aerosol and cloud properties as well as surface conditions. In this case, all retrieval parameters are consistent. However, the dimension of the problem becomes large. In addition, several factors can have a similar impact on the spectrum. In its turn, it might cause numerical instability. For instance, due to similarities in their radiometric signatures, simultaneous retrieval of cloud and aerosol properties is a very challenging task [23]. On the other hand, it is obvious that as the number of unknown parameters increases, the forward model becomes more flexible and a closer matching between the simulations and the measurements is possible. In practice, the optimal set of retrieved parameters is hard to find before making retrieval exercises, at least on synthetic measurements (i.e., simulated by using the forward model).

The simplest method to perform an inversion is the so-called fitting technique, in which the elements of the state vector $\mathbf{x}$ are manually tuned to match the output of the forward model with the measurements (see Eqs. (5.19) and (5.20)). Strictly speaking, the inversion procedure itself is avoided (as we work only with the forward model), what makes the fitting procedure robust. Since this technique involves a human, who has to tune elements of $\mathbf{x}$, the fitting procedure is efficient only when the dataset to be processed is small and the state vector has few elements. The manual fitting becomes inefficient as the number of parameters to be retrieved and to be matched increases. In this regard, this approach is suitable for small-scale problems of radiative transfer (e.g., involving one or two parameters to be retrieved) [24, 25]. In principle, the fitting procedure can be substituted by mathematical libraries for finding a solution to a nonlinear problem (e.g., [26]). However, the performance of the general algorithms designed for a wide spectrum of problems is expected to be lower than those specifically designed for a given application.

To accelerate the fitting procedure, look-up tables (LUTs) of simulated satellite signals can be precomputed for selected scenarios with different atmospheric parameters [27]. A scenario that provides the best match with the observed radiances is taken as the retrieved solution. To find a solution which is not present in LUTs, a multidimensional interpolation can be used. For instance, in [28] the golden section search method together with parabolic interpolation was employed (see [29] for a general review).

The LUT-based approach is successfully employed for interpreting POLDER [30], AVHRR [31] and MODIS [32] data. LUTs are computed offline and they replace runtime computations involving the forward model F. Thus, LUT-based approach is much faster than that involving online RTM computations. However, there are several drawbacks of the LUT-based approach. The solution of inverse problems based on LUTs is one of the cases presented in LUTs. Obviously, the true solution cannot be captured in LUTs. For a large number of parameters to be retrieved, look-up tables of corresponding observations may have larger dimensions and thus be less suitable for the operational use. As noted in [33], the efficiency of the lookup table approach is low for interpreting data from imagers with multi viewing and polarimetric capabilities (e.g., POLDER and especially 3MI [34]).

To make the inversion easier, the original forward model can be expressed by means of approximate solutions derived in a closed-form. In this case, the inversion can be performed analytically. This approach can be applied for the retrieval of cloud properties, as discussed in [35]. In particular, the usage of asymptotic equations instead of solutions expressed via the discrete ordinate method allows us to avoid the compilation of LUTs in the case of optically thick clouds and, therefore, to speed up the retrieval process.

As we will see further, if the forward model is nonlinear and analytically not invertible, it can be linearized around a guess value of **x**, and then a new guess can be found. Although such an approach looks straight-forward, it is hard to predict in advance, how accurate and robust the retrieval algorithm will be when applied to real measurements. In this regard, it is important to gain insights into the theory of inversion and do not blindly rely on mathematical libraries as black boxes. In particular, one must remember that although the direct radiative transfer problem always has a solution, this is not generally the case for an inverse radiative transfer problem.

In the general case, the retrieval of atmospheric and underlying surface parameters is performed by using specifically designed retrieval algorithms, which are based on the LUTs, various radiative transfer approximations (e.g., the asymptotic theory as in [36]), statistical estimation [7, 37] and the regularization theory [8].

5.5 Differential Optical Absorption Spectroscopy

5.5.1 General

The differential optical absorption spectroscopy (DOAS) is a widely used method to retrieve atmospheric trace gases from measurements of the solar light scattered by the Earth's atmosphere. In atmospheric science, the DOAS approach was introduced by Brewer et al. [38], Noxon [39] and Platt et al. [40–42]. A thorough review of DOAS was given by Rozanov and Rozanov [43].

DOAS can use the artificial light source (active DOAS) or scattered sunlight (passive DOAS). We will be focused on the latter one. The DOAS principle can be applied to various ground-based as well as satellite-based platforms [20]. The measurements are performed in the ultraviolet/visible spectral range (300–700 nm) depending on the target species. Usually, spectrometers are used with the spectral resolution around 0.4–1.5 nm.

In the case of ground-based measurements, the zenith viewing direction is often used. If the sensitivity to a certain trace gas is high, the measurements can be performed when the Sun is close to the zenith (high sun). If the sensitivity is not sufficient, the measurements are taken at twilight (in this case, the solar light passes a longer path). In the so-called multi axis DOAS (MAXDOAS) the measurements are performed at several viewing angles. In particular, for tropospheric observations, the MAXDOAS instruments use the horizon viewing measurement directions [44].

In the case of satellite-based observations, DOAS can be used for both down-looking and limb observations. By scanning the limb of the atmosphere, different altitudes are probed and fine vertical resolution can be obtained, although at the cost of lower horizontal resolution.

The DOAS retrieval consists of two major steps:

1. Retrieval of the slant column density (or the total slant column);
2. Conversion of the slant column to the vertical column through the air mass factor.

Let us consider these steps separately.

5.5.2 Retrieval of the Slant Column Density

Following Platt [40–42], it is assumed that the absorption cross-section of a considered trace gas can be written as a sum of two components:

$$\bar{\sigma}(\lambda) = \bar{\sigma}_s(\lambda) + \bar{\sigma}_d(\lambda), \tag{5.25}$$

where $\bar{\sigma}_s(\lambda)$ is the smooth component which varies slowly with λ, while $\bar{\sigma}_d(\lambda)$ changes rapidly with λ, and is referred to as the differential absorption cross-section (which gives the word "differential" to DOAS). Assuming that the radiance measured by the instrument is the extraterrestrial radiance weakened by gaseous absorption, Rayleigh scattering as well as scattering by cloud and aerosol particles, we apply the Bouguer law. For sun-normalized radiance we have

$$R(\lambda) = \exp\left[-\int_0^S \bar{\sigma}(\lambda) n_g ds - \int_0^S \sigma_{ext}^{Mie}(\lambda, s) ds - \int_0^S \sigma_{ext}^{Ray}(\lambda, s) ds\right], \tag{5.26}$$

where σ_{ext}^{Mie} is associated with scattering by cloud and aerosol particles,[1] while the integration is performed along the light path s. Assuming that the cross-section does not change along the light path and recalling the definition of the slant column density (SCD), we have

$$R(\lambda) = \exp\left[-\bar{\sigma}(\lambda) \text{SCD} - \int_0^S \sigma_{ext}^{Mie}(\lambda, s) ds - \int_0^S \sigma_{ext}^{Ray}(\lambda, s) ds\right]. \tag{5.27}$$

---

[1] Of course, the aerosol and cloud particles can be non-spherical. Nevertheless, we use the nomenclature "Mie" for short.

As the spectral dependence of $\bar{\sigma}_s(\lambda)$ (which is incorporated in $\bar{\sigma}(\lambda)$), $\sigma_{\text{ext}}^{\text{Mie}}(\lambda, s)$ and $\sigma_{\text{ext}}^{\text{Ray}}(\lambda, s)$ (the latter two are $\sim \lambda^{-p}, 0 < p \leq 4$), we can approximate their contributions by a low-degree polynomial. Implicitly, it is assumed that the differential part of the cross-section can be represented as a polynomial of the same degree. Thus, we have

$$R(\lambda) = \exp\left[-\bar{\sigma}_d(\lambda) \operatorname{SCD} + \sum_{k=1}^{K} b_k \lambda^k \right]. \tag{5.28}$$

Then, taking the logarithm of Eq (5.28) we obtain the following equation:

$$-\bar{\sigma}_d(\lambda) \operatorname{SCD} + \sum_{k=1}^{K} b_k \lambda^k = \ln R(\lambda), \tag{5.29}$$

in which the right-hand side is the natural logarithm of the sun-normalized radiance. This equation has $K+1$ unknown parameters, namely SCD and $\{b_k\}, k = 1, ..., K$. Considering measurements taking at N_λ wavelengths and performing a least-square fit in a selected spectral range, the unknown coefficients can be found, including SCD. This step is referred to as the DOAS fit. Using notations as in Eq. (5.21), let us rewrite Eq. (5.29) as follows:

$$-\bar{\sigma}_d(\lambda) \operatorname{SCD} = \ln \bar{R}(\lambda). \tag{5.30}$$

Equation (5.30) can be equipped with an additional term representing the rotational Raman scattering.

5.5.3   Conversion of the Slant Column Density to the Vertical Column Density

Formally, the conversion of the slant column density to the vertical column density (VCD) is performed via the air mass factor (AMF) defined as follows:

$$\operatorname{AMF} = \frac{\operatorname{SCD}}{\operatorname{VCD}}. \tag{5.31}$$

In the absence of atmospheric scattering, the light travels along straight lines, and the AMF is referred to as the geometrical AMF ($\operatorname{AMF_G}$), which is a function of the cosines of the solar zenith angle μ_0 and the viewing zenith angle μ:

$$\operatorname{AMF_G} = \frac{1}{\mu_0} + \frac{1}{\mu}. \tag{5.32}$$

Considering Eq. (5.29) for the cases with and without absorbers, the air mass factor can be expressed as follows [45]:

$$\mathrm{AMF}(\lambda) = \frac{\ln R(\lambda) - \ln \widehat{R}(\lambda)}{\bar{\sigma}_\mathrm{d}(\lambda)\,\mathrm{VCD}}, \tag{5.33}$$

where $\widehat{R}(\lambda)$ is the sun-normalized radiance as it would be in the absence of an absorber. The radiances are calculated by using the appropriate radiative transfer model.

For the vertically inhomogeneous atmosphere, we can consider a spatial discretization in N_L atmospheric layers. To retrieve the information about the trace gas concentration, we use measurements at different observation conditions. For the ith atmospheric layer and jth observation condition, the so-called box-AMF can be computed [46]:

$$\mathrm{AMF}_{i,j}(\lambda) = \frac{\ln R_{i,j}(\lambda) - \ln \widehat{R}_j(\lambda)}{\bar{\sigma}_i(\lambda)\,\mathrm{VCD}_i}, \tag{5.34}$$

where $R_{i,j}(\lambda)$ is computed neglecting absorbers everywhere expect in the ith layer. Then, Eq. (5.30) transforms into

$$-\sum_{i=1}^{N_L} \mathrm{VCD}_i \cdot \mathrm{AMF}_{i,j}(\lambda)\,\bar{\sigma}_i(\lambda) = \ln \bar{R}_j(\lambda). \tag{5.35}$$

That is the so-called modified DOAS equation [47]. In practice, for Eq. (5.34) the radiative transfer models are used, while the measurements go into the right part of Eq. (5.35). Solving system (5.35), the unknown VCD in each atmospheric layer can be retrieved.

5.5.4 Linearization and Generalized DOAS Equation

Note, that Eqs. (5.34)–(5.35) imply a linear relation between the gaseous concentration and the logarithmic radiance. For the direct solar beam, it is a consequence of the Bouguer law. However, because of the multiple scattering process, this is not true. More accurate computations are based on the linearization of the radiative transfer model.

We consider the absorption coefficient as

$$\sigma_\mathrm{abs}(\lambda, z) = \bar{\sigma}(\lambda, z)\, n(z), \tag{5.36}$$

where $n(z)$ is a gaseous concentration profile. Note that the cross-section depends on the altitude due to temperature variations. Using Eq. (5.36), the perturbation of the absorption coefficient can be written as

$$d\sigma_{abs}(\lambda, z) = d\bar{\sigma}(\lambda, z)\, n(z) + \bar{\sigma}(\lambda, z)\, dn(z),\qquad(5.37)$$

where $d\bar{\sigma}(\lambda, z)$ and $dn(z)$ are perturbations of the absorption cross-section and the number density profile, respectively. Then, for an infinitely thin atmospheric layer with the geometrical thickness δz at the altitude z, we introduce a weighting function as follows:

$$w(\lambda, z) = \left.\frac{d\ln R(\lambda)}{\delta z\, d\sigma_{abs}(\lambda, z)}\right|_{n_0(z)},\qquad(5.38)$$

where the subscript $n_0(z)$ indicates that the derivative is computed for a certain guess profile $n_0(z)$. Note also that the perturbation $d\sigma_{abs}(\lambda, z)$ is zero everywhere except for this thin layer around the altitude z. Comparing Eqs. (5.34) and (5.38) we see that the weighting function is the air mass factor computed for the layer δz. Then substituting the summation for discrete layers in Eq. (5.35) by a continuous integration from the ground to the top of the atmosphere (z_{TOA}), we obtain

$$\ln \bar{R}(\lambda, n(z)) = \ln \bar{R}(\lambda, n_0(z)) + \int_0^{z_{TOA}} w(\lambda, z')\, \sigma_{abs}(\lambda, z')\, dz'.\qquad(5.39)$$

This equation is referred to as the generalized DOAS equation. In view of Eq. (5.36), this equation provides a linear relationship between the intensity logarithm and the gaseous concentration. Thus, the gaseous profile can be fitted to match the left part of Eq. (5.39) (which is measured) with the right part of Eq. (5.39) (which is computed). Introducing

$$y(\lambda) = \ln \bar{R}(\lambda, n(z)) - \ln \bar{R}(\lambda, n_0(z)),\qquad(5.40)$$

$$x(z) = n(z),\qquad(5.41)$$

and

$$k(\lambda, z) = \bar{\sigma}(\lambda, z)\, w(\lambda, z),\qquad(5.42)$$

Eq. (5.39) is written as

$$y(\lambda) = \int_0^{z_{TOA}} k(\lambda, z')\, x(z')\, dz'.\qquad(5.43)$$

This is a Fredholm integral equation of the first kind [48], while $k(\lambda, z)$ is referred to as the kernel function.

To conclude this section, we note that the DOAS approach is considered to be fast and robust. However, the trace gas retrieval can be performed by using also the radiance model (Eq. (5.19)). In this case a direct-fitting approach is used in which spectral radiances simulated using the radiative transfer model are adjusted to measurements [49]. Although this approach is more time-consuming and somewhat complex (as the spectra have to be computed accurately), it is "more physically-based".

5.6 Temperature Profile Retrieval by Nadir Sounding

Satellite measurements can be used to retrieve the surface temperature and the temperature profile across the atmosphere [50]. For that, the spectrum in the infrared region around 10–15 μm can be used [51] (as this spectral region corresponds to the peak of the black body radiation emitted by the Earth). In particular, for obtaining information about the temperature profile, the temperature sounding generally involves the detection of radiation emitted by a gas which is uniformly distributed in the atmospheric column (e.g., CO_2 and O_2). For surface temperature retrieval, the wavelengths are chosen which are able to pass through the atmosphere.

In this regard, we may assume that the emission is a function of the temperature only. Consider a radiative transfer equation for thermal radiation:

$$\frac{dL(\lambda, z)}{dz} = -\sigma_{abs}(\lambda, z) L(\lambda, z) + \sigma_{abs}(\lambda, z) B(\lambda, z). \tag{5.44}$$

Here B is the Planck function, σ_{abs} is the absorption coefficient, while the scattering coefficient is set to zero (the scattering process is neglected). For simplicity (and in order to stay in the framework of the linear problem), we use the Rayleigh-Jeans law (which is actually valid in the microwave region, but not in the visible spectral range):

$$B(\lambda, z) = \frac{2ck_B T(z)}{\lambda^4} \tag{5.45}$$

with k_B being the Boltzmann constant and c is the speed of light. The formal solution of Eq. (5.44) is given as

$$L(\lambda, z) = L(\lambda, 0) e^{-\int_0^z \sigma_{abs}(\lambda, z')dz'} + \gamma \int_0^z T(z') e^{-\int_{z'}^z \sigma_{abs}(\lambda, z'')dz''} dz', \tag{5.46}$$

where

$$\gamma = \frac{2ck_B}{\lambda^4}. \tag{5.47}$$

It can be proved by substituting Eq. (5.46) into Eq. (5.44). Neglecting the background contribution from the surface, we set $L(\lambda, 0) = 0$. For the radiance at the top of the atmosphere ($z = z_{TOA}$), we have

$$L\left(\lambda\right) = \gamma \int_0^{z_{TOA}} T\left(z'\right) e^{-\int_{z'}^{z''} \sigma_{\mathrm{abs}}\left(\lambda, z''\right) dz''} dz'. \tag{5.48}$$

Introducing new variables, $x\left(z\right) \equiv T\left(z\right)$, $y\left(\lambda\right) \equiv L\left(\lambda\right)$, we rewrite Eq. (5.48) as Eq. (5.43), i.e.,

$$y\left(\lambda\right) = \int_0^{z_{TOA}} k\left(\lambda, z'\right) x\left(z'\right) dz', \tag{5.49}$$

where a kernel function has the following form:

$$k\left(\lambda, z'\right) = \gamma e^{-\int_{z'}^{z''} \sigma_{\mathrm{abs}}\left(\lambda, z''\right) dz''}. \tag{5.50}$$

Again we have obtained a Fredholm integral equation of the first kind [48].

If the Planck formula is used instead of the Rayleigh-Jeans law, the model becomes nonlinear providing several numerical difficulties. One way to overcome them is to retrieve, first, the vertical profile of the Planck function and then, inverting it, to restore the temperature profile. The details can be found in [50]. The description of more sophisticated retrieval techniques can be found in [52].

5.7 Inversion of the Linear Problem

5.7.1 Fredholm Equation

As we have already seen, several problems of atmospheric remote sensing, such as DOAS retrieval and temperature retrieval from nadir sounding, lead to the Fredholm equation of the first kind. It represents the mathematical model of a continuous problem, as functions x and y in Eqs. (5.43) and (5.49) are continuous functions. Fredholm equations arise in various linear models, such as those in the theory of signal processing, radiography, spectroscopy, image processing and others. The Fredholm equation tells that the output signal is a weighted sum of input parameters distributed across a given domain. The corresponding weights incorporated in the kernel function show how sensible the output signal $y\left(\lambda\right)$ at the wavelength λ is with respect to the parameter $x\left(z'\right)$ at the coordinate z'. At the same time, high values of $k\left(\lambda, z'\right)$ tell that the measurement taken at the wavelength λ is most sensible to the parameter located at the coordinate z'. For instance, in the example illustrated in Fig. 5.5, the measurement at λ_1 is most sensible to the top of the atmosphere, the measurement at λ_2 carries information mostly about the atmosphere around 60 km altitude etc. Thus, selecting the wavelengths in such a manner that the maxima of $k\left(\lambda, z\right)$ are located at different z for different values of λ, we can obtain the information about vertical distribution of the parameter of interest (e.g., the temperature profile).

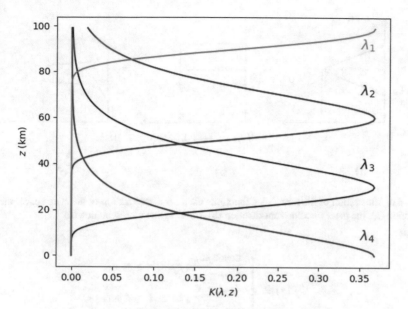

Fig. 5.5 Example of the kernel function as a function of the altitude z at different wavelengths

5.7.2 Discretization of the Linear Problem

In practice, we are confronted with a collection of discrete observations taken at specific wavelengths. Consequently, for a discrete set of measurements, Eq. (5.49) takes the following form:

$$y(\lambda_i) = \int_0^{z_{\text{TOA}}} k(\lambda_i, z') x(z') dz', \quad i = 1, \ldots, m. \tag{5.51}$$

Equation (5.51) represents a semi-discrete problem, as the discretized y-space is related to the continuous x-space. To derive a complete discrete problem, x is discretized by representing a continuous dependence $x(z)$ by a linear combination of basic functions:

$$x(z) \approx \sum_{j=1}^n \xi_j \Psi_j(z). \tag{5.52}$$

Here ξ_j are the expansion coefficients. Next, we introduce a discretization grid $\{z_j\}$, $j = 0, \ldots, n$ of the altitude interval $[0, z_{\text{TOA}}]$, where $z_0 = 0$ and $z_n = z_{\text{TOA}}$. The values of x at corresponding levels are $x_j = x(z_j)$. The simplest basic functions are the piecewise linear functions (shown in Fig. 5.6a):

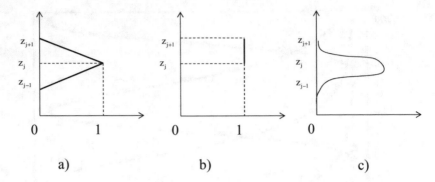

Fig. 5.6 Illustration of different basic functions used for discretization of the state space: the hat function (**a**), the piecewise constant function (**b**), the B-spline representation (**c**)

$$\Psi_j(z) = \begin{cases} \dfrac{z - z_{j-1}}{z_j - z_{j-1}}, z_{j-1} < z \leq z_j, \\ \dfrac{z_{j+1} - z}{z_{j+1} - z_j}, z_j \leq z < z_{j+1}, \\ 0, \quad \text{otherwise}. \end{cases} \tag{5.53}$$

In this case, the expansion coefficients are equal to the values of the atmospheric temperature profile at each grid point:

$$\xi_j = x_j. \tag{5.54}$$

We may also choose the piecewise constant functions (shown in Fig. 5.6b):

$$\Psi_j(z) = \begin{cases} 1, z_j < z \leq z_{j+1}, \\ 0, \text{otherwise}. \end{cases} \tag{5.55}$$

Then, the expansion coefficients are the mean values of the atmospheric profile of the retrieved quantity over each layer:

$$\xi_j = \frac{1}{2} \left(x(z_j) + x(z_{j+1}) \right). \tag{5.56}$$

For a smoother approximation, the B-spline interpolation [53] or the expansion over principal components [54] can be used. In this case, the basic functions can have a complex shape (shown in Fig. 5.6c) and the expansion coefficients may have no clear physical meaning.

For the sake of simplicity, we use the hat basic functions. Substituting Eqs. (5.52) and (5.54) into a semi-discrete problem (5.51), we obtain a fully discrete problem:

$$y\left(\lambda_i\right) = \sum_{j=1}^{n} \left[\int_0^{z_{\text{TOA}}} k\left(\lambda_i, z'\right) \psi_j\left(z'\right) dz' \right] x_j, \tag{5.57}$$

or in a compact matrix form:

$$\mathbf{y} = \mathbf{Kx}, \tag{5.58}$$

where $\mathbf{K}$ is the $m \times n$ matrix with entries

$$[\mathbf{K}]_{ij} = \int_0^{z_{\text{TOA}}} k\left(\lambda_i, z'\right) \psi_j\left(z'\right) dz'. \tag{5.59}$$

This equation can be interpreted in the following way. Since each observation is a weighted sum of x-values,

$$y_i = \sum_{j=1}^{n} [\mathbf{K}]_{ij}\, x_j \tag{5.60}$$

or in the matrix format:

$$y_i = \mathbf{k}_i^T \mathbf{x} \tag{5.61}$$

and $\mathbf{k}_i^T$ are a set of vectors in state space, the measurement y_i can be interpreted as projections of the state $\mathbf{x}$ onto $\mathbf{k}^T$. Unlike nonlinear problems, linear problems can be inverted analytically. To invert Eq. (5.58) we multiply both sides by $\mathbf{K}^T$. Obviously the matrix $\mathbf{K}^T \mathbf{K}$ is square and can be inverted. Thus, the solution to Eq. (5.58) reads as

$$\mathbf{x} = \mathbf{K}^\dagger \mathbf{y}, \tag{5.62}$$

where

$$\mathbf{K}^\dagger = \left(\mathbf{K}^T \mathbf{K}\right)^{-1} \mathbf{K}^T \tag{5.63}$$

is the pseudo-inverse (Moore–Penrose inverse [55]). Note that although the radiative transfer model is nonlinear, several inversion methods use linear iterations (e.g., Gauss-Seidel technique) approximating the original model by a linearized one at each iteration step. The details of these approaches will be considered later.

So far, we only note that the construction $\mathbf{K}^T \mathbf{K}$ is very important in the inversion theory and can be met in several contexts. For instance, the Fisher information matrix is defined as

$$\mathbf{D} = \mathbf{K}^T \mathbf{C}_\delta^{-1} \mathbf{K}, \tag{5.64}$$

where $\mathbf{C}_\delta$ is the covariance matrix of the vector $\mathbf{y}$. The Fisher matrix formalism is useful to assess how well the measurements constrain the atmospheric parameters, before doing the measurement. For instance, we can choose the most informative wavelengths and combination of viewing geometries [56, 57] by examining the Fisher matrix.

5.7.3   Concept of Ill-Posedness

Now we can formulate the criteria for ill-posed and well-posed problems. The concept of well-posed problems was defined by Courant and Hilbert [58] and by Jacques Hadamard [59]. The problem corresponding to Eq. (5.58) is considered to be well-posed if the following three conditions hold:

1. For any $\mathbf{y}$, a solution $\mathbf{x}$ exists;
2. The solution $\mathbf{x}$ is unique;
3. The solution $\mathbf{x}$ is stable with respect to perturbations in $\mathbf{y}$; that means, if $\mathbf{Kx}_0 = \mathbf{y}_0$ and $\mathbf{Kx} = \mathbf{y}$, then $\mathbf{x} \rightarrow \mathbf{x}_0$ whenever $\mathbf{y} \rightarrow \mathbf{y}_0$.

If at least one of these conditions is violated, the problem is called ill-posed.

In the original work of Hadamard [59], the third condition was not mentioned explicitly. Hadamard himself adopted the definition of Courant [58]. As discussed in [60], it might be more appropriate to attribute the well-posed problem definition to Courant rather than Hadamard. Nevertheless, in many textbooks, the phrase "well-posed in the sense of Hadamard" is used. Also, it is interesting to note that Courant and Hilbert thought the third condition should hold for any mathematical formulations that describe observable natural phenomena, which is currently not the case in many fields including remote sensing. More details about history of the concept of inverse problems can be found in [61].

5.7.4   Algebraic Interpretation of Courant-Hadamard Conditions

Thinking in terms of linear algebra, we introduce the range space of $\mathbf{K}$

$$\Re\,(\mathbf{K}) = \{\mathbf{Kx} | \mathbf{x} \in X\} \tag{5.65}$$

as all possible vectors which can be obtained by applying operator $\mathbf{K}$ to a vector $x \in X$. The null space of $\mathbf{K}$ is defined as

$$\aleph\,(\mathbf{K}) = \{\mathbf{x} \in X | \mathbf{Kx} = 0\}, \tag{5.66}$$

that is, a set of all $\mathbf{x}$ such that $\mathbf{Kx} = 0$. The rest of $\mathbb{R}^n$-space which is not $\aleph\,(\mathbf{K})$ is denoted $\aleph^\perp\,(\mathbf{K})$. Analogously, the rest of $\mathbb{R}^m$-space which is not $\Re\,(\mathbf{K})$ is $\Re^\perp\,(\mathbf{K})$, as shown in Fig. 5.7. Also, for our analysis we need to know the rank r of operator $\mathbf{K}$ [62]:

$$r = \text{rank}\,(\mathbf{K}), \tag{5.67}$$

which corresponds to the maximal number of linearly independent columns (or rows [63]) of $\mathbf{K}$. Since $\mathbf{K} \in \mathbb{R}^{m \times n}$, we have $r \leq \min\,(n, m)$. If $r < \min\,(n, m)$, then some

Fig. 5.7 Operator **K**
projects state space onto
range space which is a
subspace of measurement
space. Null-space is
projected to 0

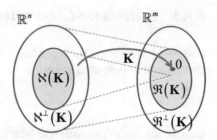

columns and rows contained in **K** are not linearly independent. The Hadamard conditions are related to the properties of the operator **K**.

The operator **K** is *surjective*, if each element of $\Re$ (**K**)-space is mapped to by at least one element of $\mathbb{R}^n$-space. For surjective operators we have $r = m$. If **K** is not surjective (i.e., $y \notin \Re$ (**K**)) then the linear equation is not solvable for all $y \in Y \not\subset \Re$ (**K**), and the first Hadamard's condition does not hold. Non-existence can occur because the forward model does not take into account important factors or because the noise distorts and corrupts the data to which the inverse operator is applied.

The operator **K** is *injective* if each element of $\Re$ (**K**)-space is mapped to by at most one element of $\mathbb{R}^n$-space. If **K** is *injective*, then $r = n$ and $\aleph$ (**K**) = 0. That implies that two different elements from $\mathbb{R}^n$-space are projected to different elements in $\Re$ (**K**)-space. If **K** is not injective (i.e., $\aleph$ (**K**) $\neq$ 0) then the linear equation may have more than one solution. Even if the initial continuous problem establishes one-to-one correspondence between input and output parameters, non-uniqueness can occur due to discretization of the problem. It is a peculiarity of the rank-deficient problems. Indeed, let

$$\mathbf{y}_0 = \mathbf{K}\mathbf{x}_0. \tag{5.68}$$

Then for any $\mathbf{x}' \in \aleph$ (**K**), we have

$$0 = \mathbf{K}\mathbf{x}'. \tag{5.69}$$

Taking a sum of Eqs. (5.68) and (5.69), we get

$$\mathbf{y}_0 = \mathbf{K}\left(\mathbf{x}_0 + \mathbf{x}'\right). \tag{5.70}$$

Thus, if $\mathbf{x}_0$ is a solution of the inverse problem, then $\mathbf{x}_0 + \mathbf{x}'$ is also a solution, i.e., the solution is not unique—that is a violation of the second Hadamard condition.

Finally, if $\mathbf{K}^{-1}$ exists but is not continuous, then the solution **x** of the linear equation does not depend continuously on the data (i.e., instability under data perturbation). Violation of the third Hadamard condition means that small errors in the data space are significantly amplified in the state space. The ill-posedness of a discrete linear problem is reflected by a huge condition number of the **K** matrix.

5.7.5 Under-/over-Constrained Problems

Following Rodgers [7], depending on m, n and r, the problem can be under-constrained or over-constrained.

If

$$r = m = n, \tag{5.71}$$

then the number of parameters to be retrieved is equal to the number of measurements, and the measurements are all independent. This is the case of well-posed problem (given that the condition number of the $\mathbf{K}$ matrix is not large). The $\mathbf{K}$ matrix is directly invertible in this case. If condition (5.71) does not hold, $\mathbf{K}$ is not invertible.

If $r = m < n$, the problem is under-constrained, i.e., there are more unknowns than measurements. Yet, since $r = m$, the measurements are independent.

If $r = n < m$, the problem is over-constrained, i.e., there are more measurements than unknown parameters of the state vector, and measurements contain information about all of them.

The problems for which $r < \min(m, n)$ are called rank-deficient problems. If $r < m$, the measurements are not independent and they can be inconsistent (they can contradict each other). If $r < n$, the information contained in the measurements is not enough to retrieve all the unknowns.

5.7.6 The Singular Value Decomposition

In this section we prove a general solution to the linear inverse problem (see Eqs. (5.62)–(5.63)). We consider a singular value decomposition for an $m \times n$ matrix $\mathbf{K}$, whose rank is r:

$$\mathbf{K} = \mathbf{U}\mathbf{\Sigma}\mathbf{V}^T, \tag{5.72}$$

where $\mathbf{U}$ and $\mathbf{V}$ are $m \times m$ and $n \times n$ unitary matrices over $\mathbf{K}$, respectively, and $\mathbf{\Sigma}$ is a diagonal $m \times n$ matrix with non-negative real numbers on the diagonal. The $\mathbf{U}$ and $\mathbf{V}$ matrices consist of sets of vectors,

$$\mathbf{U} = [\mathbf{u}_1, ..., \mathbf{u}_m], \tag{5.73}$$

$$\mathbf{V} = [\mathbf{v}_1, ..., \mathbf{v}_m], \tag{5.74}$$

which form orthonormal bases $\mathbb{R}^m = \text{span}\,\{\mathbf{u}_k\}_{k=\overline{1,m}}$ $\mathbb{R}^n = \text{span}\,\{\mathbf{v}_k\}_{k=\overline{1,n}}$ and

$$\mathbf{U}\mathbf{U}^T = \mathbf{U}^T\mathbf{U} = \mathbf{I}_m, \tag{5.75}$$

$$\mathbf{V}\mathbf{V}^T = \mathbf{V}^T\mathbf{V} = \mathbf{I}_n, \tag{5.76}$$

where $\mathbf{I}_m$ and $\mathbf{I}_n$ are the $m \times m$ and $n \times n$ identity matrices, respectively. The Σ has the following representation:

$$\Sigma = \begin{bmatrix} \sigma_1 & \cdots & 0 & \cdots & 0 \\ \vdots & \ddots & & & \vdots \\ & & \sigma_r & & \\ 0 & \cdots & 0 & \ddots & \\ \vdots & & \vdots & & 0 \\ & & & & \vdots \\ 0 & \cdots & 0 & \cdots & 0 \end{bmatrix}. \tag{5.77}$$

The numbers $\{\sigma_i\}_{i=1,\dots,r}$ are called the singular values, which can be ordered in a descending order:

$$\sigma_1 \geq \sigma_2 \geq \dots \geq \sigma_r > \sigma_{r+1} = \dots = \sigma_n = 0. \tag{5.78}$$

The singular value decomposition provides an explicit representation of the range and null space of $\mathbf{K}$. Namely, the sets $\{\mathbf{v}_i\}_{i=1,\dots,r}$ and $\{\mathbf{v}_i\}_{i=r+1,\dots,n}$ are orthonormal bases for $\aleph(\mathbf{K})^{\perp}$ and $\aleph(\mathbf{K})$, respectively, (i.e., $\aleph(\mathbf{K})^{\perp} = \text{span}\,\{\mathbf{v}_i\}_{i=1,\dots,r}$, $\aleph(\mathbf{K}) = \text{span}\,\{\mathbf{v}_i\}_{i=r+1,\dots,n}$) while the sets $\{\mathbf{u}_i\}_{i=1,\dots,r}$ and $\{\mathbf{u}_i\}_{i=r+1,\dots,m}$ are orthonormal bases of $\Re(\mathbf{K})$ and $\Re(\mathbf{K})^{\perp}$, respectively, i.e.,

$$\Re(\mathbf{K}) = \text{span}\,\{\mathbf{u}_i\}_{i=1,\dots,r}, \tag{5.79}$$

$$\Re(\mathbf{K})^{\perp} = \text{span}\,\{\mathbf{u}_i\}_{i=r+1,\dots,m}. \tag{5.80}$$

Equation (5.79) implies that

$$\mathbf{y} = \sum_{k=1}^{r} c_k \mathbf{u}_k, \tag{5.81}$$

where c_k are projections of $\mathbf{y}$ onto $\mathbf{u}_k$, i.e.,

$$c_k = <\mathbf{y}\mathbf{u}_k>, \tag{5.82}$$

and $<>$ stands for scalar product. Multiplying Eq. (5.72) by $\mathbf{V}$ from the right side, and using Eq. (5.76), we obtain:

$$\mathbf{K}\mathbf{V} = \mathbf{U}\Sigma, \tag{5.83}$$

or, in terms of singular vectors,

$$\mathbf{K}\mathbf{v}_k = \sigma_k \mathbf{u}_k. \tag{5.84}$$

Analogically, considering the transpose of Eq. (5.72), we obtain

$$\mathbf{K}^T = \mathbf{V}\Sigma^T\mathbf{U}^T \tag{5.85}$$

and then, multiplying it by $\mathbf{U}$ from the right and taking into account Eq. (5.75), we obtain

$$\mathbf{K}^T\mathbf{U} = \mathbf{V}\Sigma^T, \tag{5.86}$$

$$\mathbf{K}^T\mathbf{u}_k = \sigma_k\mathbf{v}_k. \tag{5.87}$$

Note that the condition number of the matrix $\mathbf{K}$ can be computed as follows:

$$\text{cond}\,(\mathbf{K}) = \frac{\sigma_1}{\sigma_r}. \tag{5.88}$$

In principle, the condition number is related to the third Hadamard condition, as it shows how sensible the solution is to a perturbation in the input data. Thus, a problem with a low condition number is said to be well-conditioned, while a problem with a high condition number is said to be ill-conditioned. Note that large values of condition numbers may indicate that the variables to be retrieved are highly linearly related (this case is referred to as multicollinearity) [64].

5.7.7 Solvability

To solve equation (5.58), we assume that $\mathbf{y} \in \Re\,(\mathbf{K})$. Then, substituting Eq. (5.72) into Eq. (5.58) we obtain:

$$\mathbf{U}\Sigma\mathbf{V}^T\mathbf{x} = \mathbf{y}. \tag{5.89}$$

Then, taking into account Eqs. (5.75) and (5.76), Eq. (5.89) can be solved:

$$\mathbf{x} = (\mathbf{V}\Sigma^{-1}\mathbf{U}^T)\mathbf{y} = \mathbf{K}^\dagger\mathbf{y}, \tag{5.90}$$

or written through singular vectors:

$$\mathbf{x} = \sum_{i=1}^{n} \frac{1}{\sigma_i}\left(\mathbf{u}_i^T\mathbf{y}\right)\mathbf{v}_i, \tag{5.91}$$

where the operator $\mathbf{K}^\dagger : \mathbb{R}^m \to \mathbb{R}^n$ is an $n \times m$ matrix,

$$\mathbf{K}^\dagger = \sum_{i=1}^{r} \frac{1}{\sigma_i}\mathbf{u}_i^T\mathbf{v}_i \tag{5.92}$$

maps $\mathbf{y} \in \Re\,(\mathbf{K})$ into the solution $\mathbf{x}$, and is called the generalized inverse.

Formally, multiplying Eq. (5.58) by $\mathbf{K}^T$ we get:

$$\mathbf{K}^T \mathbf{K} \mathbf{x} = \mathbf{K}^T \mathbf{y}. \tag{5.93}$$

Matrix $\mathbf{K}^T \mathbf{K}$ is square and can be inverted. Thus,

$$\mathbf{x} = \left(\mathbf{K}^T \mathbf{K}\right)^{-1} \mathbf{K}^T \mathbf{y} \tag{5.94}$$

and

$$\mathbf{K}^\dagger = \left(\mathbf{K}^T \mathbf{K}\right)^{-1} \mathbf{K}^T, \tag{5.95}$$

which proves the solution (5.62). Using Eq. (5.72) it is easy to prove that Eq. (5.95) is equivalent to Eq. (5.92).

5.7.8 Least Square Solution in the Case of Noisy Data

Next, the data vector $\mathbf{y} \in \Re(\mathbf{K})$ is considered as the exact data vector, while

$$\mathbf{x}^\dagger = \mathbf{K}^\dagger \mathbf{y} \tag{5.96}$$

is the exact solution corresponding to Eq. (5.91). The noisy data vector is defined as

$$\mathbf{y}^\delta = \mathbf{y} + \boldsymbol{\delta}, \tag{5.97}$$

where $\boldsymbol{\delta}$ is the instrumental noise. The noise can be described in either a deterministic setting or in a semi-stochastic setting. In the first case the data error is characterized by the noise level Δ, while in the second case we deal with a discrete white noise with the covariance matrix $\mathbf{C}_\delta = \sigma^2 \mathbf{I}_m$. The noise level can be estimated by using the probability distribution of the noise, i.e., $\Delta = \varepsilon\{\|\boldsymbol{\delta}\|\}$ or $\Delta^2 = \varepsilon\{\|\boldsymbol{\delta}\|^2\}$, where ε is the expected value operator. These estimates can be computed either numerically by generating randomly a sample of noise vectors and averaging, or analytically, if the explicit integrals of probability densities are available. In the case of white noise, the second criterion yields

$$\Delta^2 = m\sigma^2. \tag{5.98}$$

From a practical point of view, the Tikhonov solution does not depend on which setting the problem is treated; differences can appear in convergence rate results for different regularization parameter choice methods.

The inverse problem is to find $\mathbf{x}^\delta$ such that

$$\mathbf{K} \mathbf{x}^\delta = \mathbf{y}^\delta. \tag{5.99}$$

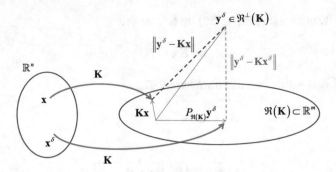

Fig. 5.8 Illustration of the least square solution

In general, $\mathbf{y}^\delta \notin \Re(\mathbf{K})$ and $\mathbf{x}^\dagger$ cannot be retrieved from $\mathbf{y}^\delta$. However, substituting $\mathbf{y}^\delta$ instead of $\mathbf{y}$ into Eq. (5.91) gives the so-called the least square solution, namely

$$\mathbf{x}^\delta = \sum_{i=1}^{r} \frac{1}{\sigma_i} \left(\mathbf{u}_i^T \mathbf{y}^\delta \right) \mathbf{v}_i. \tag{5.100}$$

Let us give a geometrical interpretation to the least-square solution. To do that, we refer to Fig. 5.8. Since $\mathbf{y}^\delta \notin \Re(\mathbf{K})$, the point corresponding to $\mathbf{y}^\delta$ does not lie on the plane corresponding to $\Re(\mathbf{K})$. Using Eqs. (5.81), (5.84) and (5.100), it can be shown that

$$\mathbf{y}^\delta - \mathbf{K}\mathbf{x}^\delta = \sum_{k=r+1}^{m} < \mathbf{y}^\delta \mathbf{u}_k > \mathbf{u}_k. \tag{5.101}$$

Multiplying Eq. (5.101) by $\mathbf{K}^T$ from the left side and using Eqs. (5.87) and (5.78) we see that

$$\mathbf{K}^T \left(\mathbf{y}^\delta - \mathbf{K}\mathbf{x}^\delta \right) = \sum_{k=r+1}^{m} < \mathbf{y}^\delta, \mathbf{u}_k > \mathbf{K}^T \mathbf{u}_k = \sum_{k=r+1}^{m} < \mathbf{y}^\delta, \mathbf{u}_k > \sigma_k \mathbf{v}_k = 0,$$
$$\tag{5.102}$$

that is, $\mathbf{y}^\delta - \mathbf{K}\mathbf{x}^\delta$ is perpendicular to $\Re(\mathbf{K})$ and for any $\mathbf{x}$ we have

$$\left\| \mathbf{y}^\delta - \mathbf{K}\mathbf{x}^\delta \right\| \le \left\| \mathbf{y}^\delta - \mathbf{K}\mathbf{x} \right\|. \tag{5.103}$$

In other words,

$$\mathbf{x}^\delta = \arg \min_{\mathbf{x}} \left\| \mathbf{y}^\delta - \mathbf{K}\mathbf{x} \right\|. \tag{5.104}$$

Thus, Eq. (5.100) solves Eq. (5.99) in the least-square sense (5.104).

5.7.9 Ill-Posedness of the Least-Square Solution

In this section we relate the condition number of the **K** matrix with the error of the least-square solution. For this, we note that

1. the number of small singular values of the **K** matrix increases with the dimension of **K**;
2. as the singular values σ_i decrease, the corresponding singular vectors $\mathbf{u}_i$ and $\mathbf{v}_i$ have more sign changes in their components. As a consequence of the oscillatory behavior of the high-order singular vectors, the norm of the least-squares solution becomes extremely large and $\mathbf{x}^\delta$ is not a reliable approximation of $\mathbf{x}^\dagger$.

Let us consider the noise in the data vector in the direction of the singular vector $\mathbf{u}_r$, corresponding to the smallest non-zero singular value σ_r:

$$\delta = \mathbf{u}_r \cdot \Delta, \tag{5.105}$$

where

$$\Delta = \left\| \mathbf{y}^\delta - \mathbf{y} \right\|. \tag{5.106}$$

Substituting Eq. (5.105) into Eq. (5.97), we obtain the following parameterization:

$$\mathbf{y}^\delta = \mathbf{y} + \mathbf{u}_r \cdot \Delta. \tag{5.107}$$

The least-squares solution (5.100) for Eq. (5.107) (taking into account Eq. (5.91) and Eq. (5.84)) is then given by

$$\mathbf{x}^\delta = \mathbf{x} + \frac{\Delta}{\sigma_r} \mathbf{v}_r. \tag{5.108}$$

From Eqs. (5.106) and (5.108), we obtain $\left\| \mathbf{x}^\delta - \mathbf{x} \right\| = \left\| \mathbf{y}^\delta - \mathbf{y} \right\| / \sigma_r$, then taking into account Eq. (5.88):

$$\left\| \mathbf{x}^\delta - \mathbf{x}^\dagger \right\| = \left\| \mathbf{y}^\delta - \mathbf{y} \right\| \mathrm{cond}(\mathbf{K}) / \sigma_1. \tag{5.109}$$

Thus, the error of the retrieval increases with the error in the data and the condition number of the matrix **K**.

5.8 Regularization of Inverse Problems

5.8.1 Regularization Principles

In order to solve the ill-posed problem, the effect of the noise in the data should be suppressed. This process is called *regularization*. To regularize the ill-posed problem, additional information about the properties of the solution can be added (e.g., its smoothness). The goal of regularization is to construct a regularized operator F_{reg}^{-1} such that

$$\mathbf{x} = F_{\mathrm{reg}}^{-1}(\mathbf{y}, \mathbf{x}_a) \tag{5.110}$$

is an acceptable solution of Eq. (5.3), while vector $\mathbf{x}_a$ is a priori state vector which expresses our previous knowledge or expectations about solution properties. In practice, it is impossible to find a single operator F_{reg}^{-1} that gives an acceptable solution for any $\mathbf{y}$. Instead, the goal is to construct a family $\{F_\alpha^{-1}, \alpha > 0\}$ of operators such that

$$\mathbf{x} = F_\alpha^{-1}(\mathbf{y}, \mathbf{x}_a), \tag{5.111}$$

where α is the regularization parameter, which tells how strong the initial problem should be regularized. The regularization parameter must be chosen in accordance with some properties of the given data vector $\mathbf{y}$.

The foundations of the regularization theory were developed by Tikhonov [65]. Tikhonov regularization is one of the most widely used techniques for regularizing ill-posed problems. In particular, Tikhonov showed that the solution for $\mathbf{y}^\delta = \mathbf{K}\mathbf{x}^\delta$ can be found as a solution to the following penalized least-squares problem:

$$\mathbf{x}_\alpha^\delta = \arg\min_{\mathbf{x}^\delta} \left\{ \left\| \mathbf{y}^\delta - \mathbf{K}\mathbf{x}^\delta \right\|^2 + \alpha \boldsymbol{\Omega}(x) \right\}, \tag{5.112}$$

where $\boldsymbol{\Omega}$ is the stabilizing functional. The expression in $\{\ \}$ is referred to as the Tikhonov function.

Usually, instability of the ill-posed problems leads to the large norm of the solution. Therefore, the simplest regularization can be done if we add a penalty term in the minimization problem (5.104), namely:

$$\mathbf{x}_\alpha^\delta = \arg\min_{\mathbf{x}} \left\{ \left\| \mathbf{y}^\delta - \mathbf{K}\mathbf{x}^\delta \right\|^2 + \alpha \left\| \mathbf{x} \right\|^2 \right\}. \tag{5.113}$$

Essentially, Eq. (5.113) means the search for $\mathbf{x}_\alpha^\delta$ providing at the same time a small residual $\left\| \mathbf{y}^\delta - \mathbf{K}\mathbf{x}^\delta \right\|$ and a moderate value of the penalty term $\alpha \left\| \mathbf{x} \right\|^2$. If α is very small, Eq. (5.113) leads to the original unstable problem; if α is very large, the penalty term dominates over the residual term $\left\| \mathbf{y}^\delta - \mathbf{K}\mathbf{x}^\delta \right\|^2$, and the solution is defined by the penalty term rather than the original problem. Due to the additional term in Eq. (5.113), solution (5.94) is updated as follows:

$$\mathbf{x}_\alpha^\delta = \left(\mathbf{K}^T \mathbf{K} + \alpha \mathbf{I} \right)^{-1} \mathbf{K}^T \mathbf{y}^\delta. \tag{5.114}$$

More general form of the Tikhonov function reads as follows:

$$\mathbf{x}_\alpha^\delta = \arg\min_{x} \left\{ \left\| \mathbf{y}^\delta - \mathbf{K}\mathbf{x}^\delta \right\|^2 + \alpha \left\| \mathbf{L}\mathbf{x} \right\|^2 \right\}, \tag{5.115}$$

where $\mathbf{L}$ is the regularization matrix. In this case, Eq. (5.115) solves the following problem:

$$\left(\mathbf{K}^T\mathbf{K} + \alpha\mathbf{L}^T\mathbf{L}\right)\mathbf{x}_\alpha^\delta = \mathbf{K}^T\mathbf{y}^\delta, \tag{5.116}$$

so that

$$\mathbf{x}_\alpha^\delta = \left(\mathbf{K}^T\mathbf{K} + \alpha\mathbf{L}^T\mathbf{L}\right)^{-1}\mathbf{K}^T\mathbf{y}^\delta. \tag{5.117}$$

From the previous discussions, it follows that the instability of solution (5.100) comes from higher-order singular vectors, which correspond to very small singular values. Since $\mathbf{L}^T\mathbf{L}$ is positive semi-definite, the singular values of the matrix $\left(\mathbf{K}^T\mathbf{K} + \alpha\mathbf{L}^T\mathbf{L}\right)$ are shifted in the positive direction and thus, the problem becomes more stable (e.g., less sensitive to noise in $\mathbf{y}^\delta$). The alternative explanation is that the condition number of $\mathbf{K}^T\mathbf{K}$ is large in the case of ill-posed inverse problems. In the framework of the Tikhonov regularization, the original $\mathbf{K}^T\mathbf{K}$ is substituted with the regularization operator $\left(\mathbf{K}^T\mathbf{K} + \alpha\mathbf{L}^T\mathbf{L}\right)$ that has a smaller condition number controlled by α.

The penalty term can be computed with respect to the residual between the solution $\mathbf{x}$ and a priori solution $\mathbf{x}_a$:

$$\mathbf{x}_\alpha^\delta = \arg\min_x \left\{ \left\|\mathbf{y}^\delta - \mathbf{K}\mathbf{x}^\delta\right\|^2 + \alpha\left\|\mathbf{L}\left(\mathbf{x} - \mathbf{x}_a\right)\right\|^2 \right\}. \tag{5.118}$$

In this case, the actual solution moves toward a priori solution when α increases.

5.8.2 Regularization Matrices

If $\mathbf{L}$ is a diagonal matrix,

$$\mathbf{L}_0 = \begin{bmatrix} \beta_1 & 0 & 0 & 0 \\ 0 & \beta_2 & 0 & 0 \\ 0 & 0 & \ddots & 0 \\ 0 & 0 & 0 & \beta_n \end{bmatrix}, \tag{5.119}$$

then the penalty term is

$$\|\mathbf{L}_0\mathbf{x}\| = \sqrt{\sum_{i=1}^n (\beta_i x_i)^2}. \tag{5.120}$$

The penalty term controls the magnitude of the solution. If $\mathbf{L}$ is the identity matrix, then Eq. (5.115) is reduced to Eq. (5.113). The matrix $\mathbf{L}_0$ can be regarded as a zero-order derivative operator.

To preserve the smoothness of the solution, the regularization matrices can be constructed on the basis of discrete approximations to derivative operators, e.g.,

$$\mathbf{L}_1 = \begin{bmatrix} -1 & 1 & \dots & 0 & 0 \\ 0 & -1 & \dots & 0 & 0 \\ \vdots & \vdots & \ddots & \vdots & \vdots \\ 0 & 0 & \dots & -1 & 1 \end{bmatrix}, \tag{5.121}$$

$$\mathbf{L}_2 = \begin{bmatrix} 1 & -2 & 1 & \dots & 0 & 0 & 0 \\ 0 & 2 & -2 & \dots & 0 & 0 & 0 \\ \vdots & \vdots & \vdots & \ddots & \vdots & \vdots & \vdots \\ 0 & 0 & 0 & \dots & 1 & -2 & 1 \end{bmatrix}, \tag{5.122}$$

where $\mathbf{L}_1$ is n-1 by n matrix, while $\mathbf{L}_2$ is n-2 by n matrix. Indeed, the norm of $\mathbf{L}_1\mathbf{x}$ is

$$\|\mathbf{L}_1\mathbf{x}\| = \sqrt{\sum_{i=1}^{n-1} (x_{i+1} - x_i)^2}. \tag{5.123}$$

By minimizing $\|\mathbf{L}_1\mathbf{x}\|$ we also minimize the difference between neighboring points in $\mathbf{x}$, thereby making the solution smoother.

Several derivative operators can be combined by using the Cholesky factorization, e.g.,

$$\mathbf{L}^T\mathbf{L} = \eta_0\mathbf{L}_0^T\mathbf{L}_0 + \eta_1\mathbf{L}_1^T\mathbf{L}_1 + \eta_2\mathbf{L}_2^T\mathbf{L}_2 + ..., \tag{5.124}$$

where the weights $\eta_0, \eta_1, ...$ satisfy the normalization condition

$$\eta_0 + \eta_1 + \eta_2 + ... = 1. \tag{5.125}$$

In this case, there is a way to control both absolute values of the solution and its smoothness.

5.8.3 Tikhonov Regularization Based on a Priori Statistical Information

The regularization matrix can also incorporate the statistical information about the solution. For instance, one can expect that the trace gas concentrations correlate with each other at different atmospheric levels within the so-called correlation length. In this case, the statistical information can be described by means of correlation matrices. So, for the vector $\mathbf{x}$ the covariance matrix $\mathbf{C}$ can be defined as follows:

$$[\mathbf{C}_x]_{ij} = \text{cov}\left[x_i, x_j\right] = M\left[(x_i - M\left[x_i\right])(x_j - M\left[x_j\right])\right], \tag{5.126}$$

where M is the expected (mean) value of the argument. Then the regularization matrix is derived using the Cholesky factorization:

$$\mathbf{C}_x^{-1} = \mathbf{L}^T \mathbf{L}. \tag{5.127}$$

Substituting Eq. (5.127) in Eq. (5.117), we obtain

$$\mathbf{x}_\alpha^\delta = \left(\mathbf{K}^T \mathbf{K} + \alpha \mathbf{C}_x^{-1}\right)^{-1} \mathbf{K}^T \mathbf{y}^\delta. \tag{5.128}$$

Typically, for trace gas profile retrieval, an exponential correlation function is used:

$$[\mathbf{C}_x]_{ij} = \bar{\sigma}_i \bar{\sigma}_j x_{ai} x_{aj} \exp\left(-2\frac{|z_i - z_j|}{l_i + l_j}\right), \quad i, j = 1, ..., n, \tag{5.129}$$

where $\bar{\sigma}_i$ are the unitless profile standard deviations, l_i are the lengths which determine the correlation between the parameters at different altitudes z_i, while x_a corresponds to a priori profile. Clearly, if the elements of $\mathbf{x}$ are not correlated, the covariance matrix is diagonal, and Eq. (5.128) simplifies to Eq. (5.114).

5.8.4 Bayesian Approach

In statistical inversion theory [7], the model F gives a relation between probability densities for $\mathbf{x}$ and $\mathbf{y}$. Bayes' theorem of inverse problems says that

$$p\left(\mathbf{x}|\mathbf{y}^\delta\right) = \frac{p\left(\mathbf{y}^\delta|\mathbf{x}\right) p\left(\mathbf{x}\right)}{p\left(\mathbf{y}^\delta\right)}, \tag{5.130}$$

where $p\left(\mathbf{x}|\mathbf{y}^\delta\right)$ is the conditional probability density of $\mathbf{x}$ given the measurements $\mathbf{y}^\delta$, $p\left(\mathbf{y}^\delta|\mathbf{x}\right)$ is the conditional probability density of $\mathbf{y}^\delta$ given the state $\mathbf{x}$, $p\left(\mathbf{x}\right)$ is the probability density of $\mathbf{x}$, while $p\left(\mathbf{y}^\delta\right)$ plays a role of normalization constant and is usually ignored. Then, assuming that the state vector $\mathbf{x}$ and the noise δ in the measurements obey a normal distribution law with covariance matrices $\mathbf{C}_x$ and $\mathbf{C}_\delta$, respectively, we have

$$p\left(\mathbf{x}\right) = \frac{1}{\sqrt{(2\pi)^n \det\left(\mathbf{C}_x\right)}} \exp\left(-\frac{1}{2}\left(\mathbf{x} - \mathbf{x}_a\right)^T \mathbf{C}_x^{-1}\left(\mathbf{x} - \mathbf{x}_a\right)\right), \tag{5.131}$$

$$p\left(\mathbf{y}^\delta|\mathbf{x}\right) = \frac{1}{\sqrt{(2\pi)^m \det\left(\mathbf{C}_\delta\right)}} \exp\left(-\frac{1}{2}\left(\mathbf{y}^\delta - F\left(\mathbf{x}\right)\right)^T \mathbf{C}_\delta^{-1}\left(\mathbf{y}^\delta - F\left(\mathbf{x}\right)\right)\right), \tag{5.132}$$

where $\mathbf{x}_a$ can be regarded as a mean state vector (that is our a priori information about the atmosphere). Substituting Eqs. (5.131) and (5.132) into Bayes' theorem (5.130), we get

$$p\left(\mathbf{x}|\mathbf{y}^\delta\right) \sim \exp\left(-\frac{1}{2}V\left(\mathbf{x}|\mathbf{y}^\delta\right)\right), \tag{5.133}$$

where $V\left(\mathbf{x}|\mathbf{y}^\delta\right)$ is the so-called a posteriori potential defined by

$$V\left(\mathbf{x}|\mathbf{y}^\delta\right) = \left(\mathbf{y}^\delta - F\left(\mathbf{x}\right)\right)^T \mathbf{C}_\delta^{-1}\left(\mathbf{y}^\delta - F\left(\mathbf{x}\right)\right) + \left(\mathbf{x} - \mathbf{x}_a\right)^T \mathbf{C}_x^{-1}\left(\mathbf{x} - \mathbf{x}_a\right). \tag{5.134}$$

The solution to inverse problem $\mathbf{x}^\delta$ maximizes the conditional probability density $p\left(\mathbf{x}|\mathbf{y}^\delta\right)$ and thus, minimizes $V\left(\mathbf{x}|\mathbf{y}^\delta\right)$:

$$\mathbf{x}^\delta = \arg\min_{\mathbf{x}} V\left(\mathbf{x}|\mathbf{y}^\delta\right). \tag{5.135}$$

For Gaussian densities with the covariance matrices

$$\mathbf{C}_\delta = \sigma^2 \mathbf{I}_m, \tag{5.136}$$

$$\mathbf{C}_x = \sigma_x^2 \left(\mathbf{L}^T\mathbf{L}\right)^{-1}, \tag{5.137}$$

and the regularization parameter

$$\alpha = \sigma^2/\sigma_x^2, \tag{5.138}$$

the a posteriori potential $V\left(\mathbf{x}|\mathbf{y}^\delta\right)$ corresponds to the Tikhonov function (5.118):

$$V\left(\mathbf{x}|\mathbf{y}^\delta\right) = \frac{1}{\sigma^2} F_\alpha\left(\mathbf{x}\right). \tag{5.139}$$

For linear problems, a posteriori potential is

$$V\left(\mathbf{x}|\mathbf{y}^\delta\right) = \left(\mathbf{y}^\delta - \mathbf{Kx}\right)^T \mathbf{C}_\delta^{-1}\left(\mathbf{y}^\delta - \mathbf{Kx}\right) + \mathbf{x}^T\mathbf{C}_x^{-1}\mathbf{x} \tag{5.140}$$

and Eq. (5.135) leads to

$$\mathbf{x}^\delta = \mathbf{Gy}^\delta, \tag{5.141}$$

where

$$\mathbf{G} = \left(\mathbf{K}^T\mathbf{C}_\delta^{-1}\mathbf{K} + \mathbf{C}_x^{-1}\right)^{-1}\mathbf{K}^T\mathbf{C}_\delta^{-1} \tag{5.142}$$

is known as the gain matrix or the contribution function matrix in statistical inversion theory [7]. As you can see, under assumptions (5.136)–(5.138) solution (5.141)–(5.142) becomes identical to solution (5.117). Note that the gain matrix incorporates the Fisher information matrix (5.64).

For these reasons, the Bayesian approach is regarded as a stochastic version of the Tikhonov regularization method with an a priori chosen regularization parameter. Note that α defined by Eq. (5.138) can be interpreted as the noise-to-signal ratio. The correspondence between the Bayesian estimation approach and the Tikhonov regularization has been discussed in numerous books (see, e.g., [8, 66, 67]).

5.8.5 Singular Value Decomposition for Regularized Solution

The regularized solution (5.115) and (5.117) can be expressed in terms of the generalized singular value decomposition (GSVD) of the matrix pair $(\mathbf{K}, \mathbf{L})$ [68]. If $\mathbf{K}$ is an $m \times n$ matrix, $\mathbf{L}$ is a $p \times n$ matrix and rank $(\mathbf{L}) = p$, then the GSVD of $(\mathbf{K}, \mathbf{L})$ reads as

$$\mathbf{K} = \mathbf{U}\Sigma_1\mathbf{W}^{-1}, \tag{5.143}$$

$$\mathbf{L} = \mathbf{V}\Sigma_2\mathbf{W}^{-1}, \tag{5.144}$$

where

$$\Sigma_1 = \begin{bmatrix} \operatorname{diag}(\sigma_i)_{p\times p} & 0 \\ 0 & E_{n-p} \\ 0 & 0 \end{bmatrix}, \tag{5.145}$$

$$\Sigma_2 = \begin{bmatrix} \operatorname{diag}(\mu_i)_{p\times p} & 0 \end{bmatrix}, \tag{5.146}$$

$\mathbf{U} = [\mathbf{u}_1, ..., \mathbf{u}_m]$ is the orthogonal $m \times m$ matrix, $\mathbf{V} = [\mathbf{v}_1, ..., \mathbf{v}_p]$ is the orthogonal $p \times p$ matrix, $\mathbf{W} = [\mathbf{w}_1, ..., \mathbf{w}_n]$ is non-singular $n \times n$ matrix, $\operatorname{diag}(\sigma_i)_{p\times p}$ and $\operatorname{diag}(\mu_i)_{p\times p}$ are $p \times p$ diagonal matrices. The following normalization condition holds:

$$\sigma_i^2 + \mu_i^2 = 1, \quad i = 1, ..., p. \tag{5.147}$$

In terms of the generalized singular systems

$$\mathbf{x}_\alpha^\delta = \sum_{k=1}^{p} \frac{\chi_k^2}{\chi_k^2 + \alpha} \frac{1}{\sigma_k} \left(\mathbf{u}_k^T \mathbf{y}^\delta\right) \mathbf{w}_k + \sum_{k=p+1}^{n} \left(\mathbf{u}_k^T \mathbf{y}^\delta\right) \mathbf{w}_k, \tag{5.148}$$

where the generalized singular values of $(\mathbf{K}, \mathbf{L})$ are

$$\chi_i = \frac{\sigma_i}{\mu_i}, \quad \chi_1 \geq ... \geq \chi_p > 0. \tag{5.149}$$

If $p = n$, Eq. (5.148) simplifies to

Fig. 5.9 Normalization
coefficient f_k as a function
of the singular value σ_k for a
given value of the
regularization parameter α

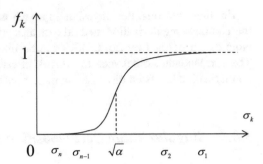

$$x_\alpha^\delta = \sum_{k=1}^n \frac{\chi_k^2}{\chi_k^2 + \alpha} \frac{1}{\sigma_k} \left(\mathbf{u}_k^T \mathbf{y}^\delta \right) \mathbf{w}_k. \tag{5.150}$$

Further, if the regularization matrix $\mathbf{L}$ is the identity matrix,

$$\mathbf{V} = \mathbf{W}, \tag{5.151}$$

and Eq. (5.150) simplifies to the following solution:

$$x_\alpha^\delta = \sum_{k=1}^n \frac{\sigma_k^2}{\sigma_k^2 + \alpha} \frac{1}{\sigma_k} \left(\mathbf{u}_k^T \mathbf{y}^\delta \right) \mathbf{v}_k. \tag{5.152}$$

Let us introduce the normalization coefficient

$$f_k = \frac{\sigma_k^2}{\sigma_k^2 + \alpha}. \tag{5.153}$$

Fig. 5.9 shows f_k as a function of the singular value σ_k. We can see that the regularization parameter α in the denominator reduces the weight of small singular values ($\sigma_k < \sqrt{\alpha}$), thereby decreasing the impact of oscillatory singular vectors.

In fact, the forward model has the effect of low-pass filters, e.g., it passes inputs with a frequency lower than a certain cutoff frequency and attenuates those with higher frequency. For instance, small variations in the ozone profile most likely do not have a significant influence on the measured spectrum. The inverse operator acts as a high-pass filter that amplifies noise in the measurements. Tikhonov regularization belongs to the spectral regularization methods which suppress the oscillatory behavior corresponding to small eigenvalues [69].

5.9 Choice of the Regularization Parameter

5.9.1 Error Characteristics in the State Space

Now a natural question arises: How shall we choose the regularization parameter? Ideally, we would like to select a regularization parameter so that the corresponding regularized solution minimizes some indicator of solution fidelity, e.g., some measure of the size of the solution error. Hence, we must perform an error analysis and estimate the accuracy of the obtained solution. To do that, we consider

1. the exact solution, i.e., the solution retrieved from the noise-free data (see Eq. (5.91)):

$$\mathbf{x}^\dagger = \mathbf{K}^\dagger \mathbf{y} = \sum_{k=1}^{n} \frac{1}{\sigma_k} \left(\mathbf{u}_k^T \mathbf{y} \right) \mathbf{v}_k, \tag{5.154}$$

2. the regularized solution obtained from the noise-free data:

$$\mathbf{x}_\alpha = \mathbf{K}_\alpha^\dagger \mathbf{y} = \sum_{k=1}^{n} \frac{\sigma_k^2}{\sigma_k^2 + \alpha} \frac{1}{\sigma_k} \left(\mathbf{u}_k^T \mathbf{y} \right) \mathbf{v}_k \tag{5.155}$$

and the retrieved regularized solution from the noisy data (see Eq. (5.152)):

$$\mathbf{x}_\alpha^\delta = \mathbf{K}_\alpha^\dagger \mathbf{y}^\delta = \sum_{k=1}^{n} \frac{\sigma_k^2}{\sigma_k^2 + \alpha} \frac{1}{\sigma_k} \left(\mathbf{u}_k^T \mathbf{y}^\delta \right) \mathbf{v}_k. \tag{5.156}$$

The total error of the solution is given by

$$\mathbf{e}_\alpha^\delta = \mathbf{x}^\dagger - \mathbf{x}_\alpha^\delta. \tag{5.157}$$

Introducing the smoothing error $\mathbf{e}_s$ and the noise error $\mathbf{e}_n$ by

$$\mathbf{e}_s = \mathbf{x}^\dagger - \mathbf{x}_\alpha, \tag{5.158}$$

$$\mathbf{e}_n = \mathbf{x}_\alpha - \mathbf{x}_\alpha^\delta, \tag{5.159}$$

respectively, we represent the total error as

$$\mathbf{e}_\alpha^\delta = \mathbf{e}_s + \mathbf{e}_n. \tag{5.160}$$

The smoothing error quantifies the loss of information due to the regularization, while the noise error quantifies the loss of information due to the incorrect data. The smoothing error can be expressed in terms of the exact solution:

$$\mathbf{e}_s = \mathbf{x}^\dagger - \mathbf{x}_\alpha = \mathbf{x}^\dagger - \mathbf{K}_\alpha^\dagger \mathbf{y} = \mathbf{x}^\dagger - \mathbf{K}_\alpha^\dagger \mathbf{K} \mathbf{x}^\dagger = (\mathbf{I} - \mathbf{A}_\alpha) \mathbf{x}^\dagger, \qquad (5.161)$$

where $\mathbf{A}_\alpha = \mathbf{K}_\alpha^\dagger \mathbf{K}$ is the averaging kernel, as well as in terms of the exact data vector:

$$\mathbf{e}_s = \left(\mathbf{K}^\dagger - \mathbf{K}_\alpha^\dagger\right) \mathbf{y} = \sum_{k=1}^n \left(1 - \frac{\sigma_k^2}{\sigma_k^2 + \alpha}\right) \frac{1}{\sigma_k} < \mathbf{u}_k, \mathbf{y} > \mathbf{v}_k. \qquad (5.162)$$

The noise error can be represented via the noise vector $\boldsymbol{\delta} = \mathbf{y}^\delta - \mathbf{y}$ as follows:

$$\mathbf{e}_n = \mathbf{x}_\alpha - \mathbf{x}_\alpha^\delta = \mathbf{K}_\alpha^\dagger \left(\mathbf{y} - \mathbf{y}^\delta\right) = -\mathbf{K}_\alpha^\dagger \boldsymbol{\delta} = -\sum_{k=1}^n \frac{\sigma_k^2}{\sigma_k^2 + \alpha} \frac{1}{\sigma_k} < \mathbf{u}_k^T, \boldsymbol{\delta} > \mathbf{v}_k. \qquad (5.163)$$

Assuming that δ has zero mean and the covariance $\sigma^2 \mathbf{I}$, we obtain

$$\varepsilon\left\{\|\mathbf{e}_n\|^2\right\} = \sigma^2 \sum_{k=1}^n \left(\frac{\sigma_k^2}{\sigma_k^2 + \alpha} \frac{1}{\sigma_k}\right)^2, \qquad (5.164)$$

where $\varepsilon\{\}$ stands for the expected value. Then, the expected value of the total error is given by

$$\varepsilon\left\{\|\mathbf{e}_\alpha^\delta\|^2\right\} = \|\mathbf{e}_s\|^2 + \varepsilon\left\{\|\mathbf{e}_n\|^2\right\}. \qquad (5.165)$$

Examining Eqs. (5.162) and (5.164) we conclude that the smoothing error $\|\mathbf{e}_s\|^2$ increases with α, while the expected value of the noise error $\varepsilon\left\{\|\mathbf{e}_n\|^2\right\}$ is a decreasing function of α. We arrive at the same conclusion just from the physical point of view. Indeed, with too little regularization, reconstructions have highly oscillatory artifacts due to noise amplification, while with too much regularization, the retrieved functions are too smooth. Hence, the expected value of the total error should have a minimum for an optimal value of α.

5.9.2   Error Characteristics in the Data Space

Another idea how to characterize the "quality" of the regularization procedure is to consider the residual

$$\mathbf{r}_\alpha^\delta = \mathbf{y}^\delta - \mathbf{K} \mathbf{x}_\alpha^\delta. \qquad (5.166)$$

The idea behind this approach is that with the retrieved solution $\mathbf{x}_\alpha^\delta$ and the model $\mathbf{K}$ we should be able to reproduce the measurements $\mathbf{y}^\delta$ with a certain accuracy. Substituting Eq. (5.152) into Eq. (5.166) and taking into account Eq. (5.84), we obtain

$$\mathbf{r}_\alpha^\delta = \mathbf{y}^\delta - \sum_{k=1}^{m} \frac{\sigma_k^2}{\sigma_k^2 + \alpha} < \mathbf{y}^\delta, \mathbf{u}_k > \mathbf{u}_k. \tag{5.167}$$

Then, making use of Eqs. (5.81) and (5.82) we obtain:

$$\mathbf{r}_\alpha^\delta = \sum_{k=1}^{n} \left(1 - \frac{\sigma_k^2}{\sigma_k^2 + \alpha} \right) < \mathbf{y}^\delta, \mathbf{u}_k > \mathbf{u}_k + \sum_{k=n+1}^{m} < \mathbf{y}^\delta, \mathbf{u}_k > \mathbf{u}_k. \tag{5.168}$$

Consequently, for the residual norm we have

$$\left\| \mathbf{r}_\alpha^\delta \right\|^2 = \sum_{k=1}^{n} \left(1 - \frac{\sigma_k^2}{\sigma_k^2 + \alpha} \right)^2 < \mathbf{y}^\delta, \mathbf{u}_k >^2 + \sum_{k=n+1}^{m} < \mathbf{y}^\delta, \mathbf{u}_k >^2. \tag{5.169}$$

The residual norm can be used for defining a criterion for choosing the optimal regularization parameter.

5.10 Regularization Parameter Choice Methods

A method to choose the regularization parameter depending only on the noise level Δ, i.e., when $\alpha = \alpha(\Delta)$, is called an a priori parameter choice method. A regularization parameter choice method depending on Δ and $\mathbf{y}^\delta$, $\alpha = \alpha(\Delta, \mathbf{y}^\delta)$, is called an a posteriori parameter choice method. A regularization parameter choice method depending only on $\mathbf{y}^\delta$, $\alpha = \alpha(\mathbf{y}^\delta)$, is called an error-free parameter choice method. Unfortunately, at the present time, there is no fail-safe regularization parameter choice method which guarantees small solution errors for any noisy data. Below we consider several methods, that are used in remote sensing applications.

5.10.1 A Priori Parameter Choice Method

In a priori parameter choice method, the regularization parameter is an explicit function of the noise level $(\alpha \sim \sigma^p)$ [70]. We define the optimal regularization parameter for error estimation as the minimizer of the expected error, namely,

$$\alpha_{\text{opt}} = \arg \min \varepsilon \left\{ \left\| \mathbf{e}_\alpha^\delta \right\|^2 \right\}. \tag{5.170}$$

The optimal regularization parameter is not a computable quantity, because the exact solution is unknown, but it is possible to design an a priori parameter choice method by combining this selection criterion with a Monte-Carlo technique. The resulting algorithm involves the following steps:

1. Perform a random exploration of a domain, in which the exact solution is supposed
   to lie, by considering a set of state vector realizations $\left\{\mathbf{x}_j^{\dagger}\right\}$, $j=1,N$, where N is
   the sample size;
2. For each $\mathbf{x}_j^{\dagger}$, compute the optimal regularization parameter for error estimation

$$\alpha_{\text{opt},j} = \arg \min \varepsilon \left\{ \left\| \mathbf{e}_\alpha^\delta \left(\mathbf{x}_j^{\dagger} \right) \right\|^2 \right\} \tag{5.171}$$

and determine the exponent

$$p_j = \frac{\log \alpha_{\text{opt},j}}{\log \sigma}, \tag{5.172}$$

1. compute the sample mean exponent

$$\bar{p} = \frac{1}{N} \sum_{j=1}^{N} p_j, \tag{5.173}$$

2. choose the regularization parameter as $\alpha_{\text{opt}} = \sigma^{\bar{p}}$.

5.10.2 Discrepancy Principle

The most popular a posteriori parameter choice method is the discrepancy principle
proposed by Morozov [71]. In this method, the regularization parameter is chosen
via a comparison between the residual norm

$$\left\| \mathbf{r}_\alpha^\delta \right\|^2 = \left\| \mathbf{y}^\delta - \mathbf{K}\mathbf{x}_\alpha^\delta \right\|^2 \tag{5.174}$$

and the assumed noise level Δ. The idea behind this approach is that it does not make
sense to ask for an approximate solution $\mathbf{x}_\alpha^\delta$ with a discrepancy $\left\| \mathbf{y}^\delta - \mathbf{K}\mathbf{x}_\alpha^\delta \right\|$ less than
Δ; a residual norm in the order of Δ is the best we should ask for. Thus, we have the
equation

$$\left\| \mathbf{r}_\alpha^\delta \right\|^2 \sim m\sigma^2. \tag{5.175}$$

Substituting Eq. (5.169) into Eq. (5.175) and taking the expected value of it, we
obtain:

$$\varepsilon \left\{ \left\| \mathbf{r}_\alpha^\delta \right\|^2 \right\} = (m - n)\sigma^2 + \sum_{k=1}^{n} \left(\frac{\alpha}{\sigma_k^2 + \alpha} \right)^2 \left[(\mathbf{u}_k^T \mathbf{y})^2 + \sigma^2 \right]. \tag{5.176}$$

Examining Eq. (5.176) we can conclude that the expected residual is an increasing function of the regularization parameter and has a plateau at $(m - n)\sigma^2$. The

discrepancy principle chooses that α at which the residual norm changes to a rapidly increasing function of the regularization parameter.

In some applications, the discrepancy principle gives too small regularization parameter and the solution is undersmoothed. In [72, 73] the generalized version of the discrepancy principle is considered in which the regularization parameter α is found from

$$\left\| \mathbf{r}_\alpha^\delta \right\|^2 - \mathbf{r}_\alpha^{\delta^T} \hat{\mathbf{A}}_\alpha \mathbf{r}_\alpha^\delta \sim m\sigma^2, \tag{5.177}$$

where $\hat{\mathbf{A}}_\alpha = \mathbf{K}\mathbf{K}_\alpha^\dagger$ is the so-called influence matrix. As $\hat{\mathbf{A}}_\alpha$ is positive definite, the left-hand side of this equation is smaller than the residual $\left\| \mathbf{r}_\alpha^\delta \right\|^2$, and hence, the regularization parameter computed from Eq. (5.177) is larger than that corresponding to the ordinary method.

5.10.3 Generalized Cross-Validation and Maximum Likelihood Estimation Methods

The generalized cross-validation method was developed by G. Wahba [74]. It does not require the knowledge of σ^2 and is a very popular error-free method for choosing the regularization parameter. The generalized cross-validation function can be derived from the "leave one out" principle. In the "leave one out" cross-validation, models that are obtained by leaving one of the m data points out of the inversion process are considered. The regularization parameter α is found as a minimizer of the generalized cross-validation function,

$$G(\alpha) = \frac{\left\| \mathbf{r}_\alpha^\delta \right\|^2}{\left[\text{trace} \left(\mathbf{I} - \hat{\mathbf{A}}_\alpha \right) \right]^2}, \tag{5.178}$$

namely,

$$\alpha = \arg\min_\alpha G(\alpha). \tag{5.179}$$

The example of the generalized cross-validation function is shown in Fig. 5.10.

The generalized cross-validation method seeks to locate the transition point where the residual norm changes from a very slowly varying function of α to a rapidly increasing function of α. But instead of working with the residual norm, the generalized cross-validation method uses the ratio of the residual norm and the degree of freedom for noise, which is a monotonically increasing function of α. The generalized cross-validation may suffer from the following drawback: the unique minimum can be very flat, thus leading to numerical difficulties in computing the regularization parameter. In addition, based on a Monte Carlo analysis [75] it has been shown that the generalized cross-validation function may not have a unique minimum. To avoid this drawback, the maximum likelihood estimation method was developed which

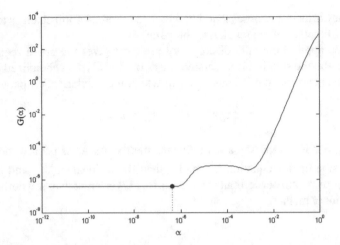

Fig. 5.10 Example of the generalized cross-validation function $G(\alpha)$ with two local minimums (one of them is flat)

is based on the maximization of the marginal likelihood function when Gaussian densities are assumed. The regularization parameter is chosen such as [76]:

$$\alpha = \arg\min_{\alpha} \frac{y^{\delta T}\left(\mathbf{I}_m - \hat{\mathbf{A}}_{\alpha}\right)y^{\delta}}{\sqrt[m]{\det\left(\mathbf{I}_m - \hat{\mathbf{A}}_{\alpha}\right)}}. \tag{5.180}$$

The maximum likelihood function has a more pronounced minimum as compared to the generalized cross-validation function.

5.10.4   L-Curve Method

The L-curve criterion has been advocated by Hansen [77]. The L-curve itself is a log–log plot of the solution norm $\left\|\mathbf{x}_{\alpha}^{\delta}\right\|$ versus the residual norm $\left\|\mathbf{r}_{\alpha}^{\delta}\right\|$ with α as the parameter (in more general case, instead of $\left\|\mathbf{x}_{\alpha}^{\delta}\right\|$, the constraint norm $\left\|\mathbf{L}\left(\mathbf{x}_{\alpha}^{\delta}\right)\right\|$ can be taken). The L-curve consists of two parts, namely a flat part where the regularization (smoothing) errors dominate and a steep part where the noise errors dominate, as shown in Fig. 5.11. The corner of the L-curve appears for regularization parameters close to the optimal parameter which balances the smoothing and the noise errors. The notion of a corner originates from a purely visual impression and it is not at all obvious how to translate this impression into mathematics. According to Hansen and O'Leary [78], the corner of the L-curve can be detected as the point of maximum curvature. Defining the L-curve components by $x(\alpha) = \log\left\|\mathbf{r}_{\alpha}^{\delta}\right\|^2$ and $y(\alpha) = \log\left\|\mathbf{x}_{\alpha}^{\delta}\right\|^2$ we select that value of α that maximizes the curvature function

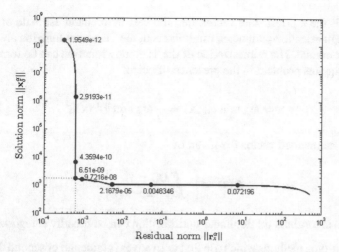

Fig. 5.11 Example of the L-curve. Values at red dots show corresponding values of the regularization parameter α

$$\alpha = \arg\min_{\alpha} \frac{\frac{d^2x(\alpha)}{d\alpha^2}\frac{dy(\alpha)}{d\alpha} - \frac{dx(\alpha)}{d\alpha}\frac{d^2y(\alpha)}{d\alpha^2}}{\left(\left(\frac{dx(\alpha)}{d\alpha}\right)^2 + \left(\frac{dy(\alpha)}{d\alpha}\right)^2\right)^{3/2}}. \tag{5.181}$$

The L-curve is more pronounced when the noise standard deviation is small.

5.11   Tikhonov Regularization for Nonlinear Problems

5.11.1   Definition of the Problem

Most of the inverse problems arising in atmospheric remote sensing are nonlinear. In this section we discuss the practical aspects of Tikhonov regularization for solving the nonlinear problem

$$F(\mathbf{x}) = \mathbf{y}^{\delta}, \tag{5.182}$$

where F is the nonlinear forward model. As in the linear case, the continuous nonlinear problem has a corresponding discrete problem.

Nonlinear problems are treated in the framework of Tikhonov regularization as linear problems. The regularized solution is a minimizer of the objective function:

$$F_{\alpha}(\mathbf{x}) = \frac{1}{2}\left[\left\|\mathbf{y}^{\delta} - F(\mathbf{x})\right\|^2 + \alpha\left\|\mathbf{L}(\mathbf{x} - \mathbf{x}_a)\right\|^2\right], \tag{5.183}$$

where $\mathbf{x}_a$ is the a priori state vector, i.e., the best beforehand estimate of $\mathbf{x}^\dagger$. For a positive regularization parameter, minimizers of the Tikhonov function always exist, but are not unique. The minimization of the Tikhonov function can be formulated as the least-squares problem in the presence of noise:

$$\mathbf{x} = \arg\min_{\mathbf{x}} F_\alpha(\mathbf{x}) = \frac{1}{2} \arg\min_{\mathbf{x}} \| f(\mathbf{x}) \|^2 , \tag{5.184}$$

where the augmented vector $\mathbf{f}$ is given by

$$\mathbf{f}(\mathbf{x}) = \begin{bmatrix} F(\mathbf{x}) - \mathbf{y}^\delta \\ \sqrt{\alpha} \mathbf{L}(\mathbf{x} - \mathbf{x}_a) \end{bmatrix} . \tag{5.185}$$

Numerical algorithms for nonlinear optimization can be broadly categorized into:

1. Newton-type methods which use first derivatives (gradients) or second derivatives (Hessians) information,
2. direct search methods, as for example genetic algorithm and simulated annealing, which do not use information on the derivatives.

5.11.2 Inversion of the Linearized Model

The key step in non linear optimization by means of the Gauss–Newton method is the linearization procedure. That means, the non linear model is linearized about a certain state vector $\mathbf{x}_0$:

$$F(\mathbf{x}) \approx F(\mathbf{x}_0) + K(\mathbf{x}_0)(\mathbf{x} - \mathbf{x}_0), \tag{5.186}$$

where $\mathbf{K}(\mathbf{x}_0)$ is the Jacobian matrix computed at $\mathbf{x}_0$:

$$\mathbf{K}(\mathbf{x}) = \begin{bmatrix} \dfrac{dy_1}{dx_1} & \dfrac{dy_1}{dx_2} & \cdots & \dfrac{dy_1}{dx_n} \\ \dfrac{dy_2}{dx_1} & \dfrac{dy_2}{dx_2} & \cdots & \dfrac{dy_2}{dx_n} \\ \vdots & \vdots & \ddots & \vdots \\ \dfrac{dy_m}{dx_1} & \dfrac{dy_m}{dx_2} & \cdots & \dfrac{dy_m}{dx_n} \end{bmatrix} \tag{5.187}$$

By doing this, we can apply methods for solving the linear optimization problem and to find the next approximation for $\mathbf{x}$. As in the linear case, the minimization of the Tikhonov function is formulated as a least-squares problem. Nonlinear optimization methods are iterative methods, which compute the new iterate $\mathbf{x}^\delta_{\alpha k+1}$ by approximating the objective function around the actual iterate $\mathbf{x}^\delta_{\alpha k}$ by a quadratic model and by imposing the so-called descent condition

$$F_\alpha \left(\mathbf{x}^\delta_{\alpha k+1} \right) < F_\alpha \left(\mathbf{x}^\delta_{\alpha k} \right). \tag{5.188}$$

The regularized solution can be computed by using optimization methods for unconstrained minimization problems. For this purpose, "step-length-based-methods" or "trust-region methods" can be employed, as described in [79]. The Gauss–Newton method for least squares problems belongs to the category of step-length methods and has an important practical interpretation which we now discuss. At the iteration step k, the forward model is linearized about the actual iterate $\mathbf{x}^\delta_{\alpha k}$, that is,

$$F \left(\mathbf{x} \right) \approx F \left(\mathbf{x}^\delta_k \right) + \mathbf{K}_k \left(\mathbf{x} - \mathbf{x}^\delta_k \right), \tag{5.189}$$

where $\mathbf{K}_k = \mathbf{K} \left(\mathbf{x}^\delta_k \right)$ is the Jacobian matrix computed at $\mathbf{x}^\delta_k$. The linearization technique allows replacing the nonlinear equation (5.182) by its linearization

$$\mathbf{K}_k \mathbf{p} = \mathbf{y}^\delta_k, \tag{5.190}$$

where

$$\mathbf{p} = \mathbf{x} - \mathbf{x}_a, \tag{5.191}$$

$$\mathbf{y}^\delta_k = \mathbf{y}^\delta - F \left(\mathbf{x}^\delta_{\alpha k} \right) + \mathbf{K}_{\alpha k} \left(\mathbf{x}^\delta_{\alpha k} - \mathbf{x} \right). \tag{5.192}$$

Because the nonlinear problem is ill-posed, its linearization is also ill-posed, and we solve the linearized equation by means of Tikhonov regularization with the penalty term $\alpha_k \|\mathbf{L}\mathbf{p}\|^2$ where α_k is the regularization parameter at iteration k. The Tikhonov function for the linearized equation takes the form

$$F_{\alpha k} \left(\mathbf{p} \right) = \left\| \mathbf{y}^\delta_k - \mathbf{K}_k \mathbf{p} \right\|^2 + \alpha_k \|\mathbf{L}\mathbf{p}\|^2. \tag{5.193}$$

Applying solution (5.117) to it, we obtain

$$\mathbf{p}^\delta_k = \mathbf{K}^\dagger_k \mathbf{y}^\delta_k \tag{5.194}$$

and

$$\mathbf{x}^\delta_{\alpha,k+1} = \mathbf{x}_a + \mathbf{K}^\dagger_k \mathbf{y}^\delta_k, \tag{5.195}$$

where

$$\mathbf{K}^\dagger_k = \left(\mathbf{K}^T_k \mathbf{K}_k + \alpha \mathbf{L}^T \mathbf{L} \right)^{-1} \mathbf{K}^T_k \tag{5.196}$$

being the generalized inverse at the iteration step k, is exactly the new iterate computed in the framework of the Gauss–Newton method (with a unit step-length). Thus, the solution of a nonlinear ill-posed problem by means of Tikhonov regularization is equivalent to the solution of a sequence of ill-posed linearizations of the forward model about the current iterate. Note that for nonlinear problems, the definition of the Fisher information matrix (Eq. (5.64)) is still valid with $\mathbf{K}$ being the Jacobian matrix.

For choosing the regularization parameter at each iteration step, the following recommendations can be taken into account:

1. At the beginning of the iterative process, large α-values should be used to avoid local minima and to get well-conditioned least-squares problems to solve;
2. During the iteration, the regularization parameter should be decreased slowly to achieve a stable solution.

Numerical experiments have shown that a brutal use of the regularization parameter computed by one of the above selection criteria may lead to an oscillation sequence of α-values. A heuristic formula that deals with this problem has been proposed by Eriksson [7]: at the iteration step k, the regularization parameter α_k is the weighted sum between the previous regularization parameter α_{k-1} and the proposed regularization parameter α, that is,

$$\alpha_k = \begin{cases} \xi\alpha_{k-1} + (1-\xi)\,\alpha, & \text{if } \alpha < \alpha_{k-1}, \\ \alpha_{k-1}, & \text{if } \alpha \geq \alpha_{k-1}. \end{cases} \tag{5.197}$$

Another option is to take regularization parameters α_k as the term of decreasing geometric sequence, namely, $\alpha_k = \xi\alpha_{k-1}$ with $0 < \xi < 1$.

Also, the maximum number of iteration steps plays the role of the regularization parameter. The iterative process should be stopped after a certain number of steps. In order to avoid an uncontrolled expansion of the errors, it is recommended to stop computations when the residual is of the same order as the noise, i.e., $\left\| r^\delta \right\|^2 \sim \Delta^2$.

5.11.3   Quadratic Model

Newton-type methods (NTM) are iterative, employing the quadratic model

$$M_k(\mathbf{p}) = F(\mathbf{x}_k) + \mathbf{g}(\mathbf{x}_k)^T \mathbf{p} + \frac{1}{2}\mathbf{p}^T \mathbf{G}(\mathbf{x}_k)\,\mathbf{p}, \tag{5.198}$$

as a reasonable approximation to the objective function in the neighborhood of the current iterate x_k [8]. In Eq. (5.198), while $\mathbf{g}$ and $\mathbf{G}$ are the gradient and the Hessian of F, that is,

$$\mathbf{g}(\mathbf{x}) = \mathbf{K}_f(\mathbf{x})^T \mathbf{f}(\mathbf{x}) \tag{5.199}$$

and

$$\mathbf{G}(\mathbf{x}) = \mathbf{K}_f(\mathbf{x})^T \mathbf{K}_f(\mathbf{x}) + \mathbf{Q}(\mathbf{x}) \tag{5.200}$$

respectively, where

$$\mathbf{K}_f(\mathbf{x}) = \left[\, \mathbf{K}(\mathbf{x})\ \sqrt{\alpha}\mathbf{L}\, \right]^T \tag{5.201}$$

is the Jacobian matrix of $\mathbf{f}(\mathbf{x})$, $\mathbf{Q}(\mathbf{x}) = \sum_{i=1}^{m} [\mathbf{f}(\mathbf{x})]_i \, \mathbf{G}_i(\mathbf{x})$ is the second-order derivative term, and $\mathbf{G}_i$ is the Hessian of $[\mathbf{f}]_i$. The objective function (5.183) can be minimized by means of step-length methods. A step-length method requires the computation of a vector $\mathbf{p}_k$ called the search direction, and the calculation of a positive scalar τ_k, the step-length, such that $F(\mathbf{x}_{k+1}) < F(\mathbf{x}_k)$, where $\mathbf{x}_{k+1} = \mathbf{x}_k + \tau_k \mathbf{p}_k$.

The Gauss–Newton method considered above can be obtained if it is assumed that the first-order term $\mathbf{K}_f^T \mathbf{K}_f$ in the expression of the Hessian (5.200) dominates the second-order term $\mathbf{Q}$. For small residual problems, the search direction solves the equation

$$\mathbf{K}_f(\mathbf{x}_k) \mathbf{K}_f(\mathbf{x}_k) \mathbf{p} = -\mathbf{K}_f(\mathbf{x}_k)^T \mathbf{f}(\mathbf{x}_k). \tag{5.202}$$

However, for large-residual problems, the term $\|\mathbf{f}(\mathbf{x}^*)\|$ is not small, and the second-order term $\mathbf{Q}$ cannot be neglected. One possible strategy for large-residual problems is to include a quasi-Newton approximation $\overline{\mathbf{Q}}$ to the second-order derivative term $\mathbf{Q}$, and to compute the search direction by solving the equation

$$\left[\mathbf{K}_f(\mathbf{x}_k)^T \mathbf{K}_f(\mathbf{x}_k) + \overline{\mathbf{Q}}(\mathbf{x}_k) \right] \mathbf{p} = -\mathbf{K}_f(\mathbf{x}_k)^T \mathbf{f}(\mathbf{x}_k). \tag{5.203}$$

Quasi-Newton methods are based on the idea of building up curvature information as the iteration proceeds using the observed behavior of the objective function and of the gradient. After $\mathbf{x}_{k+1}$ has been computed, a new approximation of $\overline{\mathbf{Q}}(\mathbf{x}_{k+1})$ is obtained by updating $\overline{\mathbf{Q}}(\mathbf{x}_k)$ to account the newly acquired curvature information. An updated formula reads as $\overline{\mathbf{Q}}(\mathbf{x}_{k+1}) = \overline{\mathbf{Q}}(\mathbf{x}_k) + \mathbf{U}_k$, where the update matrix $\mathbf{U}_k$ is usually chosen as a rank-one matrix. The standard condition for updating $\overline{\mathbf{Q}}$ is known as the quasi-Newton condition, and requires that the Hessian should approximate the curvature of the objective function along the change in $\mathbf{x}$ during the current iteration step. Then the most widely used quasi-Newton scheme, which satisfies the quasi-Newton condition and possesses the property of hereditary symmetry, is the Broyden–Fletcher–Goldfarb–Shanno (BFGS) update [80–83].

Byrd et al. [84] proposed the quasi-Newton method, which is referred to as L-BFGS method. The algorithm approximately minimizes $\mathbf{M}_k$ subject to the bounds: $l \le x \le u$, where the vectors l and u represent lower and upper bounds on the variables, respectively. This is done by first using the gradient projection method to find a set of active bounds, followed by a minimization of $\mathbf{M}_k$ treating those bounds as equality constraints. In contrast to standard BFGS methods, which store a dense $n \times n$ approximation to the inverse Hessian (n being the number of atmospheric parameters to be retrieved), L-BFGS stores only a few vectors that represent the approximation implicitly. By making use of the compact representations of limited memory matrices, the computational cost of one iteration of the algorithm can be kept to be of order n.

Newton-type methods only build up a local model of the objective function. Such algorithms will only find the local optimum, and for this reason, they are also called local optimization algorithms. Local search methods can get stuck in a local minimum, where no improving neighbors are available. A simple modification consists of iterating calls to the local search routine, each time starting from a different initial guess. Iterated local search (ILS) is based on building a sequence of locally optimal solutions by perturbing the current local minimum and by applying local search after starting from the modified solution [85]. Various versions of ILS algorithms can be found in the literature. One of the implementations is shown in pseudo-code Algorithm 1:

Algorithm 1 Iterated local search algorithm.

---
choose $\mathbf{x}_0$ and l_{max};
compute $\mathbf{x}^*$ by NTM with initial guess $\mathbf{x}_0$;
$l = 0$;
stop = .false.;
do while (.**not**.stop .**and**. $l < l_{max}$)
  $\mathbf{x}_{\pm 0} = \mathbf{x}^* \pm p$;
  compute $\mathbf{x}_{\pm}^*$ by NTM with initial guesses $\mathbf{x}_{\pm 0}$;
  set $\mathbf{x} = \arg \min \left\{ F(\mathbf{x}^*), F(\mathbf{x}_+^*), F(\mathbf{x}_-^*) \right\}$;
  if $\mathbf{x} == \mathbf{x}^*$ **then**
    stop = .true.;
  else
    $\mathbf{x}^* = \mathbf{x}$;
  end if
  $l = l + 1$;
end do

The local search is used with the initial guess $\mathbf{x}_0$ and leads to a local optimum $\mathbf{x}^*$. Two new points $\mathbf{x}_{+0} = \mathbf{x}^* + \mathbf{p}$ and $\mathbf{x}_{-0} = \mathbf{x}^* - \mathbf{p}$ are selected around $\mathbf{x}^*$, and descents from $\mathbf{x}_{+0}$ and $\mathbf{x}_{-0}$ are done with the local search routine. These lead to the new local minima $\mathbf{x}_+^*$ and $\mathbf{x}_-^*$. For $\mathbf{x} = \arg \min \left\{ F(\mathbf{x}^*), F(\mathbf{x}_+^*) F(\mathbf{x}_-^*) \right\}$, two outcomes are possible: (a) if $\mathbf{x} = \mathbf{x}^*$ one is again at the bottom of the same valley; in this case the procedure is stopped, (b) if $\mathbf{x} \neq \mathbf{x}^*$ another local optimum, better than the incumbent has been found; in this case the search is recentered around $\mathbf{x}$ and begins again with the perturbation step. Outcome (a) includes the case in which the new local minima $\mathbf{x}_+^*$ and $\mathbf{x}_-^*$ either coincide with the incumbent local minimum $\mathbf{x}^*$, or are not better than $\mathbf{x}^*$. The perturbation strength $\mathbf{p}$ has to be chosen such that a different local minimum is obtained. This parameter is crucial for the efficiency of the algorithm and depends on the problem to be solved.

5.11.4   Global Optimization (Direct Search Methods)

Direct search methods tend to converge more slowly, but can be more tolerant to the presence of noise in the objective function and constraints. Essentially, direct search methods try different state vectors. At each iteration, the state vector is chosen by using a special search strategy. In principle, this strategy can be arbitrary as long as it guarantees that (sooner or later) it is capable of checking practically the whole domain of the retrieved parameters. But ideas behind some strategies simulate a certain process in physical or biological systems.

Genetic algorithms (GAs) belong to the category of evolutionary algorithms [86–88]. An evolutionary algorithm uses mechanisms inspired by biological evolution, such as reproduction, mutation, recombination (crossover), and selection. Candidate solutions to the optimization problem play the role of individuals in a population, and the fitness function determines the quality of the solutions. GAs differ from NTM in two main ways: (a) NTM generate a single point at each iteration, and so, a sequence of points approaching an optimal solution, while GAs generate a population of points at each iteration so that the best point in the population approaches an optimal solution, and (b) NTM select the next point in the sequence by a deterministic computation, while GAs select the next population by computation which includes random number generators.

Simulated annealing (SA) is inspired by the annealing (cooling) processes of crystals that reach the lowest energy, corresponding to the perfect crystal structure, if cooled sufficiently slowly [89–91]. SA algorithms randomly generate at each iteration a candidate point in a suitable neighborhood of the current point and, through a random mechanism controlled by a parameter called temperature (in view of the analogy with the physical process), they decide whether to move to the candidate point or to stay in the current one at the next iteration.

5.12 Linearization Techniques

5.12.1 Definition

Following [92], the term "linearization" for radiative transfer models means the computations of analytic radiance derivatives (Jacobians) with respect to atmospheric and surface variables. The Jacobian matrix can be computed by finite differences:

$$[\mathbf{K}]_{ij} \approx \frac{F\left(x_j + \Delta \xi_i\right) - F\left(x_j\right)}{\Delta \xi_i}, \tag{5.204}$$

where $\Delta \xi_i$ is the ith atmospheric parameter. There are two shortcomings. First, this method is time-consuming as it requires $N+1$ forward simulations to estimate the Jacobian matrix for N atmospheric parameters for each wavelength. Secondly, there is a problem in choosing the optimal perturbation value $\Delta \xi_i$. If $\Delta \xi_i$ is too large, the accuracy of Eq. (5.204) decreases. If $\Delta \xi_i$ is very small, the rounding errors can be significant, and the computations become unstable. Therefore, in the next sections we consider alternative techniques.

5.12.2 Automatic Differentiation

The model can also be linearized using automatic differentiation algorithms [93, 94]. Essentially, all physical models are based on elementary functions and mathematical operations for which the rules of differentiation are known. The automatic differentiation algorithm works directly with the program code (listing) that corresponds to the forward problem. Each mathematical function or subprogram is differentiated according to the rule of differentiation of a complex function. As output we get the linearized code. This approach was used in [95, 96] for the line-by-line model in the infrared region. With regard to the discrete ordinate method, considered in Chap. 3, this method has two drawbacks:

1. The solution algorithm includes an eigenvalue problem, which is not elementary and direct differentiation of the code might lead to a non-optimal algorithm.
2. The automatic differentiation system is not designed in principle for generating an optimized code. Therefore, for applications in which performance is important, the resulting linearized code must be manually optimized, which may be a more laborious task than the original one.

5.12.3 Analytical Differentiation

In the last decade, programs based on analytical methods for calculating the Jacobi matrix, such as LIDORT/VLIDORT [97], SCIATRAN [98], DOME [99], have proliferated. Currently, two methods of analytical linearization are widely used. The first of these is the direct differentiation method in the context of the discrete ordinate method. In the framework of this method, the partial derivative of the input data of the model for the desired parameter is considered; further, applying the chain differentiation rule, the entire algorithm is recursively linearized. It has been shown that the method of discrete ordinates can be completely linearized analytically without invoking additional assumptions [97]. The same idea was used when creating radiation codes for the infrared range KOPRA [100] and MOLIERE [101].

The linearization of the basic matrix operations is straight-forward. The multiplication is linearized by Leibniz rule, while the linearization of the inverse matrix is performed by two matrix multiplications:

$$\frac{\partial (\mathbf{A}_1 \mathbf{A}_2)}{\partial \varsigma} = \frac{\partial \mathbf{A}_1}{\partial \varsigma} \mathbf{A}_2 + \mathbf{A}_1 \frac{\partial \mathbf{A}_2}{\partial \varsigma}, \tag{5.205}$$

$$\frac{\partial \mathbf{A}^{-1}}{\partial \varsigma} = -\mathbf{A}^{-1} \frac{\partial \mathbf{A}}{\partial \varsigma} \mathbf{A}^{-1}. \tag{5.206}$$

Here ς is an atmospheric parameter to be retrieved (for instance, the concentration of a specific gas). The linearization of the eigenvalue problem with respect to the atmospheric parameter to be retrieved is not straight-forward. As suggested in [97], we consider the eigenvalue problem related to the discrete ordinate method (see Sect. 4.7.4, Chap. 4):

$$\mathbf{A}\mathbf{v}_k = \lambda_k \mathbf{v}_k. \tag{5.207}$$

Applying the derivative operator to Eq. (5.207) and using the orthogonality condition for eigenvectors, one can get the following equation:

$$
\begin{cases}
\dfrac{d\mathbf{A}}{d\varsigma}\mathbf{v}_k + \mathbf{A}\dfrac{d\mathbf{v}_k}{d\varsigma} = \dfrac{d\lambda_k}{d\varsigma}\mathbf{v}_k + \lambda_k\dfrac{d\mathbf{v}_k}{d\varsigma} \\
\mathbf{v}_k^T \dfrac{d\mathbf{v}_k}{d\varsigma} = 0
\end{cases}
\tag{5.208}
$$

The derivatives for the initial eigenvalue problem for the matrix $\mathbf{A}$ are expressed in terms of the solution of the system (5.208), which has to be solved for each layer and each azimuth harmonics number. Then, following the rules (5.205) and (5.206), one can get the linearization of the whole radiative transfer model. It is important to note that $\mathbf{A}$, $\mathbf{v}_k$ and λ_k depend only on the parameters of the current layer.

The linearization of the matrix operator method is organized using the concept of a "stack". The stack which contains n layers is described by reflection and transmission matrices, respectively, as well as the source function J. To add a layer number $j=n+1$ to the stack with n layers, one has to perform two steps:

1. To compute the derivatives for reflection and transmission matrices and for the layer j and a source vector (note that the latter has derivatives over the parameters of upper layers);
2. Having a linearized stack of n layers, to compute the derivatives for reflection and transmission matrices for a stack of $n+1$ layers by differentiating the interaction principle equations.

Thus, the derivatives of reflection and transmission matrices should be recomputed after each adding the layer.

5.12.4 Adjoint Approach

The second method is based on the use of properties of adjoint equations (forward-adjoint perturbation theory). This technique has been proposed by Kadomtsev [102] and further developed in the theory of neutron transport [103]. In this approach, the derivative of the radiance is expressed through the derivatives of the source function and the transport equation operator. For the first time, for a plane-parallel layer, analytical expressions for partial derivatives were obtained in the work of Marchuk [104], and later in the works of Box [105], Ustinov [106], Rozanov et al. [107], Landgraf et al. [108], for the pseudo-spherical atmosphere in Walter et al. [109] and Doicu et al. [99], and for the spherical in Walter et al. [110]. The linearization algorithm through a conjugate operator does not depend on the method of solving the direct problem, but is determined only by the initial equation. The same formalism can be applied to the three-dimensional problems.

Comparing the method of direct differentiation of the method of discrete ordinates with the forward-adjoint perturbation method (or with the differentiation of the original equation), the following should be noted:

1. The adjoint method requires finding a solution inside the medium, which is not initially required in the original direct problem.
2. Relative error of the radiance field

In [99] several linearization techniques have been compared; it has been shown that the forward-adjoint perturbation method is about three times faster than the analytical differentiation approach.

5.13   Dimensionality Reduction Based Retrievals

5.13.1   *Basic Concept of Dimensionality Reduction*

In simple words, dimensionality reduction means representing the initial data set with less number of parameters than it is initially represented. It can be considered as one of the lossy data compression paradigms [111]. Dimensionality reduction is crucial for stable and high-performance processing of spectral measurements. It excludes redundant information from the initial dataset, reduces the number of independent parameters and improves the efficiency of machine learning.

There is a distinction between linear and non linear techniques for dimensionality reduction. A more detailed review can be found in [112–115] and references therein. Linear and non linear techniques have been inter-compared in [116]. Results of these numerical experiments reveal that non linear techniques perform well on selected artificial tasks. However, they hardly outperform the principal component analysis technique on real-world tasks. Bearing in mind that no obviously superior method has emerged in the benchmarking studies (increasingly time-consuming and sophisticated dimensionality reduction techniques lead to more accurate results, and vice versa), our discussion given below is limited to the classical principal component analysis (PCA).

5.13.2   *Principal Component Analysis*

The sizes of matrices are specified using the notation $\in \mathbb{R}^{\text{rows} \times \text{columns}}$. Let $\mathbf{y} = (y(\lambda_1), y(\lambda_2), ..., y(\lambda_W))$, $\mathbf{y} \in \mathbb{R}^{1 \times W}$, be a row-vector of atmospheric radiances at W wavelengths $\{\lambda_w\}_{w=1,...,W}$. A set of S spectra are assembled into a matrix

$\mathbf{Y} \in \mathbb{R}^{S \times W}$ whose ith row is $\mathbf{y}_i$. Then, $\mathbf{y}_i$ can be represented in a new basis system as follows:

$$\mathbf{y}_i = \bar{\mathbf{y}} + \sum_{k=1}^{W} t_{ik} \mathbf{f}_k. \tag{5.209}$$

Here, $\bar{\mathbf{y}} = \sum_{i=1}^{S} \mathbf{y}_i / S$, $\bar{\mathbf{y}} \in \mathbb{R}^{1 \times W}$ is the sample mean of the spectra (the average spectrum), t_{ik} is the kth coordinate of the vector $\mathbf{y}_i$ in the new basis system and $\mathbf{f}_k = (f_k(\lambda_1), f_k(\lambda_2), ..., f_k(\lambda_W)) \in \mathbb{R}^{1 \times W}$ is the kth basis vector. Noting that high-dimensional real data are often situated on or near a lower-dimensional manifold, the spectrum $\mathbf{y}_i$ can be projected onto the K-dimensional subspace ($K < W$) as follows:

$$\mathbf{y}_i \approx \bar{\mathbf{y}} + \sum_{k=1}^{K} t_{ik} \mathbf{f}_k, \tag{5.210}$$

or in matrix form for the initial dataset:

$$\mathbf{Y} \approx \bar{\mathbf{Y}} + \mathbf{TF}, \tag{5.211}$$

where $\bar{\mathbf{Y}} = \{\bar{\mathbf{y}}, ..., \bar{\mathbf{y}}\} \in \mathbb{R}^{S \times W}$, $\mathbf{F} = \{\mathbf{f}_1, \mathbf{f}_2, ..., \mathbf{f}_K\}^T \in \mathbb{R}^{K \times W}$, $\mathbf{T} \in \mathbb{R}^{S \times K}$ is the matrix whose entries are $\{t_{ik}\}_{i=1,...,S}^{k=1,...,K}$. Hereinafter the superscript ^{T} stands for "transpose". The transformation (5.210) can be done using dimensionality reduction techniques, such as PCA [54]. In the latter, basic vectors $\mathbf{f}_k$ in (5.210) are referred to as "principal components" (PCs) or empirical orthogonal functions (EOFs) and are taken as K eigenvectors related to the K most significant eigenvalues of the covariance matrix $\text{cov}(\mathbf{Y}, \mathbf{Y}) \in \mathbb{R}^{W \times W}$. The coordinates t_{ik} in the new coordinate system and the corresponding matrix $\mathbf{T}$ are called "principal component scores".

5.13.3 Principal Component Regression Models

In the retrieval algorithms based on linear regression, the following representation for the retrieved parameter x is exploited:

$$x = c + \sum_{w=1}^{W} l_w y(\lambda_w), \tag{5.212}$$

where c is the linear offset and l_w are the regression coefficients. In order to capture the essential features of the simulated data and to avoid "over-dimensionality" (the so-called Hughes effect [117]), the simulated spectral data can be compressed using an appropriate dimensionality-reduction technique. The principal component regression (PCR) method employs the linear regression model between x and the principal component scores of the spectral radiance:

$$x = c + \sum_{k=1}^{K} l_k t_k.$$

(5.213)

As $K << W$, the dimension of the linear regression model (and the corresponding inverse problem) is reduced. Moreover, since the instrument noise does not affect PC scores of low order, the whole inversion scheme is more stable.

For noisy data, the set of eigenvectors F must be computed for the matrix $C_Y + C_e$, rather than for C_Y, where C_e is the noise covariance matrix. In this case, the PC scores for the noisy data are correlated and are therefore called "projected principal components" [118]. If the statistics of the noise is unknown, the noise covariance matrix can be estimated by making some assumptions (e.g., Gaussian noise) or by using the following approximation $C_e \approx \alpha I$, where I is the identity matrix and α is the regularization parameter. This procedure reduces the impact of high-order principal components.

The kernel ridge regression (KRR) algorithm [119] generalizes the PCR method; KRR has been used for predicting atmospheric profiles from the IASI (the infrared atmospheric sounding interferometer) instrument [120]. One drawback of the PCR and KRR models is that the basis vectors F characterize the measurements Y, while the information contained in X is not taken into account. An alternative model that gets a round this drawback is the partial least-squares regression (PLSR) [121]. In [122, 123], it was shown that PLSR leads to model-fitting with fewer PCs than required with PCR. In its turn, the PLSR approach can be generalized to the case when we are retrieving a set of correlated parameters (e.g., the temperature profile) rather than a single variable x. The corresponding method is then referred to as canonical correlations [124]. The use of canonical correlations in atmospheric sciences applications is summarized in [125].

The approach based on the PCR has been successfully applied for solving the problems of volcanic plume-height retrieval from GOME-2 [126] and CO_2 retrieval from GOSAT measurements [127]. This approach can be extended by substituting the linear regression by the artificial neural network to find a better relation between the parameter to be retrieved and principal components. This method was used for retrieving the volcanic plume height from TROPOMI measurements [128], surface properties [129] and other parameters.

5.13.4 Principal Component Analysis in the Framework of DOAS

In this section we consider the application of the PCA in the framework of DOAS. In [130] the modification of the DOAS approach was proposed for retrieving the SO_2 total column. The PCA is applied for the measured spectra in regions with no significant SO_2, e.g., the equatorial Pacific:

$$\ln y (\lambda) = \overline{\ln y} (\lambda) + \sum_{i=1}^{K} t_k f_k (\lambda). \qquad (5.214)$$

In this way the EOFs capture the variability of the data caused by physical processes (i.e., Rayleigh and Raman scattering and ozone absorption). In addition, the features of the instrument (e.g., the instrumental degradation, the slit function and measurement artifacts) are implicitly accounted for by EOFs. That is the training phase. Then, for polluted regions with SO_2, representation (5.214) will produce a residual which is presumably associated with the SO_2 content. Thus,

$$\ln y (\lambda) = \overline{\ln y} (\lambda) + \sum_{i=1}^{K} t_k f_k (\lambda) + \Omega_{SO_2} \frac{\partial \ln y (\lambda)}{\partial \Omega_{SO_2}}, \qquad (5.215)$$

where Ω_{SO_2} is the SO_2 vertical column density. The derivative in the last term can be estimated either by finite differences or by using linearized radiative transfer models [97, 99]. Then, from Eq. (5.215) Ω_{SO_2} can be readily retrieved.

This method has been applied to the ozone monitoring instrument (OMI) [131] data in the spectral interval 310–340 nm. As the high-order principal components represent the noise rather than a useful signal, the truncation over the principal components also acts as a filter. To reconstruct the spectral radiances, at least 20–30 principal components are required while in the presence of relatively strong SO_2 signals that number could be reduced to 8. Authors claim that the noise in the data was decreased by a factor of 2, thereby providing greater sensitivity to anthropogenic sources of SO_2.

So far, there are no reports of applying a similar approach to other trace gases. One reason for that is the difficulty of obtaining the system of EOFs for regions without a certain trace gas. The second reason is that, strictly speaking, representation (5.215) is an approximate one. For SO_2, it works correctly and the residual is associated with the SO_2 signal. For other trace gases representation (5.215) might not be valid, so more elaborated approach is required.

5.13.5 Retrieval in the Reduced Input Space

Since the atmospheric retrieval problem is severely ill-posed, a physically correct result can be obtained only by using a regularization procedure. The latter takes into account some a priori information. In this context, dimensionality reduction of the input data space can be regarded as a special type of regularization, i.e., the retrieved parameters should obey a certain dependency reproduced by a chosen set of EOFs.

Timofeyev et al. [132] applied the dimensionality reduction technique to parameterize the aerosol extinction coefficient for incorporation into the inversion algorithm, in which the corresponding PC scores rather than aerosol extinction dependence were retrieved. The system of EOFs was defined for a dataset of aerosol extinction coefficients computed on the basis of Mie theory and algorithms for particle ensembles.

Finally, in [133] the dimensionality reduction is performed in the input (temperature and humidity profiles) and output spaces (spectral radiances), while artificial neural networks are used to establish the interdependency between corresponding PC scores. Since the number of independent parameters is reduced, such a scheme is considered to be more robust and efficient than the conventional approach.

References

1. C. Bohren, D. Huffman, *Absorption and Scattering of Light by Small Particles* (Wiley, 1998). https://doi.org/10.1002/9783527618156
2. G.B. Airy, Account of some circumstances historically connected with the discovery of the planet exterior to Uranus. Astronomische Nachrichten **25**(10), 131–148 (1847). https://doi.org/10.1002/asna.18470251002
3. Mon Challis, Account of observations at the Cambridge observatory for detecting the planet exterior to Uranus. Monthly Not. R. Astron. Soc. **7**(9), 145 (1846). https://doi.org/10.1093/mnras/7.9.145
4. J. Burrows, P. Borrell, U. Platt (eds.), *The Remote Sensing of Tropospheric Composition from Space* (Springer, Berlin, 2011). http://orcid.org/10.1007/978-3-642-14791-3
5. S. Liang, *Comprehensive Remote Sensing* (Elsevier, Amsterdam, 2018)
6. M. Coldewey-Egbers, S. Slijkhuis, B. Aberle, D. Loyola, Long-term analysis of GOME in-flight calibration parameters and instrument degradation. Appl. Opt. **47**(26), 4749 (2008). https://doi.org/10.1364/ao.47.004749
7. C. Rodgers, *Inverse Methods for Atmospheric Sounding: Theory and Practice* (Wolrd Scientific Publishing, 2000)
8. A. Doicu, T. Trautmann, F. Schreier, *Numerical Regularization for Atmospheric Inverse Problems* (Springer, Berlin, 2010)
9. W. Decbski, in *Advances in Geophysics* (Elsevier, 2010), pp. 1–102. https://doi.org/10.1016/s0065-2687(10)52001-6
10. A. Kokhanovsky, V.V. Rozanov, Droplet vertical sizing in warm clouds using passiveoptical measurements from a satellite. Atmosp. Meas. Tech. **5**(3), 517–528 (2012). https://doi.org/10.5194/amt-5-517-2012
11. A. Loew, W. Bell, L. Brocca, C. Bulgin, J. Burdanowitz, X. Calbet, R. Donner, D. Ghent, A. Gruber, T. Kaminski, J. Kinzel, C. Klepp, J.C. Lambert, G. Schaepman-Strub, M. Schröder, T. Verhoelst, Validation practices for satellite based Earth observation data across communities. Rev. Geophys. **55**(3), 779–817 (2017). https://doi.org/10.1002/2017rg000562
12. S. Wang, X. Li, Y. Ge, R. Jin, M. Ma, Q. Liu, J. Wen, S. Liu, Validation of regional-scale remote sensing products in China: From site to network. Remote Sens. **8**(12), 980 (2016). https://doi.org/10.3390/rs8120980
13. G.M.B. Dobson, Forty years' research on atmospheric ozone at Oxford: a history. Appl. Opt. **7**(3), 387 (1968). https://doi.org/10.1364/ao.7.000387
14. F. Götz, A. Meetham, G. Dobson, The vertical distribution of ozone in the atmosphere. Proc. R. Soc. Lond. Ser. A Contain. Pap. Math. Phys. Character **145**(855), 416–446 (1934). http://orcid.org/10.1098/rspa.1934.0109
15. C. Pekeris, On the interpretation of the Umkehr effect in atmospheric ozone measurements. Publ. Oslo Obs. **1**, Ii–I31 (1933)
16. C. Mateer. A study of the information content of Umkehr observations (Ph.D.), University of Michigan (1964)
17. I. Petropavlovskikh, New Umkehr ozone profile retrieval algorithm optimized for climatological studies. Geophys. Res. Lett. **32**(16) (2005). https://doi.org/10.1029/2005gl023323
18. S.R. Aliwell, Analysis for BrO in zenith-sky spectra: An intercomparison exercise for analysis improvement. J. Geophys. Res. **107**(D14) (2002). https://doi.org/10.1029/2001jd000329

19. C. Sioris, C. Haley, C. McLinden, C. von Savigny, I. McDade, J. McConnell, W.J. Evans, N. Lloyd, E. Llewellyn, K. Chance, T. Kurosu, D. Murtagh, U. Frisk, K. Pfeilsticker, H. Bösch, F. Weidner, K. Strong, J. Stegman, G. Mégie, Stratospheric profiles of nitrogen dioxide observed by optical spectrograph and infrared imager system on the Odin satellite. J. Geophys. Res.: Atmosp. **108**, D7 (2003). https://doi.org/10.1029/2002JD002672

20. U. Platt, J. Stutz, *Differential Optical Absorption Spectroscopy* (Springer, Berlin, 2008). http://orcid.org/10.1007/978-3-540-75776-4

21. M. Seah, I. Gilmore, S. Spencer, Background subtraction: Ii. general behaviour of reels and the tougaard universal cross section in the removal of backgrounds in aes and xps. Surf. Sci. **461**(1–3), 1 (2000). https://doi.org/10.1016/s0039-6028(00)00373-3

22. A.V. Lubenchenko, A.A. Batrakov, A.B. Pavolotsky, O. Lubenchenko, D.A. Ivanov, XPS study of multilayer multicomponent films. Appl. Surf. Sci. **427**, 711–721 (2018). http://orcid.org/10.1016/j.apsusc.2017.07.256

23. O.P. Hasekamp, Capability of multi-viewing-angle photo-polarimetric measurements for the simultaneous retrieval of aerosol and cloud properties. Atmosp. Meas. Tech. **3**(4), 839 (2010). https://doi.org/10.5194/amt-3-839-2010

24. V.P. Afanasyev, D.S. Efremenko, A.V. Lubenchenko, M. Vos, M.R. Went, Extraction of cross-sections of inelastic scattering from energy spectra of reflected atomic particles. Bull. Russ. Acad. Sci.: Phys. **74**(2), 170–174 (2010). https://doi.org/10.3103/s1062873810020152

25. V. Afanas'ev, A. Gryazev, D. Efremenko, P. Kaplya, Dierential inverse inelastic mean free path and dierential surface excitation probability retrieval from electron energy loss spectra. Vacuum **136**, 146–155 (2017). http://orcid.org/10.1016/j.vacuum.2016.10.021

26. W. Press, *Numerical Recipes 3rd Edition: The Art of Scientific Computing* (Cambridge University Press, Cambridge, 2007)

27. J. Vicent, J. Verrelst, N. Sabater, L. Alonso, J. Rivera-Caicedo, L. Martino, J. Muñoz-Marí, J. Moreno, Comparative analysis of atmospheric radiative transfer models using the atmospheric look-up table generator (ALG) toolbox (version2.0). Geosci. Model Dev. **13**(4), 1945–1957 (2020). https://doi.org/10.5194/gmd-13-1945-2020

28. O. Zawadzka, K. Markowicz, Retrieval of aerosol optical depth from optimal interpolation approach applied to SEVIRI data. Remote Sens. **6**(8), 7182 (2014). https://doi.org/10.3390/rs6087182

29. R. Brent, *Algorithms for Minimization Without Derivatives* (Prentice-Hall Inc, New Jersey, 1973)

30. D. Tanré, F.M. Bréon, J.L. Deuzé, O. Dubovik, F. Ducos, P. François, P. Goloub, M. Herman, A. Lifermann, F. Waquet, Remote sensing of aerosols by using polarized, directional and spectral measurements within the A-Train: the PARASOL mission. Atmosp. Meas. Tech. **4**(7), 1383 (2011). https://doi.org/10.5194/amt-4-1383-2011

31. M. Mishchenko, I. Geogdzhayev, B. Cairns, W. Rossow, A. Lacis, Aerosol retrieval sover the ocean by use of channels 1 and 2 AVHRR data: sensitivity analysis and preliminary results. Appl. Opt. **38**(36), 7325 (1999). https://doi.org/10.1364/ao.38.007325

32. L.A. Remer, Y.J. Kaufman, D. Tanré, S. Mattoo, D.A. Chu, J.V. Martins, R.R. Li, C. Ichoku, R.C. Levy, R.G. Kleidman, T.F. Eck, E. Vermote, B.N. Holben, The MODIS aerosol algorithm, products, and validation. J. Atmosp. Sci. **62**(4), 947–973 (2005). https://doi.org/10.1175/jas3385.1

33. O. Dubovik, M. Herman, A. Holdak, T. Lapyonok, D. Tanré, J.L. Deuzé, F. Ducos, A. Sinyuk, A. Lopatin, Statistically optimized inversion algorithm for enhanced retrieval of aerosol properties from spectral multi-angle polarimetric satellite observations. Atmosp. Meas. Tech. **4**(5), 975–1018 (2011). https://doi.org/10.5194/amt-4-975-2011

34. 3MI - EUMETSAT. https://www.eumetsat.int/website/home/Satellites/FutureSatellites/EUMETSATPolarSystemSecondGeneration/3MI/index.html

35. A. Kokhanovsky, in *Cloud Optics. Atmospheric and Oceanographic Sciences Library*, vol. 34 (Springer Netherlands, 2006), pp. 207–258. https://doi.org/10.1007/1-4020-4020-2_4

36. A.A. Kokhanovsky, V. Rozanov, E. Zege, H. Bovensmann, J. Burrows, A semi analytical cloud retrieval algorithm using backscattered radiation in 0.4-2.4 μm spectral region. J. Geophys. Res. **108**, D1 (2003). https://doi.org/10.1029/2001jd001543

37. W. Edie, D. Dryard, F. James, M. Roos, B. Sadoulet, *Statistical Methods in Experimental Physics* (North-Holland, New York, 1971)
38. A.W. Brewer, C.T. Mcelroy, J.B. Kerr, Nitrogen dioxide concentrations in the atmosphere. Nature **246**(5429), 129–133 (1973). https://doi.org/10.1038/246129a0
39. J.F. Noxon, Nitrogen dioxide in the stratosphere and troposphere measured byground-based absorption spectroscopy. Science **189**(4202), 547–549 (1975). https://doi.org/10.1126/science.189.4202.547
40. U. Platt, D. Perner, H.W. Pätz, Simultaneous measurement of atmospheric CH2O, O3, and NO2 by differential optical absorption. J. Geophys. Res. **84**(C10), 6329 (1979). https://doi.org/10.1029/jc084ic10p06329
41. U. Platt, D. Perner, A.M. Winer, G.W. Harris, J. Pitts, Detection of NO3 in the polluted troposphere by differential optical absorption. Geophys. Res. Lett. **7**(1), 89–92 (1980). https://doi.org/10.1029/gl007i001p00089
42. U. Platt, D. Perner, G.W. Harris, A.M. Winer, J.N. Pitts, Observations of nitrous acid in an urban atmosphere by differential optical absorption. Nature **285**(5763), 312–314 (1980). https://doi.org/10.1038/285312a0
43. V.V. Rozanov, A.V. Rozanov, Differential optical absorption spectroscopy (DOAS) and air mass factor concept for a multiply scattering vertically in homogeneous medium: theoretical consideration. Atmosp. Meas. Tech. **3**(3), 751–780 (2010). https://doi.org/10.5194/amt-3-751-2010
44. F. Wittrock, A. Richter, H. Oetjen, J. Burrows, M. Kanakidou, S. Myriokefalitakis, R. Volkamer, S. Beirle, U. Platt, T. Wagner, Simultaneous global observations of glyoxal and formaldehyde from space. Geophys. Res. Lett. **33**, 16 (2006). https://doi.org/10.1029/2006gl026310
45. L. Perliski, S. Solomon, On the evaluation of air mass factors for atmospheric near ultraviolet and visible absorption spectroscopy. J. Geophys. Res. **98**(D6), 10363 (1993). https://doi.org/10.1029/93jd00465
46. T. Wagner, J.P. Burrows, T. Deutschmann, B. Dix, C. von Friedeburg, U. Frieß, F. Hendrick, K.P. Heue, H. Irie, H. Iwabuchi, Y. Kanaya, J. Keller, C.A. McLinden, H. Oetjen, E. Palazzi, A. Petritoli, U. Platt, O. Postylyakov, J. Pukite, A. Richter, M. van Roozendael, A. Rozanov, V. Rozanov, R. Sinreich, S. Sanghavi, F. Wittrock, Comparison of box-air-mass-factors and radiances for multiple-axis differential optical absorption spectroscopy (MAX-DOAS) geometries calculated from different UV/visible radiative transfer models. Atmosp. Chem. Phys. **7**(7), 1809 (2007). https://doi.org/10.5194/acp-7-1809-2007
47. D.J. Fish, R.L. Jones, E.K. Strong, Midlatitude observations of the diurnal variation of stratospheric BrO. J. Geophys. Res. **100**(D9), 18863 (1995). https://doi.org/10.1029/95jd01944
48. A.M. Wazwaz, The regularization method for Fredholm integral equations of the first kind. Comput. Math. Appl. **61**(10), 2981–2986 (2011). https://doi.org/10.1016/j.camwa.2011.03.083
49. C. Lerot, M. van Roozendael, J.C. Lambert, J. Granville, J. van Gent, D. Loyola, R. Spurr, The GODFIT algorithm: a direct fitting approach to improve the accuracy of total ozone measurements from GOME. Int. J. Remote Sens. **31**(2), 543–550 (2010). https://doi.org/10.1080/01431160902893576
50. P.K. Rao, W. Smith, R. Koffler, Global sea-surface temperature distribution determined from an environmental satellite. Mon. Weather Rev. **100**(1), 10–14 (1972). https://doi.org/10.1175/1520-0493(1972)100<0010:gstddf>2.3.co;2
51. W.L. Smith, H.M. Woolf, H.E. Fleming, Retrieval of atmospheric temperature profiles from satellite measurements for dynamical forecasting. J. Appl. Meteorol. **11**(1), 113–122 (1972). https://doi.org/10.1175/1520-0450(1972)011<0113:roatpf>2.0.co;2
52. S.P. Ho, W. Smith, H.L. Huang, Retrieval of atmospheric-temperature and water-vapor profiles by use of combined satellite and ground-based infrared spectral radiance measurements. Appl. Opt. **41**(20), 4057 (2002). https://doi.org/10.1364/ao.41.004057
53. I.J. Schoenberg, 2. the basis property of B-splines, in *Cardinal Spline Interpolation* (Society for Industrial and Applied Mathematics, 1973), pp. 11–19. https://doi.org/10.1137/1.9781611970555.ch2

54. K. Pearson, On lines and planes of closest fit to systems of points in space. Phil. Mag. **2**(6), 559 (1901)
55. R. Penrose, A generalized inverse for matrices. Math. Proc. Camb. Philos. Soc. **51**(3), 406–413 (1955). https://doi.org/10.1017/s0305004100030401
56. D. Wittman. Fisher matrix for beginners. http://wittman.physics.ucdavis.edu/Fisher-matrix-guide.pdf
57. O. Dubovik, Optimization of numerical inversion in photopolarimetric remote sensing, in *Photopolarimetry in Remote Sensing* (Kluwer Academic Publishers, 2004), pp. 65–106. https://doi.org/10.1007/1-4020-2368-5_3
58. R. Courant, D. Hilbert, *Methods of Mathematical Physics* (Wiley, 1989). http://orcid.org/10.1002/9783527617210
59. J. Hadamard, Sur les problèmes aux dérivées partielles et leursignification physique. Princeton Univ. Bull. **13**, 49–52 (1902). [in French]
60. M. Gockenbach, *Linear inverse problems and Tikhonov regularization*. The Carus mathematical Monographs (The Mathematical Association of America, 2016)
61. V. Mazýa, T. Shaposhnikova, *Jacques Hadamard, a Universal Mathematician* (American mathematical Society, 1998)
62. R. Bapat, *Linear Algebra and Linear Models* (Springer, London, 2012). http://orcid.org/10.1007/978-1-4471-2739-0
63. S. Banerjee, *Linear Algebra and Matrix Analysis for Statistics* (Chapman and Hall/CRC, 2014). https://doi.org/10.1201/b17040
64. D. Belsley, E. Kuh, R. Welsch, *Regression Diagnostics: Identifying Influential Data and Sources of Collinearity* (Wiley, New York, 1980)
65. A. Tikhonov, V. Arsenin, *Solutions of Ill-posed Problems*. Scripta Series in Mathematics (Winston & Sons, Washington, D.C., 1977)
66. J. Kaipio, E. Somersalo, *Statistical and Computational Inverse Problems* (Springer, 2005). http://orcid.org/10.1007/b138659
67. C. Vogel, *Computational Methods for Inverse Problems* (Society for Industrial and Applied Mathematics, 2002). https://doi.org/10.1137/1.9780898717570
68. P. Hansen, *Rank-Deficient and Discrete Ill-Posed Problems* (Society for Industrial and Applied Mathematics, 1998). https://doi.org/10.1137/1.9780898719697
69. L.L. Gerfo, L. Rosasco, F. Odone, E.D. Vito, A. Verri, Spectral algorithms for supervised learning. Neural Comput. **20**(7), 1873–1897 (2008). https://doi.org/10.1162/neco.2008.05-07-517
70. H. Engl, M. Hanke, A. Neubauer, *Regularization of Inverse Problems, Mathematics and Its Applications* (Springer, Netherlands, 2000)
71. V.A. Morozov, On the solution of functional equations by the method of regularization. Dokl. Akad. Nauk SSSR **167**(3), 510–512 (1966)
72. H. Gfrerer, An a posteriori parameter choice for ordinary and iterated Tikhonov regularization of ill-posed problems leading to optimal convergence rates. Math. Comput. **49**(180), 507 (1987). https://doi.org/10.1090/s0025-5718-1987-0906185-4
73. M. Lukas, Asymptotic behaviour of the minimum bound method for choosing the regularization parameter. Inverse Probl. **14**(1), 149–159 (1998). https://doi.org/10.1088/0266-5611/14/1/013
74. G. Wahba, Practical approximate solutions to linear operator equations when the data are noisy. SIAM, J. Numer. Anal. **14**(4), 651–667 (1977). https://doi.org/10.2307/2156485
75. A. Thompson, J. Kay, D. Titterington, A cautionary note about crossvalidatory choice. J. Stat. Comput. Simul. **33**(4), 199–216 (1989). https://doi.org/10.1080/00949658908811198
76. G. Kitagawa, W. Gersch, A smoothness priors long AR model method for spectral estimation. IEEE Trans. Autom. Control **30**(1), 57–65 (1985). https://doi.org/10.1109/tac.1985.1103786
77. P. Hansen, Analysis of discrete ill-posed problems by means of the L-curve. SIAM Rev. **34**(4), 561–580 (1992). https://doi.org/10.1137/1034115
78. P. Hansen, D. O'Leary, The use of the L-curve in the regularization of discrete ill-posed problems. SIAM, J. Sci. Comput. **14**(6), 1487–1503 (1993). https://doi.org/10.1137/0914086

79. P.E. Gill, W. Murray, M.H. Wright, *Practical Optimization* (Academic, London, 1981)
80. C.G. Broyden, The convergence of a class of double-rank minimization algorithms1. General considerations. IMA, J. Appl. Math. **6**(1), 76–90 (1970). https://doi.org/10.1093/imamat/6.1.76
81. R. Fletcher, A new approach to variable metric algorithms. Comput. J. **13**(3), 317–322 (1970). https://doi.org/10.1093/comjnl/13.3.317
82. D. Goldfarb, A family of variable-metric methods derived by variational means. Math. Comput. **24**(109), 23–23 (1970). https://doi.org/10.1090/s0025-5718-1970-0258249-6
83. D.F. Shanno, Conditioning of quasi-newton methods for function minimization. Math. Comput. **24**(111), 647–647 (1970). https://doi.org/10.1090/s0025-5718-1970-0274029-x
84. R. Byrd, P. Lu, J. Nocedal, C. Zhu, A limited memory algorithm for bound constrained optimization. SIAM, J. Sci. Comput. **16**(5), 1190–1208 (1995). https://doi.org/10.1137/0916069
85. H.R. Lourenço, O.C. Martin, T. Stützle, Iterated local search: Framework and applications, in *Handbook of Metaheuristics* (Springer Science + Business Media, 2010), pp. 363–397. https://doi.org/10.1007/978-1-4419-1665-5_12
86. D. Goldberg, *Genetic Algorithms in Search, Optimization and Machine Learning*, 1st edn. (Addison-Wesley Longman Publishing Co. Inc., Boston, 1989)
87. J. Holland, *Adaptation in Natural and Artificial Systems* (MIT Press, Cambridge, 1992)
88. Z. Michalewicz, *Genetic Algorithms + Data Structures = Evolution Programs* (Springer Science + Business Media, 1996). https://doi.org/10.1007/978-3-662-03315-9
89. N. Metropolis, A.W. Rosenbluth, M. Rosenbluth, A.H. Teller, E. Teller, Equation of state calculations by fast computing machines. J. Chem. Phys. **21**(6), 1087 (1953). https://doi.org/10.1063/1.1699114
90. S. Kirkpatrick, C. Gelatt, M. Vecchi, Optimization by simulated annealing. Science **220**(4598), 671–680 (1983)
91. V. Cerny, Thermodynamical approach to the traveling salesman problem: An efficient simulation algorithm. J. Optim. Theory Appl. **45**(1), 41–51 (1985). https://doi.org/10.1007/bf00940812
92. R. Spurr, T. Kurosu, K. Chance, A linearized discrete ordinate radiative transfer model for atmospheric remote-sensing retrieval. J. Quant. Spectrosc. Radiat. Transf. **68**(6), 689–735 (2001). https://doi.org/10.1016/S0022-4073(00)00055-8
93. A. Griewank, *On automatic differentiation. Preprint ANL/MCS-P10-1088* (Argonne National Laboratory. Mathematics and Computer Science Division., 1989). http://softlib.rice.edu/pub/CRPC-TRs/reports/CRPC-TR89003.pdf
94. A. Griewank, *Evaluating Derivatives: Principles and Techniques of Algorithmic Differentiation* (SIAM, Philadelphia, 2000)
95. F. Schreier, B. Schimpf, A new ecient line-by-line code for high resolution atmosphericradiation computations incl. derivatives, in *IRS, Current Problems in Atmospheric Radiation*, ed. by W.L. Smith. Y. Timofeyev, vol. 2001, 381–384 (2000)
96. F. Schreier, U. Boettger, MIRART, a line-by-line code for infrared atmospheric radiation computations incl. derivatives. Atmos. Ocean Opt. **16**, 262–268 (2003)
97. R. Spurr, LIDORT and VLIDORT. Linearized pseudo-spherical scalar and vector discrete ordinate radiative transfer models for use in remote sensing retrieval problems, in *Light Scattering Reviews*, vol. 3, ed. by A. Kokhanovsky (Springer, 2008), pp. 229–275. https://doi.org/10.1007/978-3-540-48546-9_7
98. V. Rozanov, A. Rozanov, A. Kokhanovsky, J. Burrows, Radiative transfer through terrestrial atmosphere and ocean: Software package SCIATRAN. J. Quant. Spectrosc. Radiat. Transf. **133**, 13–71 (2014). http://orcid.org/10.1016/j.jqsrt.2013.07.004
99. A. Doicu, T. Trautmann, Two linearization methods for atmospheric remote sensing. J. Quant. Spectrosc. Radiat. Transf. **110**(8), 477–490 (2009). https://doi.org/10.1016/j.jqsrt.2009.02.001
100. G. Stiller, T. Clarmann, B. Funke, N. Glatthor, F. Hase, M. Hopfner, A. Linden, Sensitivity of trace gas abundances retrievals from infrared limb emission spectra to simplifying approximations in radiative transfer modelling. J. Quant. Spectrosc. Radiat. Transf. **72**(3), 249–280 (2002). https://doi.org/10.1016/S0022-4073(01)00123-6

101. J. Urban, P. Baron, N. Lautie, N. Schneider, K. Dassas, P. Ricaud, J. Noe, Moliere (v5): A versatile forward- and inversion model for the millimeter and sub-millimeter wavelength range. J. Quant. Spectrosc. Radiat. Transf. **83**(3–4), 529–554 (2004). https://doi.org/10.1016/S0022-4073(03)00104-3

102. B. Kodomtsev, About inuence function in radiative trasnfer theory. DAN USSR **113**(3), 541–543 (1957). [in Russian]

103. G. Bell, S. Glasstone, *Nuclear Reactor Theory* (Van Nostrand Reinholt, New York, 1970)

104. G. Marchuk, Equation for value of information from meteorological sattelites andformulation of inverse problems. Kossmicheskie issledovaniya **2**(3), 462–477 (1964). [in Russian]

105. M. Box, Radiative perturbation theory: A review. Environ. Modell. Softw. **17**(1), 95–106 (2002). https://doi.org/10.1016/S1364-8152(01)00056-1

106. E. Ustinov, Atmospheric weighting functions and surface partial derivatives for remote sensing of scattering planetary atmospheres in thermal spectral region: General adjoint approach. J. Quant. Spectrosc. Radiat. Transf. **92**(3), 351–371 (2005). https://doi.org/10.1016/j.jqsrt.2004.08.003

107. V. Rozanov, A. Rozanov, Relationship between different approaches to derive weighting functions related to atmospheric remote sensing problems. J. Quant. Spectrosc. Radiat. Transf. **105**(2), 217–242 (2007). https://doi.org/10.1016/j.jqsrt.2006.12.006

108. J. Landgraf, O. Hasekamp, M. Box, T. Trautmann, A linearized radiative transfer model for ozone profile retrieval using the analytical forward-adjoint perturbation theory approach. J. Geophys. Res.: Atmosp. **106**(D21), 27291–27305 (2001). https://doi.org/10.1029/2001JD000636

109. H. Walter, J. Landgraf, O. Hasekamp, Linearization of a pseudo-spherical vector radiative transfer model. J. Quant. Spectrosc. Radiat. Transf. **85**(3–4), 251–283 (2004). https://doi.org/10.1016/S0022-4073(03)00228-0

110. H. Walter, J. Landgraf, Linearization of radiative transfer in spherical geometry: anapplication of the forward-adjoint perturbation theory, in *Light Scattering Reviews*, vol. 5, ed. by A. Kokhanovsky (Springer, Berlin, 2010), pp. 105–146. https://doi.org/10.1007/978-3-642-10336-0_4

111. S. Najmabadi, P. Offenhäuser, M. Hamann, G. Jajnabalkya, F. Hempert, C. Glass, S. Simon, Analyzing the effect and performance of lossy compression on aeroacoustic simulation of gas injector. Computation **5**(4), 24 (2017). https://doi.org/10.3390/computation5020024

112. S. Roweis, L. Saul, Nonlinear dimensionality reduction by locally linear embedding. Science **290**(22), 2323–2326 (2000). https://doi.org/10.1126/science.290.5500.2323

113. M.A. Kramer, Nonlinear principal component analysis using auto associative neural networks. AIChE J. **37**(2), 233–243 (1991). https://doi.org/10.1002/aic.690370209

114. A. Gorban, B. Kégl, D. Wunsch, A. Zinovyev (eds.), *Principal Manifolds for Data Visualization and Dimension Reduction* (Springer, Berlin, 2008). http://orcid.org/10.1007/978-3-540-73750-6

115. I. Fodor, A survey of dimension reduction techniques. Technical report, Lawrence Livermore National Lab., CA (US) (2002). https://doi.org/10.2172/15002155

116. L. van der Maaten, E. Postma, H. van den Herik, Dimensionality reduction: A comparative review. Technical report, ticc-tr 2009-005. Tech. rep., Tilburg University (2009). http://lvdmaaten.github.io/publications/papers/TR_Dimensionality_Reduction_Review_2009.pdf

117. G. Hughes, On the mean accuracy of statistical pattern recognizers. IEEE Trans. Inform. Theory **14**(1), 55–63 (1968). http://orcid.org/10.1109/tit.1968.1054102

118. W. Blackwell, F. Chen, *Neural Networks in Atmospheric Remote Sensing* (Lexington, 2009)

119. J. Shawe-Taylor, N. Cristianini, *Kernel Methods for Pattern Analysis* (Cambridge University Press, New York, NY, USA, 2004)

120. G. Camps-Valls, J. Munoz-Mari, L. Gomez-Chova, L. Guanter, X. Calbet, Nonlinear statistical retrieval of atmospheric profiles from MetOp-IASI and MTG-IRS Infrared Sounding Data. IEEE Trans. Geosci. Remote Sens. **50**(5), 1759 (2012). https://doi.org/10.1109/TGRS.2011.2168963

121. R. Rosipal, N. Krämer, Overview and recent advances in partial least squares, in *Subspace, Latent Structure and Feature Selection* (Springer Science + Business Media, 2006), pp. 34–51. https://doi.org/10.1007/11752790_2

122. P. Wentzell, L.V. Montoto, Comparison of principal components regression and partial least squares regression through generic simulations of complex mixtures. Chem. Intell. Laborat. Syst. **65**(2), 257–279 (2003). https://doi.org/10.1016/s0169-7439(02)00138-7

123. S. Maitra, J. Yan, Principal component analysis and partial least squares: Two dimensionreduction techniques for regression, in *Discussion Papers: 2008 Discussion Paper Program - Applying Multivariate Statistical Models* (Casualty actuarial society, 2008), pp. 79–90. https://www.casact.org/pubs/dpp/dpp08/08dpp76.pdf

124. H. Hotelling, The most predictable criterion. J. Educ. Psychol. **26**, 139 (1935). http://orcid.org/10.1037/h0058165

125. D. Wilks, *Statistical Methods in the Atmospheric Sciences*, 3rd edn. (Elsevier, New York, 2011)

126. D. Efremenko, D. Loyola, P. Hedelt, R. Spurr, Volcanic SO2 plume height retrieval from UV sensors using a full-physics inverse learning machine algorithm. Int. J. Remote Sens. **38**(sup1), 1 (2017). https://doi.org/10.1080/01431161.2017.1348644

127. M. Kataev, A. Lukyanov, Empirical orthogonal functions and its modification in the task of retrieving of the total amount CO2 and CH4 with help of satellite fourier transform spectrometer GOSAT (TANSO-FTS), in *22nd International Symposium on Atmospheric and Ocean Optics: Atmospheric Physics*, ed. by G. Matvienko, O. Romanovskii (SPIE, 2016). https://doi.org/10.1117/12.2249360

128. P. Hedelt, D. Efremenko, D. Loyola, R. Spurr, L. Clarisse, Sulfur dioxide layer height retrieval from Sentinel-5 Precursor/TROPOMI using FP_ILM. Atmosp. Meas. Tech. **12**(10), 5503–5517 (2019). https://doi.org/10.5194/amt-12-5503-2019

129. D. Loyola, J. Xu, K.P. Heue, W. Zimmer, Applying FP_ILM to the retrieval of geometry-dependent reflective lambertian equivalent reflectivity (GE_LER) daily maps from UVN satellite measurements. Atmosp. Meas. Tech. **13**(2), 985–999 (2020). https://doi.org/10.5194/amt-13-985-2020

130. C. Li, J. Joiner, N. Krotkov, P. Bhartia, A fast and sensitive new satellite SO2 retrieval algorithm based on principal component analysis: Application to the ozone monitoring instrument. Geophys. Res. Lett. **40**(23), 6314–6318 (2013). https://doi.org/10.1002/2013GL058134

131. P. Levelt, G. van den Oord, M. Dobber, A. Malkki, H. Visser, J. de Vries, P. Stammes, J. Lundell, H. Saari, The ozone monitoring instrument. IEEE Trans. Geosci. Remote Sens. **44**(5), 1093–1101 (2006). https://doi.org/10.1109/tgrs.2006.872333

132. Y. Timofeyev, A. Polyakov, H. Steele, M. Newchurch, Optimal eigen analysis for the treatment of aerosols in the retrieval of atmospheric composition from transmission measurements. Appl. Opt. **42**(15), 2635 (2003). https://doi.org/10.1364/ao.42.002635

133. A.V. Polyakov, Y.M. Timofeev, Y.A. Virolainen, Izvestiya, Using artificial neural networks inthe temperature and humidity sounding of the atmosphere. Atmosp. Ocean. Phys. **50**(3), 330–336 (2014). http://orcid.org/10.1134/s0001433814030104

Useful links

We would like to conclude our book with links which the readers may find useful.

- https://hitran.org—The most extensive database of the absorption cross-sections and spectroscopic parameters; the website provides also a tutorial;
- https://hitran.org/hapi/—python interface to the HITRAN database;
- https://hitran.iao.ru—The Information System HITRAN on the Web (HotW). It provides a web interface information about the spectral line parameters of atmospheric gases and pollutants. It is mostly based on HITRAN;
- https://scattport.org—Light Scattering Information Portal for the light scattering community. It provides codes for computing single-scattering phase functions. The website contains a lot of materials about light scattering;
- https://www.giss.nasa.gov/staff/mmishchenko/—a website devoted to light scattering; it contains computer codes, datasets and publications. Initially created by M. Mishchenko;
- https://atmos.eoc.dlr.de/tools/Py4CAtS/—Library Py4CAtS. It allows to read HITRAN data and perform line-by-line computations;
- http://www.libradtran.org—Library for radiative transfer. It is a collection of C and Fortran functions and programs (including DISORT model) for calculation of solar and thermal radiation in the Earth's atmosphere;
- http://157.82.240.167/~clastr/—Open Clustered Libraries for Atmospheric Science and Transfer of Radiation. In particular, it contains the vector radiative transfer model PSTAR;
- https://coloradolinux.com/shdom/—Radiative transfer model SHDOM—a widely used deterministic three-dimensional radiative transfer model based on Spherical Harmonic Discrete Ordinate Method (SHDOM) ;
- https://github.com/korkins/SORD_JQSRT_2017—Radiative transfer model SORD;
- https://github.com/korkins/IPOL—Radiative transfer model IPOL;
- http://6s.ltdri.org—Radiative trasnfer model 6S;
- https://artmotoolbox.com—a software package that provides convenient tools for running the radiative trasnfer models;

© Springer Nature Switzerland AG 2021 293
D. Efremenko and A. Kokhanovsky, *Foundations of Atmospheric Remote Sensing*,
https://doi.org/10.1007/978-3-030-66745-0

- https://atmos.eoc.dlr.de—the ATMOS webserver operated by The department Atmospheric Processors (ATP) of DLR's Remote Sensing Technology Institute (IMF);
- https://www.wmo-sat.info/oscar—provides detailed information on all Earth observation satellites and instruments;
- https://s5phub.copernicus.eu/dhus/—Sentinel-5P Pre-Operations Data Hub;
- https://ceres.larc.nasa.gov/data/—CERES data products;
- https://modis.gsfc.nasa.gov/data/—MODIS data products;
- https://worldview.earthdata.nasa.gov/—NASA Worldview portal;
- https://www.unidata.ucar.edu/software/netcdf/examples/programs/—Some sample netCDF programs;
- http://www.ncl.ucar.edu/—The NCAR Command Language designed for scientific data analysis and visualization;
- https://scitools.org.uk/cartopy/docs/latest/—Python package designed for geospatial data processing in order to produce maps and other geospatial data analyses

Index

© Springer Nature Switzerland AG 2021
D. Efremenko and A. Kokhanovsky, *Foundations of Atmospheric Remote Sensing*,
https://doi.org/10.1007/978-3-030-66745-0

Printed in the United States
by Baker & Taylor Publisher Services